GUIDED EVOLUTION OF SOCIETY

A Systems View

Contemporary Systems Thinking

Series Editor: Robert L. Flood
Monash University
Australia

DESIGNING SOCIAL SYSTEMS IN A CHANGING WORLD
Bela H. Banathy

GUIDED EVOLUTION OF SOCIETY
A Systems View
Bela H. Banathy

LIBERATING SYSTEMS THEORY
Robert L. Flood

OPERATIONAL RESEARCH AND SYSTEMS
The Systemic Nature of Operational Research
Paul Keys

POWER, IDEOLOGY, AND CONTROL
John C. Oliga

SELF-PRODUCING SYSTEMS
Implications and Applications of Autopoiesis
John Mingers

SOCIOPOLITICAL ECOLOGY
Human Systems and Ecological Fields
Frederick L. Bates

SYSTEMIC INTERVENTION
Philosophy, Methodology, and Practice
Gerald Midgley

SYSTEMS METHODOLOGY FOR THE MANAGEMENT SCIENCES
Michael C. Jackson

A Continuation Order Plan is available for this series. A continuation order will bring delivery of each new volume immediately upon publication. Volumes are billed only upon actual shipment. For further information please contact the publisher.

GUIDED EVOLUTION OF SOCIETY

A Systems View

Bela H. Banathy
Saybrook Graduate School and Research Center
San Francisco, California and
International Systems Institute
Carmel, California

Kluwer Academic / Plenum Publishers
New York, Boston, Dordrecht, London, Moscow

Library of Congress Cataloging-in-Publication Data

Banathy, Bela H.
Guided evolution of society: a systems view/Bela H. Banathy.
p. cm. — (Contemporary systems thinking)
Includes bibliographical references and index.
ISBN 0-306-46382-2
1. Social evolution. 2. Human evolution. I. Title. II. Series.

GN360 .B36 2000
303.4—dc21

00-033113

ISBN: 0-306-46382-2

233 Spring Street, New York, New York 10013

http://www.wkap.nl

10 9 8 7 6 5 4 3 2 1

A C.I.P. record for this book is available from the Library of Congress

Printed in the United States of America

To Eva

With Loving Remembrance of Our
First Meeting Sixty Years Ago

Up-to-now, our societal way of life has been the result of random chain of changing events. And now we suddenly realize that unless we take things at hand, the process of change will continue under the sway of relentless change, entirely blind to human dreams and if there is a central task for humankind in the next millennium, it is to start on the right track in its efforts to control the direction of evolution.

Csikszentmihalyi

Preface

The Agoras of the city-states of the classical Greek period were "public spheres" where true democracy was lived by citizens, who made collective decisions about issues affecting their daily lives. Inspired by their story, in this book, I developed the idea of "New Agoras of the twenty-first century." This idea is a metaphor of social context in which people can make collective decisions about their future. I see the social contexts of the New Agoras to be our families, our neighborhoods, and the many community groups and organizations in which we live and work. People in these settings would collectively envision their ideal future and re-create themselves as evolutionary designing communities in which they would steer themselves toward their aspired future. The many New Agoras would link up with each other to bring to life the *Guided Evolution of the Society.* This work sets forth a possible approach toward establishing New Agoras for the twenty-first century.

The evolutionary inquiry idea developed here is grounded in my public philosophy, which holds that: *The right of people to take part directly in making decisions that affect their lives and guide their own destiny is a* ***fundamental human right.*** *If the life of the society is so organized that people can learn how to exercise this right then they can develop competence that enables them to guide their own evolution toward a desired future.*

Today, however, "the right of people to take part in decisions affecting their life"—constantly quoted by political establishments—is just empty rhetoric. We never had that right. That right is exercised by those who "represent" us or possess authority over us. We never had the opportunity to learn how to exercise that right. So, to be true to my own philosophy, one of my core missions in life became to do something about this. With others, I have spent years to design and develop learning resources that enable people to become competent, to become the best they can, and fulfill their potential. In the course of the last decade, however, I have become increasingly convinced that even if people fully develop their individual potential, they cannot give direction to their lives, they cannot forge their individual and collective destiny unless they understand how evolution works and develop evolutionary competence by evolutionary learning. Only then can they enter into the evolutionary design space and engage in the conscious, guided evolution of their lives and the systems in which they live and work.

I offer this work as a resource that individuals, groups, communities, and organizations of all kind might use to learn to create a common ground, collectively define values and qualities they seek to realize, envision ideal images of a desired future, and bring those images to life by design.

The ideas and propositions presented in this work have been influenced by many. What I have learned about systems, design, and evolutionary inquiries I have learned with—and from—many others. And I am grateful to them. I wish to acknowledge the special help of a few friends and colleagues who reviewed the manuscript and offered advice and guidance. The help of Robert Artigiani, Aleco Christakis, Allan Combs, Duane Elgin, Stanley Krippner, Ervin Laszlo, and Gordon Rowland is much treasured and appreciated. Furthermore, I am most grateful to Ken Derham, a Director and Managing Editor of Kluwer Academic/Plenum Publishers for embracing the idea of this work and for his guidance of the production of the book. I wish to extend my appreciation to Robert Maged and Robert Wheeler of the editorial staff.

Contents

PART III. SELF-GUIDED EVOLUTION

GUIDED EVOLUTION OF SOCIETY

A Systems View

1

Introduction

In the course of the evolutionary journey of our species, there have been three seminal events. The first occurred some seven million years ago, when our humanoid ancestors silently entered on the evolutionary scene. Their journey toward the second crucial event lasted more than six million years, when—as the greatest event of our evolutionary history—*Homo sapiens sapiens*, started *the human revolution,* the revolutionary process of *cultural evolution.* Today, we have arrived at the threshold of the third revolution: the *revolution of conscious evolution,* when it becomes our responsibility to enter into the evolutionary design space and guide the evolutionary journey of our species.

I tell the story of the first six million years of the journey in just enough detail to understand how evolution had worked in times when it was primarily biological, driven by natural selection. With the human revolution some fifty thousand years ago, with the emergence of self-reflective consciousness, the evolutionary process transformed from biological to cultural. From that point on we follow the journey with detailed attention, because we want to learn how cultural evolution works.

Since the middle of the last century we became increasingly conscious of evolution. The science of evolution became the context for much of what science is about. Our attainment of evolutionary consciousness provides the springboard of conscious purposeful evolution, marking the third crucial event: the *revolution of conscious evolution.* The focus of this work is to contemplate the unfolding of this crucial event.

In what follows, I address five questions: Why this work? What is it about? How was it developed? Who might benefit from it? How can it be used as a learning resource?

1.1 Why this Work: Rationale and Underlying Assumptions

Today there is an ever increasing awareness of the massive societal changes and transformations that reflect the new realities of our current era. These changes touch the lives of every individual, family, community, nation, and the global society. As we are faced with these massive changes, the following questions arise: Are we only spectators of these changes? Are we destined to be their victims? Do we have to relegate decisions affecting our lives to others? Are we at the mercy of "social engineers" and politicians who design systems and laws for us? In this work I respond with a *resounding no* to these questions.

1.1.1 Underlying Assumptions

The right of people to guide their own destiny, to create authentic, nurturing, sustainable communities, to control their resources, to govern themselves and guide their own evolution is a most fundamental human right. If people learn *how* to exercise this right, then they shall have the power to create a truly democratic civil society: a society that is motivated by voluntary commitment, in which people can design and organize their lives for the development of their full potential, and individually and collectively serve the common good.

1.2 What It Is About: The Content

The learning journey commences with an exposition of a brief history of the evolutionary idea through time as I review the science of general evolution and specifically sociocultural evolution. Next, the journey unfolds the "evolutionary story" of our species from the time when the first humanoid entered the evolutionary scene to our current era. The journey continues as I develop a systems view of evolution, explore evolutionary consciousness, and set forth the key conditions of conscious, self-guided evolution. I suggest that a core condition of engaging in evolutionary inquiry is the acquisition of evolutionary competence through evolutionary learning. Evolutionary competence empowers us to create an image of guided evolution and develop an approach by which evolutionary guidance systems can be designed by families, organizations, social systems, communities, and societies and eventually by humanity as a whole.

The twelve chapters of the book are organized into three parts.

1.2.1 Part I: The Evolutionary Journey of Humankind

Part I establishes the foundation of the work. Several evolutionary ideas, position, and perspectives are explored and synthesized to create a comprehensive knowledge base for understanding human and societal evolution. In this part, I address the following questions: What is evolution in general? How have ideas about human and societal evolution evolved through time? What does all this mean to us today?

In ***Chapter 2: The Evolution of the Evolutionary Idea,*** I start with exploring historical perspectives of evolution, "sweeping in" the views and positions of several "students of evolution" since the time of the classical Greek scholars. Then, I focus on the nineteenth century, when scientifically based evolutionary scholarship emerged, producing a general theory of evolution. Next, I search for perspectives that draw upon the insights of contemporary evolutionary scholars who have examined the logic of evolution and its laws and principles.

In ***Chapter 3: The Story of Human and Societal Evolution: The Evolutionary, Journey of Our Ancestors,*** I take on a journey that starts at the point where the humanoid species split off from the common ancestral line shared with the primates some seven million years ago. The story from then on follows the evolutionary journey of the various humanoid and *Homo* species until the emergence of *Homo sapiens sapiens*: modern man.

In ***Chapter 4: The Story of the Evolution of Homo sapiens sapiens (HSS)***, we meet the *First Generation of HSS*, the Cro-Magnons. Their emergence represents a giant leap forward in the evolution of our species. As we continue our journey, we next meet the *Second Generation of HSS,* the first *farmer*s who brought about the agricultural revolution and the agriculture-based ancient civilizations. The Third Generation HSS emerged with the Renaissance as the era of the scientific–industrial revolution.

1.2.2 Part II: The Journey from Evolutionary Consciousness to Conscious Evolution

In this part, we jointly explore the questions: What have we learned in the course of our journey about how evolution works? What does the attainment of evolutionary consciousness mean to us? And, How can we respond to the challenge of conscious evolution?

In ***Chapter 5: What Have We Learned from the Evolutionary Story?,*** I organize the core ideas of human and societal evolution into an evolutionary model. Then, in the context of this model, I propose that we come to understand evolution as: (1) a phenomenon of constantly expanding boundaries of space and time dimensions of the human experience; (2) the emergence of new modes and technologies of communication; (3) the continuous unfolding of new relationships among human systems, leading to their reorganization at ever higher levels of complexity; (4) the emergence of new beliefs, and world views; (5) the emergence of new paradigms of knowledge organization and utilization; and (6) from the integration of all the above, the emergence of new images of mankind at a higher level of consciousness.

In ***Chapter 6: From Evolutionary Consciousness to Conscious Evolution,*** I suggest that we find ourselves today in the "twilight zone" between two evolutionary stages. With one foot we are still in the Third Generation of HSS and with the other we seek a foothold in the Fourth. Our current state is one of confusion, uncertainty, chaos, and disorganization as well as hope, aspiration, and determination. Today we are faced with an evolutionary predicament. Its primary source is the speed and intensity of changes and transformations at the now emerging stage of evolution. It creates a twofold evolutionary gap: a gap in collective consciousness and a gap between technological intelligence and sociocultural intelligence.

In ***Chapter 7: Calling for Conscious Evolution,*** I revisit evolutionary scholars who earlier made remarks on conscious evolution and consult with other evolutionary scientists and philosophers who developed rather comprehensive statements about conscious and guided evolution in recent years.

1.2.3 Part III: Self-Guided Evolution

In this part, I set forth the conditions of conscious evolution; define evolutionary guidance, set forth its purpose, and develop the concept of evolutionary guidance systems (EGSs). Then, I propose an approach to designing EGSs and explore a set of issues that enhance—and add value to—the design of EGSs. In conclusion, I ask the question: What if the idea of EGS would become reality by the purposeful design of EGSs at all levels of the society?

In ***Chapter 8: Organizing Perspectives and Conditions of Engaging in Self-Guided Evolution,*** I introduce perspectives that will guide our thinking about the process of conscious evolution, and propose that a precondition of acting upon conscious evolution is the individual and collective acquisition of evolutionary competence and a willingness to engage in conscious evolution.

In ***Chapter 9: Self-Guided Societal Evolution,*** I present the four domains of evolutionary inquiry: philosophy, theory, epistemology, and application. Then, I review philosophical and theoretical knowledge bases that are relevant to evolutionary inquiry. Finally, I ask: Who should be the designers of evolutionary systems?

In ***Chapter 10: An Epistemology of Conscious, Self-Guided Evolution,*** I set forth the major strategies of evolutionary epistemology:

transcending the existing state, envisioning an ideal image of our evolutionary future, and transforming our systems by bringing the evolutionary image to life by purposeful design.

In ***Chapter 11: The Design of Evolutionary Guidance Systems,*** I introduce the approach to the design of evolutionary systems and explore ways and means by which we can develop design competence by acquiring a design culture.

In ***Chapter 12: The Agoras of the Twenty-First Century,*** I ask: How can we create evolutionary designing communities in the twenty-first century? How would they work? What would happen if our families, the systems in which we live and work, in our communities would engage in the design of their own evolutionary systems? Then, I will propose sets or qualities we should abide by in evolutionary inquiry and seek to realize in our evolutionary systems. In closing, I propose a set of characteristics that would add value to the work of our evolutionary designing communities.

1.3 How Was the Book Developed?

In writing the book, I have used historical and contemporary evolutionary literature and consulted sources from the systems and design sciences. Working with the evolutionary literature, however, I found that the existing evolutionary knowledge base is rather fragmented. From the aspect of the theme of this work, while several sources refer to conscious evolution as an emerging and important evolutionary issue, their authors do not elaborate on how to engage in and how to conduct evolutionary inquiry. I consider the work to be a learning resource. I included activities that enable the user to create his or her own meaning and apply what has been learned in the functional context of the learner's choice. As the book is a learning resource, I also revisit content items from different perspectives that might enrich understanding and provide deeper insight to the learner.

1.4 Who Will Benefit?

There is today a mounting interest in the idea of evolutionary consciousness, conscious evolution, self-determination, and taking responsibility individually and collectively for shaping our future. While the interest is high, practical approaches to conscious evolution and evolutionary guidance are not yet explored or defined and are not yet operationalized. This book focuses on these issues, and therefore should capture the interest not only of the evolutionary scholarship community, but more pointedly people in the public arena, the helping professions, learning and human development systems, and community and volunteer agencies. More significantly, the work is a useful guide of the many who have already attained evolutionary consciousness and are ready to enter into the evolutionary design space as creators of the future.

1.5 The Learning Strategy

The work is not expository but learning focused. Sets of activities are integrated in the chapters. The activities invite the learner to create his or her own understanding and meaning from the content, capture the core ideas and perspectives, and apply what is being learned in real-life functional contexts of interest to the learner. The text is particularly useful, and it has been tested in at-a-distance educational settings.

The work, if used as a textbook, is a resource for evolutionary learning and developing functional competence in the design of evolutionary guidance systems. The activities, integrated into the text, offer the "personal learning space" of the learner. In this space, the learner works with the text and carries out the proposed activities. In so doing, the learner: (1) develops an understanding of human and societal evolution, (2) attains evolutionary consciousness, (3) generates competence to engage in conscious evolution, and (4) applies the competence in the functional contexts of the learner's interest. Keep in mind that *learning that is not applied in a real-life context is learning not completed.*

As a guide to the use of the material as a learning resource I suggest that: the richer, the more extended, and the more inclusive are the sets of "core ideas" generated by the learner, the more powerful and the richer will be his or her understanding of evolution and the acquisition of evolutionary consciousness and conscious evolution.

We can now proceed with the learning journey. In the course of this journey we will understand societal evolution, develop evolutionary consciousness, generate conscious evolution, engage in evolutionary learning, and discover for ourselves the marvel of what the acquisition of evolutionary competence can offer for the enrichment of our own lives and the lives of the systems in which we live and work. We shall become the creators of our future.

I

The Evolutionary Journey of Humankind

> We shall not obtain the best insight into things until we actually see things growing from the very beginning.
>
> *Aristotle*

The aim of this work is to develop an understanding of our evolutionary history, attain evolutionary consciousness, embrace the idea of conscious evolution, capture a vision of guided evolution, and bring that vision to life by learning how to design evolutionary guidance systems. It is clear that we have a long journey ahead of us. Following Aristotle's advice, we start at the very beginning, which means addressing a twofold task.

The first part of the task is to capture the meaning of evolution itself as we explore the progression of the idea from the time at which early scholars defined it, to its scientific definition in the nineteenth century, through its contemporary characterization. This exploration is presented in Chapter 2.

Second, we unfold the evolutionary story of humankind, starting at the point at which the first humanoids emerged from their ancestral lineage. From there, we follow the evolutionary trail that will lead us to the present time. We tell the story in two chapters.

In Chapter 3, we describe the various evolutionary experiments of the humanoid species, during the several millions of years when biological evolution was the dominant force in selecting out and shaping our species in the evolutionary design space.

In Chapter 4, we reach the threshold of cultural evolution as we describe the emergence of modern man and listen to the story of the Cro-Magnons, the First Generation of *Homo sapiens sapiens* (HSS). Then, we move on to meet the farmers in the small agricultural villages, and visit with agriculture-based civilizations of the Second Generation HSS. With the Renaissance, we enter the scientific–industrial period of the Third Generation. Finally, we follow the evolutionary trail of that generation and arrive to our present age.

2

The Evolution of the Evolutionary Idea

Who am I? Where am I? Who are you? Why are we here? These are some of the existential questions humans have asked since leaving the blissful state of undifferentiated subconsciousness of "oneness with nature" and "oneness with wholeness." At the dawn of consciousness, *Homo sapiens sapiens* (HSS) took the first step into the world of self-reflection about the "I and thou," "we and others," and "we and nature." This separation from nature and from "oneness with wholeness" marked the critical point of transcending "biological evolution" and entering the path of "cultural evolution." After dozing in a faint dawn of preconsciousness for several million years, "we humans finally awakened into the stark sunlight of life and discovered that we are here—'the flat fact'—of our existence. Although still shaking loose from the ancient slumber, early humans used the power of reflective consciousness to achieve an unprecedented scale of social cooperation and creativity" (Elgin, 1993, p. 49). Stepping through the threshold of consciousness marks the true miracle of the evolutionary story of our species.

Although I could begin writing the story of societal evolution now (a story that will begin to unfold in the next chapters), it is useful to create a context for the evolutionary story by painting a larger picture of overall evolution. The larger picture will show similarities as well as differences between biological and cultural/societal evolution. Making such a distinction will help us to better understand the potential of evolutionary consciousness, conscious evolution, as well as the possibility of guided evolution.

Following some initial definitions, I first trace the history of the development of the idea of evolution from its emergence up to the nineteenth century. Then—recognizing that the nineteenth century and the first part of the twentieth were the watershed era of evolutionary scholarship—I review the ideas of several scholars who devoted their life work to the study of evolution, as they created evolutionary science and established its philosophical foundations. They were true pioneers of the disciplined inquiry of evolutionary science. Next, I review the contributions of a group of evolutionary scholars, who during recent decades—building upon the work of the pioneers—have made significant contributions to advancing evolutionary thought.

This chapter includes several activities. The first task of each activity calls for capturing and describing key evolutionary ideas discussed in the text. Other tasks ask the reader to use the core ideas in a creative way. Those who use this work as a textbook are asked to enter their findings in a workbook.

2.1 *Initial Definitions*

In the broadest sense, evolution is the core process of the cosmos. The pattern of evolution can be perceived in its space and time structure. Showing the arrow of time of cosmic evolution, Chaisson (1988) places the formation of matter at fifteen billion years ago. The Milky Way was formed some twelve billion years ago, and the physical/chemical evolution on Earth reached some stability about five billion years ago. The origin of life is placed at three and a half billion years ago.

We call this dynamic cosmic process *evolution*. Today, it is generally held that the process consists of four phases: the energy, the material, the biological, and the cultural. On Earth, material evolution stabilized about five billion years ago, when our solar system had formed its systemic structure which was about the same as what we have today. Boulding (1981) noted that "Curiously enough, in our part of the universe physical evolution, under the impact of human knowledge, has now resumed a fairly rapid rate with production of transuranium elements and chemical compounds previously unknown, such as nylon and Freon. The consequences may be large and unfortunate" (p. 29).

The process that led to the beginning of biological evolution is not known for certain, despite much plausible speculation. When this process went into "high gear" it produced hosts of increasingly more complex new species. It was this trend that eventually led to the emergence of *Homo sapiens*, who—through a continuing process of evolution and the various stages of consciousness—attained the current state of development. In the general evolutionary sequence material evolution had to reach a point where biological evolution became possible, and biological evolution had to produce a human species with the potential to develop consciousness and the capability of developing complex images and creating language.

2.2 *A Review of Early Evolutionary Ideas*

This section provides a cursory account of early evolutionary ideas of philosophers and scientists of the last three millennia, leading up to the nineteenth century, when the science of evolution was established. Beyond offering a picture of early evolutionary thinking, the paragraphs that follow also reveal the evolution of philosophy and science in that early period.

In sociocultural evolution it was only when "deep consciousness" emerged that humans attained the state of cognitive and intellectual development that enabled them to reflect upon the origin of life and the evolution of their species. The review that follows is a manifestation of such a state of development. The formal expression of the evolutionary idea goes back to Hermes Trismegistos. It is not known for certain in which period he lived, "presumably long before the days of Moses," according to Jantsch (1975, p. 22). Later, his evolutionary ideas were represented in "Corpus Hermeticus." He projected a succession of universes that followed the course of birth, evolution, and death. What he called the "All" is the One who created the evolving universes by first conceiving, building, and perfecting a "plan" that subsequently is implemented in the evolutionary process in reality. What we now call "hermetic philosophy" influenced evolutionary thinking in the Asian and Mediterranean cultures for a millennium.

About 3000 years ago, some of the Greek natural philosophers suggested that all forms of life had a common origin. Anaximenes, in the sixth century B.C., believed that an originally liquid earth, on drying, produced living matter. His contemporary, Thales, claimed that water was the source of life. Among the classical Greek philosophers Heraclites held that "change is all" and change is the core principle of evolution. Aristotle described a "chain" of living things from inanimate objects to plants, to animals, and on to humans. He held that evolution unfolds from the least complex and least perfect to the most complex and most perfect.

For many in the Western world, during the first and second millennia—and for some even today—the literal interpretation of the story of creation in the Bible was the basis of ideas on the origin of life and evolution. This account of evolution is called *creationism*.

In the fourth century, Julian the Apostate said that soil and climate modified men. In the fifth century, St. Augustine and St. Gregory held that the powers of creation resided in matter as exercised throughout the ages. In the tenth century, the Muslim philosopher Ibn al-Masudi explained evolution as a transition from mineral to plant, from plant to animal, and from animal to human. Mailet, in the seventeenth century, conceived of evolution of terrestrial animals from marine animals. In the eighteenth century, Buffon attributed evolutionary change to climate and environment, as did Erasmus Darwin, Charles Darwin's grandfather.

2.3 *The Emergence of Evolutionary Science*

The nineteenth century saw the emergence of evolutionary scholarship through the life work of four scientists: Lamarck, Darwin, Wallace, and Spencer, the pioneers of evolutionary science. Here, their work is discussed briefly, with special attention to human and societal evolution.

2.3.1 *Jean Lamarck (1744–1829): The Pioneer of Cultural Evolution*

Jean de Monte Lamarck, a French natural scientist of the late eighteenth and early nineteenth centuries, set forth an evolutionary theory known as "inheritance of acquired characteristics." The theory accounts for evolution in terms the attempts of species to adjust to environmental changes. These adjustments become evident during the life of the species by use and disuse of organs. Lamarck postulated that these adjustments can be directly inherited. This theory was accepted by Darwin, who considered it as supportive of his theory of natural selection. However—as Gould (1996) points out—Lamarck's theory cannot be considered to be such a booster. The Lamarckian style of inheritance gives our evolutionary history a direct and cumulative character that Darwinian evolution does not possess. Lamarck's theory influenced several evolutionary scholars, among them Spencer. But by the end of the nineteenth century, new discoveries led to a rejection of Lamarck's theory. It was revitalized later, however, by Lamarck's followers, who did not believe that natural selection was the sole or most important explanation for evolution. This trend became know as "neo-Lamarckism." Today, evolutionary scholars accept the relevance of Lamarck's theory to cultural evolution.

2.3.2 *Charles Darwin (1809–1882): The Pioneer of General Evolution*

First, I introduce Darwin's core ideas of evolution as articulated in his two major works. Then, I interpret some of his ideas from the aspects of sociocultural evolution.

2.3.2.1 Darwin's Evolutionary Idea in a Nutshell

Although the idea of natural selection was expressed earlier, it was Charles Darwin, who, on a voyage around the world, discovered the variability of animals. Based on his observations he conceptualized that variations were caused by natural selection, and change in species was determined by the reinforcement of some variations and the elimination of others. His theory of natural selection was published in *The Origin of Species by Means of Natural Selection* (1859). To recapitulate his massive work, he says that

> species have been modified during a long course of descent. This has been effected chiefly through natural selection of numerous successive, slight, favorable variation; aided in an important manner by the inherited effects of the use and disuse of parts; and in an unimportant manner, that is in relation to adaptive structures, whether past or present, by the direct action of external conditions, and by variations which seem to us in our ignorance to arise spontaneously. (p. 421)

In the *Origin*, Darwin avoided addressing sociocultural evolution. But in a later edition he mentioned that his work might shed light on the origin of man. In his *The Descent of Man* (1871), however, Darwin devoted several chapters to man's descent and development from lower forms. Commenting on this, Leakey and Lewin (1977) state that he "erected two pillars in the theoretical structure of anthropology" (p. 1). The first addressed the "where" of the origin of man. In this aspect, Darwin correctly identified Africa as the place of origin. His second conclusion was that the "important distinguishing features of humans—bipedalism, technology, and an enlarged brain—evolved in concert" (p. 3). This conclusion—called by Leakey the "Darwinian package"—placed considerable distance in time between humans and apes. This proposition was very attractive to many who had strong views of human distinctiveness. Darwin's "when" scenario placed the time of separation back to fifteen to thirty million years ago (depending on two alternative anthropological theories). Darwin's view prevailed up to just over a decade ago, when the experiments of Allan Wilson and Vincent Sarich were used as a "molecular clock" in calculating the mutation rate of humans and African apes. They established that an evolutionary event some five plus million years ago caused a split between humans and chimpanzees. The split between chimpanzees and gorillas occurred some eight million years ago. After a long and heated debate, it was generally accepted that the split between humans and chimpanzees took place about seven million years ago.

In three chapters of *Descent* (1871), Darwin compares the mental faculties of man to those of some "lower forms," and the development of man's intellectual and moral faculties. These aspects of human evolution are relevant to our present interest and invite some quotations from *Descent*.

> There can be no doubt that the difference between the mind of the lowest man and that of the highest animal is immense . . . The thought of fashioning a stone into a tool was quite beyond the scope of the highest animal. Still less could it follow metaphysical reasoning, or solve a mathematical problem, or reflect on God, or admire a grand natural scene (p. 127). The difference between man and the lower animals, the moral sense and conscience is far the most important (p. 98). Man is enabled through his mental faculties to keep with an unchanged body in harmony with the changing universe. He has great power of adapting his habits to new conditions of life. He invents weapons, tools, and various stratagems to procure food and to defend himself. When he migrates into a colder climate he uses clothes, builds sheds, and makes fires: and by the aid of a fire cooks food otherwise indigestible. He aids his fellow men in many ways, and anticipates the future. Even in a remote period he practiced some division of labor. Ultimately, our moral sense or conscience becomes a highly complex sentiment—originating in the social instinct, largely guided by the approbation of our fellow-men, ruled by reason, self interest, and in later times by deep religious feeling, and confirmed by instruction and habit. (p. 129)

In this quote, I attempted to capture vignettes from chapters that elaborate Darwin's views of human evolution. It is interesting to note that throughout the book Darwin constantly compares man's evolutionary faculties with those of animals, which is understandable inasmuch as his primary interest was biological evolution.

In this chapter, my intent is to paint an overall picture of the evolutionary landscape with only broad strokes. A detailed picture of societal and cultural evolution can then emerge within this context. Thus, given our interest, the discussion above should be an adequate introduction to Darwin's work. In what follows, I explore and interpret a few "core" ideas of Darwin that have relevance to this work.

2.3.2.2 Exploring Relevance

In the sections that follow, I explore the relevance of Darwin's ideas to human and social evolution.

2.3.2.2.a Confronting the Creation Hypothesis. Thomas Henry Huxley, a biologist and contemporary of Darwin, called himself a "general agent" of Darwin's evolutionary theory. (Huxley was a towering natural scientist in nineteenth century England. He made exceptional contributions in such areas as physical anthropology, marine biology, paleontology, and anatomy.) Huxley's promotion of Darwin's work was a main factor in establishing Darwinism as the most important contribution to evolutionary scholarship. He suggested that the *Origin* provided us with a working hypothesis and did an immense service in freeing us forever from the "creation hypothesis."

The last sentence of *Origin* (1859) clarifies Darwin's position of creation. There is a grandeur in a

> view of life, with its various power, having originally breathed by the Creator into a few forms or into one, and that, whilst this planet has gone cycling on according to the fixed law of gravity, from so simple a beginning endless forms most beautiful and most wonderful have been, and are being evolved. (p. 429)

2.3.2.2.b Gradualism vs. Sudden Change. In Darwin's view the two crucial factors of evolution are individual differences—which he called variability—and environmental selection. Within this overall conceptual context, he suggested that a slow accumulation of minor changes produces evolution. Using a contemporary term, today Darwin would be labeled a "gradualist." Around the turn of the century, the idea of gradualism was challenged based on Mendelian genetics and the discovery of mutant genes. A heated debate developed between gradualists and Mendelists. Mutation was a manifestation of a sudden change of traits. Gradualism explained evolution through the accumulation of small steps. It turned out not to be an "either–or" issue, but to involve both. Mendel's work with genetics, the discovery of the phenomenon of mutation—linked with the idea of sudden inherited changes—offered a much needed clarification in evolutionary scholarship. Discussing human and societal evolution in the next chapter, we shall see the manifestation of both gradual changes within a specific evolutionary stage and sudden changes between evolutionary stages.

2.3.2.2.c Telos–Order–Design. At the dawn of science, Aristotle identified "telos" as purpose that guides our inquiry about anything of interest. Pursuing telos we ask: What is its purpose? What is its reason for existing? In the context of evolution: Does the cosmos exist for a reason? Does purpose play a role in evolution? Does it have telos? In the *Origin of Species*, Darwin defined the process of natural selection as a mindless, purposeless mechanical process. Darwin asked for *order* and his mindless and purposeless algorithm gives us *design*. Dennett (1996, p. 65) states Darwin's position as follows: "Let me start with regularity—the mere purposeless, mindless, pointless regularity of physics—and I will show you a process that eventually will lead to products that show not only regularity but purposive design." But the question arises: Is the purposeless mindless regularity of evolutionary mechanisms the process of sociocultural evolution? The physicist Paul Davis (1992) responded to the same question at the end of his book: "Through conscious beings the universe has generated self-awareness. This can be no trivial detail, no minor by-product of mindless purposeless forces. We are truly meant to be here"

(p. 232). Even this brief discussion shows us the discontinuity and contrast between biological and cultural evolution. Once we attain evolutionary consciousness, our evolution is guided by telos, by our conscious choices.

2.3.2.2.d Natural Selection vs. Conscious Selection. Darwin's core idea was not evolution as such but rather "natural selection." The process of the evolution of species produces diversity. It brings forth and tests alternative life forms in the evolutionary design space, and through many small steps of trial over a long period of time eventually selects the one alternative that "best fits" the environment and proves to be the most successful. Thus, evolution is a process of divergence and convergence. Darwin called this process "natural selection" or the "survival of the fittest" and defined it as mindless and purposeless. This process is the algorithm of biological evolution, but not of cultural evolution. Still, in one aspect it seems to be similar to it. In both biological and societal evolution, "selection" is involved but the process by which it is attained is very different. Humans create physical products, mental constructs, belief systems, and social systems. The unfolding story of these creations is the story of our evolution. The process of these creations is conscious selection among defined alternatives. We bring forth alternative forms and ideas, we test them, try them out, and select the one that has "a goodness of fit"with our purpose as well as the environment in which we create the alternatives by design and then converge by making conscious selection.

2.3.2.3 Confronting Difference

Building upon the preceding discussion, we can now elaborate on the difference between biological and cultural evolution. Understanding the difference and making distinctions between the two modes of evolution helps us to better understand both. Gould (1996) notes that human/cultural evolution is radically different from biological evolution. "Using the same term—evolution—for both natural and cultural history obfuscates far more than enlightens. The common designation of 'evolution' then leads to the most frequent and portentous errors in our analysis of human life and history—the overly reductionist assumption that the Darwinian paradigm will fully encompass our social and technological history as well" (p. 219).

2.3.2.3.a The Phasing Out of Biological Evolution in Our Species. We have no evidence of progressive and continuous change of the bodies and brains of humans during the past 100,000 years. Genetic change, and evolution, is often confused with cultural change and evolution. An example, discussed by Dennett (1996), is the height of humans, which has greatly increased during the last several centuries. This change in body height, he says, has not much to do—if anything at all—with genetic changes but rather with cultural changes, such as improvements in public health measures, living conditions, diet, and education. "Cultural transmission operates many orders of magnitude faster than genetic evolution and this is part of its role in making our species special" (p. 339).

2.3.2.3.b Biological vs. Cultural Change. There is a crucial difference between Darwinian and cultural evolution, says Gould (1996, p. 220). "It lies in the enormous capacity that culture holds—and nature lacks—for explosive rapidity and cumulative directionality. In an unmeasurable blink of geological eyelash, human cultural change has transformed the surface of the planet as no event of natural change has ever accomplished at Darwinian scales of myriad generations." Cultural evolution operates under principles that are radically different from those governing biological evolution. They allow for purposeful direction and progress. There are numerous manifestations of these differences, two of which are described next.

2.3.2.3.c Distinct Separation vs. Free Integration. Once a species has evolved and separated from its ancestors it remains distinct from then on (Gould, 1996). Furthermore, its members are unable to reproduce with those of other species. Species interact with each other in ecosystems but do not join. Natural evolution is a process of ever ongoing separation and distinction. In contrast with the rule of separation of species, cultural evolution is powerfully enriched by the free exchange and integration of cultural products, mental and social constructions, and cultural ideas. For example, as I write this work, I am constantly informed by others' ideas, internalize them, synthesize them with my ideas and—by so doing—expect to create a new "idea species," a new proposition about guided societal evolution.

Gould (1996) gives another example. Back in time, a traveler saw a wheel (unknown to him before) and observed how it worked. He imported the invention back home and changed its culture—the way of life of his people—forever. The helpful or harmful impact of shared cultural artifacts, mental constructions, or cultural evolution brings forth cultural change by a process unknown in biological evolution.

2.3.2.3.d Passive Selection vs. Active Learning and Design. The natural selection of biological evolution is a process by which the variety of a species is tested by external conditions (forces in a given environment), eliminating variations that fail the test by not being able to adapt to the environment. This is a long process of divergence of variations and convergence to preserve the "fittest," encompassing numerous generations. In contrast, cultural evolution is a highly accelerated process. Changes and adaptations happen as each generation passes on cultural changes to the next generation, which learns through the process of enculturation. This is the Lamarckian concept of inheritance of acquired characteristics. Furthermore, people, individually as well as collectively, can engage in the design of artifacts, technologies, mental constructs, human and social systems of all kind, that are passed on to others and become shared property. Such a "cultural selection" is the active process of societal evolution.

2.3.2.3.e Local Environment vs. No Boundaries. The Darwinian biological evolution theory limits natural selection and adaptation within the boundaries of a specific local environment. Cultural and social evolution at the dawn of consciousness was limited to small social groups, but in the course of our evolutionary journey we have continuously reorganized the human experience: first in tribes, then in agricultural communities, city-states, empires, nations, and multinational federations. We have created ever larger social entities and extended and even exploded the boundaries of human experience to the point where today the global system defines it, and now we go even farther beyond it, and see no boundaries.

2.3.2.4 What Have We Learned?

In the previous section, we reviewed Darwin's work, explored its relevance to cultural evolution, and contrasted biological evolution with cultural evolution. As a result we understand the following:

1. *Biological evolution* became the evolutionary process for our species at the point when some seven million years ago—splitting from the ape family—it came into its *own* as the human species. The slow process of natural selection shaped our species for several millions of years, until some 35,000 years ago, when—at the dawn of human consciousness—we phased out from biological evolution and entered our own cultural evolutionary path.
2. *Biological evolution* and its process, natural selection, are fundamentally different from cultural evolution. The algorithm of biological evolution is not applicable to cultural evolution. The time rate of change and the design space of cultural evolution are also fundamentally different from those for biological evolution.

3. *The primary shaper* of biological evolution comes from without, as a result of the external conditions and forces that operate in the environment. Cultural evolution primarily unfolds from within and is guided by self-reflective and creative consciousness.

Activity 1

Review the text of Section 2.1. Identify and note core ideas and work with them so that you can construct your own meaning of those ideas. Now, you are asked to attend to two tasks.

Task 1. Describe the nature and characteristics of biological evolution.

Task 2. Go to your public or institutional library and check out Darwin's *Descent*. Review Chapters 3 to 6, identify Darwin's core ideas relevant to human evolution, and describe the characteristics of human evolution as you understand them now.

2.3.3 Alfred Russell Wallace (1823–1914)

"If it had not been for the mere chance that Alfred Russell Wallace chose to dispatch the account of his discovery to Darwin, we might today be acclaiming him as the founder of modern biology" (Eiseley, 1979, p. 14). Wallace was eventually given the designation of Darwin's co-discoverer of natural selection, even though he discovered it first. In the following paragraphs I describe how this happened and then highlight some of Wallace's achievements. (Unfortunately, in the current evolutionary literature little mention is made of Wallace's work. I personally consider him the most sympathetic—albeit still the forgotten—hero of evolutionary scholarship.)

After engaging in paleontological and anthropological research for many years in various parts of the world, Wallace was in New Guinea, where in a flash of insight, the many years of study and understanding of evolution fell into place. It became clear to him that "the best fitted live." His mind leaped forward as he wrote: "Considering the amount of individual variation that my experience as a collector had shown me to exist, then it followed that all changes necessary for the adaptation of species to changing conditions would be brought about. And as great changes in the environment are always slow, there would be ample time for the change to be affected by the survival of the fittest in every generation." (From a paper to Darwin, quoted by Eiseley, 1979, p. 26.) In two evenings, Wallace elaborated this theme and dispatched it to Darwin (whom he knew well as a friend) hoping—as he said—that "the idea would be as new to him as it was to me."

The paper reached Darwin in June of 1858. Wallace's elation was to be Darwin's despair as he said in a note to his mentor Charles Lyell: "All my originality, whatever it may amount to, will be smashed—I never saw a more striking coincidence." In genuine agony he sought the advice of his friends, wondering if he could now publish honorably. He said: " I would far rather burn the whole book than he or any other man would think that I behaved in a paltry spirit." (Quotations are from Eiseley's book, p. 26.) His friends persuaded him to go ahead and publish. It was decided that Wallace's paper would be read with Darwin's unpublished sketch (which he wrote to Asa Gray at Harvard College) at the meeting of the Linnean Society on July 1, 1958. "This was the dramatic prelude to the great intellectual storm which would shake the later half of the nineteenth Century" (Eiseley, 1979, p. 26).

In 1864, Wallace wrote a paper that presented the view that it was the evolution of the human brain by which *Homo sapiens* escaped the need for specialization of parts. While animals had to develop special parts in their bodies to deal with the changing environments and the demands of survival, *Homo sapiens* was able to address those external changes and demands by designing appropriate technology, such as making clothing and implements designed for various activities. For *Homo sapiens*, biological evolution was replaced by cultural evolution. Darwin strongly disagreed with Wallace's idea. But as we shall see, Wallace's idea prevailed. Darwin differed even more strongly with one of Wallace's later suggestions that a higher intelligence played a role in the evolution of humankind.

During the rest of the nineteenth century, numerous events extolled Darwin's achievement. Even though Wallace was present at most of these events not much mention was made of his contribution. Eiseley (1979, p. 14) remarked that "Moreover, Wallace, by nature modest and retiring, never thrust himself forward and his contributions have in some cases passed into the body of scientific thought without acknowledgment." Many who claim to be knowledgeable about evolutionary science do not even know his name. In reviewing the contemporary evolutionary literature one can find little mention of Wallace except in a line or two or a brief paragraph designating Wallace as the co-discoverer of the concept of natural selection. [In his more than 500-page treatment of evolution and the meaning of life, Dennett (1996) has a only a brief paragraph on Wallace's discovery.]

2.3.4 Herbert Spencer (1820–1903): The Pioneer of Social Evolution

Spencer, the leading philosopher of nineteenth century England, was also a first-rate natural and social scientist. He was the first scholar who developed a comprehensive conception of social evolution. Even though social scientists of the second part of the twentieth century practically dismissed his contribution, during his lifetime he was widely admired and his influence was overpowering. Darwin considered him his superior. He was a prolific writer. In what follows, I capture his key ideas for the obvious reason that he was the first evolutionary scholar who comprehensively addressed societal evolution, which is the topic of this work. (As we saw earlier, Darwin basically stayed away from societal evolution.)

2.3.4.1 Spencer's Key Evolutionary Ideas

Spencer suggested two fundamental evolutionary laws. One is the process of individuation: a move from indefinite homogeneity to a coherent heterogeneity. The second law defines how social entities constitute a cooperative system, when these entities are integrated and live by and for one another. He viewed society as an organism and suggested that social morality is an aspect of evolution.

His famous article, "The Development Hypothesis," was published in 1852. In it, he rejected special creation (creationism) as an explanation of the diversity of life and set forth the concept of organic evolution through successive modification. The first volume of his life's work: *Principles of Sociology*, was published in 1874 and revised in 1896. In the *Principles* he defined the process of general evolution as follows: "Evolution is a change from a state of relatively indefinite, incoherent, homogeneity to a state of relatively definite, coherent, heterogeneity" (Spencer, 1898, p. 353). Spencer (1874) was the first to set forth the idea of evolution as a cosmic process as he said:

> The advance from simple to the complex, through a process of successive differentiation, is seen alike in the earliest changes of the Universe to which we can reason our way back; And in the earliest changes we can inductively establish; it is seen in the geologic and climatic evolution of the Earth, and of every single organism on its surface; it is seen in the evolution of Humanity, whether contemplated in the civilized individual or in the aggregation of races; it is seen in the evolution of Society . . . and in the evolution of all the endless concrete and abstract products of human activity. (p. 465)

2.3.4.2 Spencer's Ideas on Social/Societal Issues

Spencer saw the emergence of social structures from the functions that a social system is expected to carry out. He held that it is necessary to understand how an organization originated and developed in order to understand the need served by it. His general law of organizations implied that difference in functions implies differentiation among parts (of the system) that perform the functions. Much of the *Principles* was devoted to evolutionary exploration of increased specialization of functions and the resulting differentiation of structures.

Coining the word "superorganic," Spencer referred to something beyond the biological. To Spencer the term meant "social." Spencer held that social behavior was dependent in part on the nature of the individual and in part by the environment, the external forces to which individuals are subjected. Later in his life, Spencer recognized culture as being a distinct phenomenon of reality. In the final issue of *Principles* (1896), commenting on societal evolution, he says: "As fast as societies become large and highly organized, they acquire such separation from individual efforts as to give them a character of their own" (pp. 13–14). Furthermore, Spencer saw the values and attitudes of a society as reflecting the society. "For every society, and for each stage of its evolution, there is an appropriate mode of feeling and thinking" (p. 356).

2.3.4.3 Spencer and "Social Darwinism"

Spencer was an early advocate of "Social Darwinism." The term refers to the application of natural selection and "survival of the fittest" in the social arena. It is not a single doctrine but several. One of its forms holds that the elimination of "unfit" individuals from the society will benefit the human species biologically. Thus the state should not try to relieve the condition of the "less fit," the poor and needy. The second form suggests that the economic system works best when individuals are allowed to pursue their own private interests. In this climate, only the fittest enterprises will survive. "This is the familiar doctrine of "laissez faire." Spencer was an advocate of this form of Social Darwinism. A third form, which operated on the societal level—which was set forth by Spencer—was the role of wars as means of natural selection. Societies best able to endure this mode will survive and flourish. While holding a Social-Darwinian position he also professed that the welfare of citizens cannot and should not be sacrificed to the benefit of the State. Furthermore, in a human society the corporate life must be subservient to the general good of the society, instead of the society being subservient to the corporate life.

2.3.4.4 Spencer Was an Early System Thinker

In *Principles of Sociology*, in a chapter titled "A Society Is an Organism," Spencer claimed that he effectively established the mutual dependence of parts that constitute the organization, which is to be considered a living whole. This "whole" undergoes continuous growth and as it does, its parts exhibit increased structure and differentiation. Their differences are so related that they mutually support and aid each other—in a mutually dependent relationship—as they live by and for each other.

2.3.4.5 Spencer on Societal Progress

In his closing chapter of *Principles*, Spencer provides a summary characterization of the progress of social evolution. The key markers of this characterization include integration, heterogeneity, coherence, multiformity, and definiteness. He suggests that societies progress toward greater size through integration. Societies change from homogeneity to heterogeneity, up from a simple tribe (whose members are alike) to civilized nations, with unlike structural and functional characteristics. Progressive integration and heterogeneity bring forth increased coherence. We can observe multiformity emerging through time. Wandering groups had no bonds to hold them together; tribes had social bonds that connected them into a whole. Tribes then clustered under chiefs and subchiefs, and this went on up to civilized nations that consolidated enough to write their history for a thousand years. Furthermore, there has been continuous progress toward definiteness. The definition of social organizations was first vague, but advancement brought more and more settled arrangements. Customs passed into laws that became more specific in their application to a variety of situations and activities as institutions emerged and developed with increasingly and distinctively different component structures. Thus we have witnessed in a grand scheme of things the continuous progress of social evolution; and increased complexity by integration; and increased heterogeneity, coherence, multiformity, and definiteness.

2.3.4.6 A Brief Appraisal

Spencer's scientific contribution to evolution was not made in the context of a single discipline. He applied the principles of evolution to science in general in a transdisciplinary way. Prior to Spencer, no other scholar, with the exception of Aristotle, had such broad scientific knowledge. Working from such a wide knowledge base, he created his evolutionary model as a rigorous holistic system.

Activity 2
Here, we explore statements of cultural and societal evolution, which is the focus of this work. Review the preceding text discussing the work of

Lamarck, Wallace, and Spencer. Identify core ideas that represent their views on cultural evolution. Enter those in your workbook for use later on as we discuss evolutionary consciousness and conscious evolution.

2.4 *Bergson and Teilhard*

I introduce Bergson and Teilhard in the same section because what Bergson proposed as the work of the inner force as the prime mover of evolution in a subtle and opaque way, Teilhard refined, developed further, and made transparent. Teilhard's thinking on this inner force "was very clear, and is as interesting today as when he first conceived it. Teilhard de Chardin, like Bergson, not only believed that *consciousness ascends to greater richness and depth as the physical organism achieves greater complexity, but that this was the very purpose of evolution*" (Combs, 1996, p. 85).

2.4.1 Henri Bergson (1859–1941): The Pioneer of Creative Evolution

The French philosopher Bergson held the view that ultimate reality is a "vital impulse" (*elan vital*) that can be captured only by intuition. The essence of reality is change and it cannot be known by reason. (This position placed Bergson squarely in opposition to the French positivist movement.) His principal work, *Creative Evolution* (1907/1983), sets forth his theory of evolution. He denied and replaced the evolutionary idea of persistence of forms through variation, with the force of the initial creative energy which is released in divergent directions. He could not accept the notion that random variation can lead to the emergence of such a highly complex organism as the human being. An evolutionary scholar noted that a chance evolution of an organism such as the human being would be as likely as a wind sweeping though a junk yard and assembling an airplane.

In his *Creative Evolution*, Bergson strongly objected to the mechanistic, mindless, purposeless conception of evolution. He held that evolution is a continuous creative process in which a vital impulse working from inside strives for achieving perfection and freedom. This vital impulse is the source of evolutionary change. Obtainable only through intuition, this vital impulse enables us to creatively guide evolution. The principal objection to Bergson's theory has been that it uses the idea of the inner force of the mysterious vital impulse and gives prominence to intuition over intellect.

He held that the purpose of evolution is to free human consciousness from the constraints of organic evolution. The key to such was the development of complex and flexible nervous systems, including a large and complex brain, which was attained by biological evolution, at which point consciousness has reached its guiding expression in human beings. Bergson believed that the human being is the highest manifestation of the evolutionary ascent. Further, the appearance of the human species is the reason for life on earth. Explaining Bergson's evolutionary idea, Allan Combs (1996) notes that Bergson's

> zoological position may seem offensive to many today, but we must realize hat such statements were very much in keeping with the heady intellectual atmosphere of pre-war Europe, one in which leading thinkers were intoxicated with the success of the industrial revolution and, along with it, the apparent supremacy of human reason over nature. At the same time, however, they were confronted with the Darwinian reality of human's intimate connection with the world of animals and nature. The result, in Bergson case, and many others, was an evolutionary philosophy that placed humankind at the pinnacle of nature. (p. 80)

Even though Bergson's evolutionary position was dismissed later, his way of thinking about evolution continued to influence evolutionary scholars, most notably his famous countryman, Teilhard.

2.4.2 Pierre Teilhard de Chardin (1881–1955): The Pioneer of Spiritual Evolution

The *Phenomenon of Man* is a remarkable work by a remarkable individual, says Julian Huxley in his introduction to Teilhard's major work (1955/59/65/75). (I use the 1975 edition as a source.) In his book, Teilhard developed a fourfold synthesis: synthesizing (1) biological evolution with cognitive and spiritual evolution, (2) the evolutionary past with the evolutionary future, (3) variety with unity, and (4) the many with the one in human evolution. He envisioned knowable reality not as static but as a process. He was seeking to understand human significance in relationship to such an enduring and comprehensive process. Huxley, appraising Teilhard's work, suggests that:

> Pierre Teilhard starts from the position that mankind in its totality is a phenomenon to be described and analyzed like any other phenomenon: in all its manifestations, including human history and human values. His second and perhaps most fundamental point is the absolute necessity of adopting an evolutionary point of view. The different branches of science combine to demonstrate that the universe in its entirety must be regarded as one gigantic process, a process of becoming, of attaining new levels of existence and organization, which can be properly called a genesis of evolution. For this reason, he uses the term of noogenesis, to mean the gradual evolution of mind and mental properties . . . Similarly, he likes to use a pregnant term like homanization, to denote the process by which the original proto-human stock became (and is

> still becoming) more truly human, the process by which potential man realized more and more of his possibilities. Indeed, he extends this evolutionary terminology by employing such terms as ultra-homanization to denote the deductible future stage of the process in which man will have so far transcended himself as to demand some new appellation. (Huxley, 1946, pp. 202–203)

2.4.2.1 Contemplating Some of Teilhard's Core Ideas

These core ideas are captured from Teilhard"s *Phenomenon of Man* (1975) and are offered for reflection and contemplation.

2.4.2.1.a Two Trends in Personal Evolution. Teilhard saw the evolution of personality emerging from the integration of two trends. One of the trends is a movement toward achieving and developing ever more individuation, increasing self-awareness, self-organization, and self-direction. The other trend is the attainment of ever more other-awareness, interrelation, cooperation, and a state of transcending individuality in order to engage in social participation.

2.4.2.1.b Evolving the Noosystem. While noogenesis is the evolutionary purpose of the individual, Teilhard saw the evolution of humankind as becoming a single noosystem developing a common and shared thinking. He considers the noosystem becoming the mind of the human community. This evolutionary idea of Teilhard implies that "we should consider interthinking of humanity as a new type of organism, whose destiny is to realize new possibilities for evolving on this planet" (1975, p. 20).

2.4.2.1.c Science and Spirituality. Teilhard, integrating his broad scientific knowledge and deep spiritual convictions and values, created a unique vision of human evolution. Exploring the relationship of science and spirituality, he held that no concept is more familiar to us than spirituality. At the same time, this concept is the one that is so opaque to science. The difficulty of holding together spirit and matter is harshly revealed in evolutionary scholarship. Yet, nowhere is the need more urgent to build a bridge—to close the gap—than between the material and the spiritual. Teilhard held that it is no longer possible to maintain science and religion apart, operating in tightly separated compartments or having the two concern separate sectors of life. Teilhard says that "they are both relevant to the whole of human existence. The religiously-minded can no longer turn their backs upon the natural world; nor can the materialistic-minded deny importance to spiritual experiences and religious feelings" (1975, p. 26). He persuaded theologians to embrace the new perspective of evolution and scientists to recognize the spiritual implications of their science.

2.4.2.1.d Wholeness by Convergence. Our field of vision—says Teilhard—should extend to the whole; "like the meridians—as they approach the poles—science, philosophy, and religion are bound to converge as they draw nearer to the whole" (p. 30). Without such convergence Teilhard did not see how is it possible to give a full and comprehensive account of the phenomenon of evolution and the phenomenon of man. From such convergence will emerge under our eyes "the organic whole in which no element can any longer be separated from those surrounding it" (p. 74).

2.4.2.1.e The Evolutionary Potential. "Nothing could burst forth as final across the different thresholds successively traversed by evolution (however critical that might be) which has not already existed in an obscure and primordial way" (p. 71.). This idea brings forth the notion of the enfolded and unfolding evolutionary potential. In the evolution of matter (geogenesis) over one billion plus years the potential of the *biogenesis* of life was enfolded. *Biogenesis* evolved through another three billion plus years, enfolding the potential of *noogenesis*. Then, noogenesis was unfolded into the evolution of human beings. And now we contemplate: What evolutionary potential is enfolded in noogenesis, and what evolutionary potential will unfold at the next stage of evolution? (Teilhard saw the unfolding of evolution toward the "Omega Point.")

2.4.2.1.f "Groping." "Evolution is an ever on-going and unfolding 'groping.' Groping (p. 110) is the fundamental technique of "all expanding multitudes." It combines the trust for large numbers with "the precise orientation of a specific target." It is not mere chance (as it was seen by Darwin and his followers). It is "directed chance." "It means pervading everything so as to try everything, and trying everything as to find everything." It is precisely this process that nature had to use in order to have recourse to profusion.

2.4.2.1.g "Hominization." (Term coined by Teilhard.) Some biologists and psychologists are still debating if the difference between humans and some animals is a difference in "degree" rather than in "kind." Since Darwin, much has been said about "continuity" and the intelligence of animals as a way of questioning the "evolutionary leap" of the human specie and the breach of continuity. Rather than entering this debate, Teilhard resolutely focused on the central phenomenon of "reflection." (Attaining "reflection" means the evolutionary event of the human species stepping through the threshold of reflective consciousness.) It was "the power acquired by consciousness to turn in upon itself, to take possession of itself; no longer merely to know, but know oneself, no longer to know, but to know that one knows "(p. 165). This continuously unfolding process is what Teilhard meant by the term "hominization." With hominization, in spite of the insignificance of the anatomical leap, we have the beginning of a new age. "The earth 'gets a new skin', Better still, it finds its soul" (pp. 182–183).

2.4.2.1.h Discontinuity in Continuity. The advocates of spiritual explanation of social and cultural evolution hold the notion of transcendence and the superiority of man over the rest of nature. On the other hand, the materialists profess that man is just one further "term" in the evolution of life forms. As in many other appositional perspectives, Teilhard sees a resolution of this dichotomy if we are willing to "emphasize the highly natural phenomenon of the change of state. From the cell to the thinking animal a single process goes on in the same direction, without interruption. But it is inevitable that certain leaps suddenly transform the subject of the operation. "Discontinuity in continuity: that is how the birth of thought, like that of life, presents itself and defines itself" (p. 169).

2.4.2.1.i The Noospere. In the course of evolution on this planet, geogenesis turned into biogenesis, which turned into psychogenesis, which led to man, engendering the development of mind—in one word: noogenesis. The first time a living being perceived itself in the mirror of its mind, our species took a giant step forward. "The biological change of state, terminating in the awakening of thought, does not represent merely a critical point that the individual or even a species must pass through. Vaster than that, it affects life itself and its organic totality, and consequently it marks a transformation affecting the state of the entire planet" (p. 181). The first sparkle of conscious reflection in man brought forth the evolutionary leap of noogensis, developing into a new layer: the thinking layer, which since its germination has spread over and above the world. "In other words, outside and above the biosphere there is a noonsphere" (p. 182).

2.4.2.1.j Unity of Movement. Through all times man has lived and experienced the rise and expansion of his consciousness. And we can ask, Where does this rise and expansion lead? "Man is not the center of the universe as once we thought in our simplicity, but much more wonderful—the arrow pointing the way to the final unification of the world in terms of life. Man alone constitutes the last-born, the freshest, the most complex, the most subtle of all successive layers of life. This is nothing else than the fundamental vision and I shall leave it at that" (p. 224). I leave the further characterization of Teilhard's evolutionary ideas at this point. We shall visit with Teilhard again in the next chapters.

Activity 3

This activity concludes our review of the work of the six "pioneers" of evolutionary scholarship. The activity has two tasks.

Task 1. Review the text that accounts for the works of Bergson and Teilhard, the two scholars who focused on the cognitive and spiritual aspects of the evolution of human consciousness. Write a brief essay that interprets their evolutionary ideas.

Task 2. Review the text that introduced the work of Lamarck, Darwin, Wallace, and Spencer. Select evolutionary ideas from these scholars that seem to "fit in" with the ideas of Bergson and Teilhard explicated in the essay you developed in Task 1. Rewrite the essay by including/ integrating the selected ideas of the four evolutionary scholars. Enter your findings in your workbook.

2.5 *Evolutionary Scholarship of the Last Several Decades*

In Section 2.4, we visited with the pioneers of evolutionary science who developed the conceptual foundations and the knowledge bases of evolutionary scholarship. Lamarck, Darwin, and Wallace presented to us the panorama of general—primarily biological—evolution. Spencer explored sociocultural evolution. Bergsen and Teilhard focused on social, cognitive, and spiritual evolution. In the section that follows, I review the work of several evolutionary scholars of recent decades who, in their scholarship, represent various orientations. In view of the overall theme of this book, I selected from their work ideas that portray sociocultural evolution. Furthermore, my selection is biased toward reviewing the work of those who represent a systems science orientation.

2.5.1 Vilmos Csanyi

In his *Evolutionary Systems and Society: A General Theory* (1989), Csanyi discusses culture and cultural evolution, the emergence of group society, technological evolution, and the emergence of the global system.

2.5.1.1 A Definition of Culture

The first comprehensive theory of cultural evolution was developed by Spencer, who defined sociocultural evolution as a continuation of biological evolution, calling it "superorganic" organization. Cultural evolution is a generally accepted designation in evolutionary scholarship, even though we do not have a commonly accepted definition for it. Csanyi, referring to Osgood's (1959) social, material, and mental categories of culture, suggests that *social* culture evolved from interrelationship between people, *material* culture is the production and use of artifacts, and *mental* culture comprises those ideas that are not parts of the two others. Studying cultural evolution, there is a tendency to seek the smallest "units" of culture, which we call" cultural traits," that form organically interconnected cultural complexes.

2.5.1.2 Capacity for Culture

Csanyi suggests that the evolution of cultural traits has been the consequence of the emergence of genes that were responsible for producing the "capacity of culture." This means that:

> culture is adaptive independently of its particular form, and genetic selection is required only to ensure the evolution of physiological mechanism; such as well developed learning ability, large memory, social attraction, communication, ability to imitate, etc., necessary to acquire culture. If these mechanisms exist, some kind of culture will develop that will be adaptive, organization-enhancing, with reproductive-success. This idea would also account for the great variety of cultures. The emerged culture would also influence the group culture and reproductive mechanism and feed back to the genes; that is a co-evolutionary process begins. In this co-evolutionary process, natural selection is replaced by cultural selection; that is individuals best adapted to the cultural environment will have the greatest reproductive success. (p. 153)

2.5.1.3 Cultural Machines

Csanyi finds Mumford's idea (1967) of cultural complexes as "cultural machines" compelling. These are "second-order organisms," assemblies of human individuals that are created by training and discipline. They are teams of experts that can perform tasks according to a plan, for example, organizations producing artifacts, the military, corporations, factories, and so forth. Csanyi differentiates two types: the *autopoetic* cultural machine, which produces itself, and the *allopoetic* machine, such as a traditional university in which students are produced "in a controlled and planned teaching process by which the most standardized products are favored" (p. 157). "Cultural machines are characterized as operating by idea schemes, working on prearranged tasks. Their 'parts,' humans, can be gradually replaced. A new 'part' can operate properly only after acquiring the idea network that characterizes the given machine" (p. 158).

2.5.1.4 The Emergence of Cultural Man

Often, the emergence of cultural man is attributed to a single function, for example, using language, making tools, using tools, and hunting in groups. Recent comparative evolutionary studies have shown that all these functions depend on one another. "They are components of a functionally organized adaptive co-evolutionary process in which they promote each other's development" (p. 170). Furthermore, it has become increasingly apparent that the extremely high intelligence of man had developed primarily through interaction within the social environment.

2.5.1.5 The Emergence of Group Society

The replication of cultural ideas is essential to the development and maintenance of shared culture. The precondition of this is "that the human population of a given culture should not be too large so that ideas can be spread by verbal communication without any constraint of time and space, and all members of the population should be able to participate in the formation of common culture" (p. 180). The result of identical replication by verbal communication is nothing other than a cultural group that can be regarded as "group society." If the afore mentioned condition is not fulfilled, the result will be separated and isolated compartments, from which new types of cultures and new and different group societies emerge. A further outcome of the culture forming evolutionary process described here is the linguistic and cultural isolation of a "group society," which created group consciousness. The groups of the greatest cultural cohesion and shared consciousness were able to conquer the best territories and defend themselves against others.

2.5.1.6 Technical Evolution

"The early phases of cultural and technical evolution coincide. In the course of cultural evolution a 'creative technique-space' evolves, in which an artifact producing mechanism is acting" (1989, p. 184). In the early phase of article production, objects are produced piece by piece. Their multiplication is slow and the accuracy of their replication is low. With the appearance of machinery, the identical replication of artifacts begins. Suddenly, the number of different articles and their identical replication increase. Now, "the technical evolution is presently in the process of separating from cultural evolution" (p. 185). This separation has three phases. The first is the design of the artifact itself. After it is created, the artifact is separated from the mental template of the design. The second phase starts with the design and development of the machines that can produce countless identical copies of the artifacts. In the third phase, the mental template can be completely eliminated by computerized production. "The replication of the artifact becomes a self-maintaining process, not requiring human participation" (p. 186).

2.5.1.7 Autogenesis of the Global Systems

This review of Csanyi's work on cultural evolution concludes with a quotation from one of his most visionary ideas: the evolution of the global system. "The highest organizational level of the evolutionary process on Earth is the global level" (p. 187). At this level exists a single entity, the "global biocultural system." Its components are the various human societies. Within the global system, open societies have allopoetic machines that are interconnected across societies. Their interconnection may continue to lead to the emergence of higher level organizations, attaining—eventually—an operative global system. But for this to happen, we need to create new regulatory mechanisms "that help to form the coordinate temporal replication of the 'whole global system" (p. 189). The formation of the global system has been recognized already. Global models have been created. Global problems have been identified and efforts to solve them are being initiated. But these problems "cannot be understood and solved within the framework of former social paradigms" (p. 189). Csanyi holds that the evoking of systems thinking and the development of a new systems theory will be necessary as a new paradigm for global application. Furthermore, the design and introduction of new learning content and learning strategies are inevitable if we are to transform the idea systems of present societies into the idea system of global humanity. (Such a new paradigm and the learning content and approach to "evolutionary learning" will be introduced in later chapters.)

2.5.2 *Stephen Jay Gould*

Here, I introduce Gould's idea system of cultural evolution from the last chapter of his book, *Full House* (1996). I selected five core themes that are relevant to the basic theme of this work.

2.5.2.1 The Two Walls of Evolution

To understand Gould's idea of cultural evolution, I must first introduce a core theme of his work, represented by the metaphor of "left and right walls" of the evolution of life. It is at the left wall where life emerged some three and one-half billion years ago. The right wall stands for the possible limits of how far evolution can go. Life at the left wall is of minimal possible complexity, represented by prokaryotes, loosely called bacteria. "More than half the history of life is a tale of bacteria only" (p. 170). Bacteria are still here with us unchanged, as it ever shall be. Space is available for new forms of life only away from the left wall, in the space not occupied by bacteria, and it moves toward the direction of ever increasing complexity. But as it moves, it runs into limits (right walls). As Goethe said: "It is ordained that trees cannot grow to heaven." Things created by nature—or artifacts created by man—have their limits of height. Based on this central theme of his book, Gould in his last chapter discusses the case for worrying about the right walls in the evolution of human life.

2.5.2.2 The Saga of Cultural Change through Time

In *Full House,* Gould discussed the issue of why the basic character of Darwinian evolution—which causes only local adaptation and not general progress—engenders only a passive trend toward increasing complexity in the form of a small right tail. "In this context, the issue of right walls hardly comes up—because they exist in some far uncharted distance, and life as a whole has not yet impacted by them—the tree cannot reach the heaven" (Gould, 1996).

> But human cultural change is an entirely distinct process operating under radically different principles that do allow for the strong possibility of a driven trend to what we may legitimately call "progress." In this sense, I deeply regret that common usage refers to the history of artifacts and social organizations as "cultural evolution." Using the same term—evolution—for both natural and cultural history obfuscates far more than enlightens. Unfortunately, when we speak of "cultural evolution," we unwittingly imply that the process shares essential similarity with natural or Darwinian change. The common designation of 'evolution' then leads to one of the most frequent and portentous errors in our analysis of human life and history—the overly reductionist assumptions that the Darwinian natural paradigm will fully encompass our social and technological history as well. (pp. 218–220)

Gould wishes that the term "cultural evolution" would drop from use and be replaced by a term like "cultural change."

2.5.2.3 The Difference between Darwinian and Cultural Evolution

The difference between Darwinian and cultural evolution is in the explosive and cumulative capacity of culture for directional and rapid change. "In an unmeasurable blink of geological eyelash, human cultural change has transformed the surface of the planet as no event of natural evolution could ever accomplish at Darwinian scales of myriad generations" (p. 220). The most impressive contrast between natural and cultural evolution is that we have no evidence of change in the modal form of human bodies and brains in the past 100,000 years, while during this time we experienced explosive cultural changes. "The Cro-Magnon people, who painted the caves of Laxaus and Altamira some fifteen thousand years ago, are 'us.' One look at the incredible richness and beauty of this work convinces us, in the most immediate and visceral way, that Picasso held no edge to mental sophistication over these ancestors. Anything we have accomplished since—from the origin of agriculture to the entire human civilization of our days—has been built upon the capacities of an unaltered brain" (p. 220).

2.5.2.4 Topology and Mechanism of Inheritance

The topology and mechanism of inheritance stand out among the many differences between natural and cultural evolution as the motor of cultural rapidity and directionality.

2.5.2.4.a Topology. In Darwinian evolution, once a species becomes separate from its ancestral line, it becomes unable to reproduce with members of any other species; it remains distinct for ever. It interacts with others in an ecological niche but it does not amalgamate with them. "Natural selection is a process of constant separation and distinctions" (p. 221). Any given culture—on the other hand—receives a powerful boost from amalgamating with other cultures. Gould gives us an example of a traveler, who observing how a wheel works in another country, brings the invention home and changes his local culture forever.

2.5.2.4.b Mechanism of Inheritance. The mechanism of inheritance in natural selection is indirect and inefficient. It is a "negative force that can make nothing by itself" (p. 221). It eliminates most variants, preserving those best adapting to changes in local environments. The mechanism of cultural change is inheritance of acquired characteristics—the Lamarckian notion. "Any cultural change acquired in a generation can be directly passed on to the next by what we call, in a most noble world, education" (p. 222). The cultural mechanism of inheritance gives cultural evolution a "directional and cumulative" character that no natural selection can provide. It accumulates favorable innovation by the amalgamation of cultural features. It allows a culture to choose from the most useful innovations of other cultures.

2.5.2.5 Back to the Right Wall

In his closing remarks, Gould brings back to us his core notion of the *right wall*. The question is: How far can we progress—how high can we go—before we run into a right wall of evolution? He says, "Cultural change that operates by mechanism that can validate a general and driven trend to technological progress—is so very different from the minor and passive trend that Darwinian processes permit in the realm of natural evolution. And once you start to operate by general and driven trends, you can move very deliberately and very fast. With directed motion of this sort, you start to running into right walls" (p. 224). While natural selection's slow motion is rarely bounded by being encountered by right walls, our cultural idea complexes and institutions are "frequently shaped and troubled by right walls" (p. 224).

Enlightened by Gould's idea system of cultural evolution, for me the question remains: Can our culture, our directional progress, our vision—the power of our consciousness and thought—can our imagination and creativity shape the right walls of sociocultural evolution? Developing an answer to this question is a central task of this work.

2.5.3 *Eric Chaisson*

In this section, I draw upon sections of three of Chaisson's works: *Cosmic Dawn* (1981), *The Life Era* (1987), and *Universe* (1988).

2.5.3.1 A Grand View

Evolutionary science stipulates that "steady change has been the hallmark for the development of all things" (1981, p. ix). We have now a good understanding how life came to be as the consequence of the evolution of matter. "There now seem to be a clear tread linking the evolution of simple atom into galaxies and stars, the evolution of stars into heavy elements, the evolution of those into molecular building blocks of life, of those molecules into life itself, of life into intelligence, and of intelligent life into cultured and technological civilization" (1981, p. ix).

2.5.3.2 Revisiting Complexity: Our Brain

Living beings are exposed to a tremendous amount of information through their senses. The extent to which they can process this information depends upon their "complexity." "Organisms manifest this complexity in terms of an exquisite piece of matter—the brain. The brain is the central clearinghouse of all behavior" (1987, p. 199). The human brain is the most complex matter known, the most versatile of all creations, the "ultimate example of the extent to which matter has evolved in the known Universe" (1981, p. 200). It is our brain that enables us to be creative, visionary, and adventurous. The brain is the organ of our mind, and it is the mind that manifests itself in thoughts, visions, creations: creating what we call culture.

2.5.3.3 The Evolution of Culture

It is hard to pinpoint the onset of time of use of technology by our ancestors. It was probably a million years ago. "The beginning of cultural, as opposed to utilitarian, activities may be that old, for brightly colored mineral pigments have been found alongside skeletons of the earliest of the true humans" (1981, p. 234). Changes were very slow at first and it was only within the past 100,000 years or so that they have markedly accelerated. It was only around the end of the Stone Age that we find more sophisticated technological advances, such as the discovery of the wheel some 50,000 years ago and the bow and arrow some 10,000 years ago. Deliberate burials some 70,000 years ago manifested cultural advances. As mentioned earlier, prehistoric art was created some 35,000 years ago. It was a uniquely human product of *Homo sapiens*. "Whatever truly made us humans involved the creative products of emotions and imagination—qualities virtually impossible to define scientifically" (p. 235).

2.5.3.4 Making Us Cultured

The development of language—meaningful communication—was a crucial factor in making us cultured. It enhanced group hunting and conveyed the skills of tool making and use. It enabled us to store experience in memory and pass it on from one generation to another. Small statuets and bone markings of about 50,000 years ago were not accidental or decorative: they were symbolic. Nearly six thousand years ago the Sumerians developed an intricate system of numerals and abstract symbols. Several thousands of their uncovered clay tables show a basic vocabulary of some fifteen hundred separate signs. These tables represent an advanced stage in the evolution of the art of writing. They also speak for the search for truth.

> The epitome of cultures is the search for truth, the need to understand ourselves and the world around us. The desire and the ability to undertake this search truly identify us as humans, distinguishing us from all known life

> forms. To be sure, understanding is the major goal of science, though science shares this goal with other disciplines. Religion, the arts, philosophy, among others, represent alternative efforts to appreciate who we are and how we came to be. For thousands of years, humans have recognized that the best way to dispel mystery is to understand it. (1981, p. 244)

Sumerian tablets of five thousand years ago attempt to explain how gods created humans to be their slaves. Their system provided for the needs of priestly households, it distinguished managers from the managed, and priests from plebeians. "Sumerian poetry clearly documents how Sumerian religion incorporated a systematic theology of human, worldly, and cosmic phenomena" (1981, p. 245).

2.5.3.5 Consciousness

But what is it that enables us—even drives us—to search for truth and ask fundamental questions of our existence? The answer, says Chaisson in *Universe* (1988), is consciousness, that part of human nature that permits us to wonder, to introspect, to abstract, to explain—the ability to step back, perceive the big question, and examine how our existence relates to the existence of all things. "How did consciousness emerge? When did humans become aware of themselves? Is consciousness a natural consequence of neurological evolution? Some researchers think so, but they cannot prove it. Others object, suggesting that some specific, perhaps unlikely mechanism is required for its development; they argue that the capacity for imagery and imagination represent something more than just a continued accumulation of neurons" (p. 526).

2.5.3.6 Evolution Guided by Evolutionary Ethics

I conclude my review of Eric Chaisson by capturing one of his core ideas of evolution. In his *Life Era* (1987) he says "If our species is to survive to enjoy the future, then we must make synonymous the words 'future' and 'ethical,' thus terming our next grand evolutionary epoch 'ethical evolution'" (p. 201). The escalating rate and intensity of change at which we evolve is our own making and it is dominated by technological changes that are often beyond our control as technological intelligence is increasingly outdistancing our sociocultural intelligence. Thus, our challenge is to guide social, technological, scientific, political, and economic changes and become "the agents of change." Once we understand this we can guide evolution so that it is beneficial to humankind. "Of all the implications for the Life Era concept, the most important to my view is that we, as the dominant species on Earth, develop—evolve if you will, and quickly too—a global culture. We need to identify and embrace a form of planetary ethics that will guide our attitude and behavior toward what is best for all humankind" (p. 176).

2.5.4 *Richard Dawkins*

Dawkins is the author of *The Selfish Gene* (1989). A Neo-Darwinian, he suggests that his book is not intended to be a general advocacy of Darwinism, but is an exploration of the biology of selfishness and altruism.

2.5.4.1 Memes: The New Replicators

Dawkins asks: Are there good reasons to suppose that the human species is unique? I believe the answer is yes, he says, and he continues: Most of what is unusual about our species can be summed up in one word: "culture" and cultural transmission. He suggests that cultural transmission is analogous to genetic transmission, in that it gives rise to a new form of evolution. "For an understanding of the evolution of modern man, we must begin by throwing out the gene as the sole basis of our idea of evolution" (p. 191). What is special about genes is that they are replicators in biological evolution.

A new kind of replicator emerged in evolution, "achieving evolutionary change at a rate that leaves the old gene panting far behind" (p. 192). Dawkins names this new duplicator as the unit of cultural transmission: "meme." (From the Greek root of *mimene* meaning imitation or related to memory from the French word *meme*.) Examples of memes include musical tunes, ideas of fashion, ideas of making instruments and building a structure, ceremonies and customs, ideas of technology and art, and the idea of God. Scientists pass on ideas to colleagues, in lectures to students, in articles. "Just as genes propagate themselves in the gene pool by leaping from body to body via sperms and eggs, so memes propagate themselves in the meme pool by leaping from brain to brain via a process which, in the broad sense, can be called imitation" (p. 192). "An idea meme might be defined as an entity that is capable of being transmitted from one brain to the other" (p. 196). In the evolution of our species, the gene-selected evolution, by making brains, provided the "evolutionary soup" in which the first memes arose. "Once self-copying memes had arisen, their much faster kind of evolution took off. We biologists have assimilated the idea of genetic evolution so deeply that we tend to forget it is only one of many possible evolutions" (p. 194).

2.5.4.2 Selfishness and Altruism

Dawkin's key argument is that we are created by our genes, and the dominant quality to be expected in a successful gene is a ruthless selfishness that rises in the behavior of the individual. Dawkin's own feeling is that "a human society based simply on the gene's law of universal ruthless selfishness would be a very nasty society in which to live" (p. 3). He also says that if we wish to have a society in which people cooperate unselfishly toward a common goal, we should understand what our "selfish genes are up to" so that we might have a chance to upset their design. This means that we are to develop a culture in which we have altruistic idea systems—a system of memes, and we can teach and learn to become unselfish in the service of the common good. Dawkin also notes that it is a very common fallacy to suppose that genetically inherited traits are fixed and unmodifiable. Genes may instruct us to be selfish, but we are not compelled to obey them. Of course, it may be less difficult to be altruistic if we were genetically programmed to be altruistic. "Some say that culture is so important that genes, whether selfish or not, are virtually irrelevant to the understanding of human nature. It all depends on where you stand in the debate over nature versus nurture, as a determinant of human attributes" (p. 3). Dawkin closes his discussion on "memes: the new replicators" by suggesting that the capacity for genuine altruism may be another of our qualities. We can defy the selfish genes as well as the selfish memes of indoctrination. "We alone on earth can rebel against the tyranny of the selfish replicators" (p. 201).

2.5.5 Daniel C. Dennett

I summarize here some of Dennett's ideas from his recent book, *Darwin's Dangerous Idea: Evolution and the Meaning of Life* (1996). I start with his chapter on "The Cranes of Culture" that comments on "memes." One of his metaphors differentiates between cranes that "lift" and skyhooks that "pull." I use a delightful quote on this from his book: "Reductionists think everything in nature cane explained without skyhooks; greedy reductionists think it can all be explained without cranes" (p. 394).

2.5.5.1 The Crane and the Meme of Culture

Human beings are vastly different from all other species: Our culture makes us different. "The primary difference between our species and all others is our reliance on the cultural transmission of information, and hence cultural evolution" (p. 331). Using the crane metaphor Dennett says that human culture is not just a crane made of cranes, but a "crane making crane." Cultural evolution made our species special; it works many orders of magnitude faster than genetic evolution. "Anyone who worries about 'genetic determinism' should be reminded that virtually all the differences between the people, of say, Plato's day and the people living today—their physical talents, proclivities, attitudes, prospects—must be due to cultural changes" (p. 338). Following Dawkin's designation, Dennett refers to the unit of cultural evolution as "meme," which, he says, has a powerful role to play in the analysis of the human sphere. Our minds are "meme nests," in which memes interact with—and became adjusted to—each other, and swiftly change to fit circumstances. As the new meme emerges it becomes replicated as the mind broadcasts the new meme. The result of the interaction is adjustment-based synthesis.

Dennett closes his discussion of memes by saying that "The invasion of human brains by culture, in the form of memes, has created human minds, which alone among animal minds can conceive of things distant in the future and formulate alternative goals. It is the shaping of our minds by memes that gives us the autonomy to transcend our selfish genes" (p. 369).

2.5.5.2 The Origin of Morality

Dennett asks: What brought morality—something altogether so new—on the face of the Earth? Answering the question, Dennett tells the story that took us from a time when there was no right or wrong to the time when the ethical perspective emerged. "Once upon a time there was no morality at all," tells the story (p. 454). There were human beings, they had language and memes for good bows and bad ones, good hunting and bad, but no concept of good—moral person or moral act. "They had no concept of right or wrong because these are qualities that relate to men in Society not in solitude" (p. 454). Life was raw competition, full of conflicts, brutish, nasty, and short. Then, one day—says the story—when another conflict arose, instead of persisting in selfishness and distrust, "These particular lucky competitors hit upon a new idea of cooperation for mutual benefit. They formed a 'social contract.' Whereas before there had been families, or herds, or tribes, this was the birth of a different kind of group, a society. This was the birth of civilization. And the rest, as one says, is history" (p. 454).

In life on Earth, the step taken by human beings from persons without morality to responsible citizens was a giant one indeed. Morality was "an emergent product of a major innovation. It has been achieved by just one species, *Homo sapiens*" (p. 455). Dennett closes his discussion on the origin of morality by saying: "We, unlike the cells that compose us, are not ballistic trajectories, we are guided missiles, capable of altering course at any point, abandoning goals. For us, it is always decision time, and because we live in a world of memes, no consideration is alien to us, or a foregone conclusion" (p. 460).

2.5.6 *Sir Julian Huxley*

Sir Julian authored over a dozen books, including *Essays of a Humanist* (1946), which is reviewed here as his contribution to understanding evolution. My summary of the review of his work discusses his ideas on what he called "psychosocial" evolution and his notion of the "humanist frame" of evolution.

2.5.6.1 Psychosocial Evolution

"Evolution in the most general terms is a natural process of irreversible change, which generates novelty, variety, and increase in organization, and all reality can be regarded in one aspect as evolution" (p. 29). The total process of evolution has an inorganic or cosmic phase, a biological, and a post-biological, human phase. "In the broadest terms, the biological phase of evolution stems from the new invention of self-producing matter; the human phase from that of self-producing mind" (p. 32). Each phase proceeds with its own tempo, has its own characteristics and method of operation, and possesses its own possibilities and results. According to Huxley, the psychosocial phase has four subphases. The first generates variety within and between cultures. The second produces improvement in technological methods; in economic, political, education systems; in creative expressions; in science; in morality; and in social and international organizations. The third limits progress and preserves old systems, attitudes, and thinking. The fourth diffuses ideas and techniques between individuals, communities, and cultures. "This is tending to give unity to the world; but we must see to it that it does not also impose uniformity and destroy desirable variety" (p. 32). Referring to our role in evolution, Huxley suggests that "man's true destiny emerges in a startling new form. It is to be the chief agent for the future of evolution on this planet. Only in and through man can any further major advance be achieved" (p. 32). Huxley's conclusion confirms—as the conclusions of authors reviewed earlier—the major theme of this work: that it is our role to guide evolution.

2.5.6.2 The Humanist Frame of Evolution

Huxley suggests that today we must embark on the psychosocial stage of evolution to create a major advance, a radical change in the dominant idea systems. If our situation is not to lead to chaos, despair, or escapism, we must reunify our life within the framework of a new idea system. This idea system—whose birth we are witnessing—is *humanism*, says Huxley. And this radical change process toward the new idea system of humanism is *evolutionary humanism*. It is unitary rather than dualistic, it unifies mind and body, it affirms man's continuity with life, and life with the rest of the universe. It seeks the unity of the spiritual with the material and seeks the unity of mankind. It thinks of directional processes rather than static mechanisms. It affirms that knowledge and understanding can be advanced. It holds that social conduct and society can be improved, and that more desirable directions for individual and social development can be envisioned and implemented. We shall visit with Huxley again when we explore the agenda of evolutionary learning.

2.5.7 Allan Combs

Here I review Combs's recent book, *The Radiance of Being: Complexity, Chaos and the Evolution of Consciousness* (1996).

2.5.7.1 Perspectives on Evolution

Combs (1996) sees evolution in three distinct ways: (1) Biological evolution projects a gradual change and diversification of species. The principles and processes of biological evolution have been subject to intensive discussion among evolutionary scholars since Darwin. The contemporary view emphasizes the principles of genetics and the process of how population genetics change over time. (2) Historical evolution implies the idea of growth, maturation, and improvement over time. From the point of view of consciousness—which is Combs' main interest—evolution suggests "a progressive development of the inner person towards greater freedom, self-expression, and in particular toward a natural and harmonious way of being in the cosmos" (p. 183). This view of evolution is one that is embraced by the "wisdom traditions." (3) The third, the "grand evolutionary theory," emerged from systems thinking and the sciences of complexity. The theory accounts for the self-organizing and self-creating properties of complex system. It encompasses the entire range of evolution, from matter to life, from the origin of life upward to include "increasingly complex hierarchical systems all the way through natural ecologies and human societies, including multi-national political and economic structure" (pp. 182–183).

Combs—reflecting upon the self-creating properties of systems—brings forth the evolutionary limits of variability, within which self-creation may take place. He refers to the work of Goodwin (1994), who suggested that the variety in evolution, while rich, is limited in a sense because only certain patterns of life forms are viable, and even among those, some are more viable than others. Goodwin shows that "of the two hundred fifty thousand species of higher plants, around the stems, only three basic distribution of leaves are seen and all but twenty percent of these are of spiral forms. The same limitation is manifested in the formation of bones. These likenesses are not just the results of genetics but they represent basic patterns in the growth process of life forms, implying that only certain life forms are viable" (p. 148). Goodwin's idea demonstrates, says Combs, that nature does not take an infinite number of forms, but assumes only certain stable patters, or attractors. We can contemplate: What could be the implications of the notion of "viable patterns" in cultural and societal evolution?

2.5.7.2 On the Evolution of Consciousness

Earlier, I discussed the evolutionary theories of Bergson and Teilhard, the two pioneers of spiritual evolution and the evolution of consciousness. Building upon their work, says Combs, Jean Gebser, a cultural philosopher, unfolded a broad vision of the evolution of consciousness, which he called integral consciousness. In his exploration of the historical evolution of consciousness, Gebser discovered "several historically older structures of consciousness, forming an evolutionary progression down to the consciousness of modern human kind" (Combs, 1996, p. 89). Combs, building upon Gebser's work (1949, 1986), and using his structures of consciousness, further develops the ideas of spiritual evolution and the evolution of consciousness. The five major structures—archaic, magical, mythical, mental, and integral—represent successively dominant, but overlapping patterns of experience during human history. Each structure represents a unique view of the world, a change in the perception of space and time.

2.5.7.2.a The Archaic Structure. The archaic structure is essentially prehuman. It is the mythological state of "life in the Garden of Eden." It was a time when our humanoid ancestors, living in prehistory as protohumans, lived in a unity with nature, and did not have recognizable human culture. They lived in a "state of deep dreamless sleep." Their perception was zero-dimensional, a state of being completely embedded in nature. These prehuman ancestors could include *Australopithecus*, who lived on the African savanna from five to one million years ago; *Homo habilis* dating from three to one and one-half million years ago; and *Homo erectus*, who lived from two million to seventy-five thousand years ago. *Homo erectus*, however, might represent a transition to the magical structure.

2.5.7.2.b The Magical Structure. We relate to the world in terms of logic; the ancient human related to it through magic. In the magical state we find our ancestors already in a human mode of consciousness, separated from nature, living the one dimension of "a point" but not yet in the dimension of time and place. They did not yet have complete identity. For them, through magic, one thing could be substituted for another, or a part of a thing for the whole.

Feuerstein (1987) defined five principal features of magic consciousness: egolessness, identity with the group, one-dimensional perception (space–time has not unfolded in consciousness), life richly interwoven with nature, and magic as the main vehicle of asserting oneself. The Neanderthal man already lived magical consciousness. But it was the Cro-Magnon, the first *Homo sapiens sapiens*, who brought magical consciousness into full bloom and then transformed it to the mythical consciousness. (This shows an example of "overlap" between structures of consciousness.) Coming to the scene between forty and seventy thousand years ago, the Cro-Magnon man was definitely our direct ancestor. He was human and possessed language.

2.5.7.2.c The Mythical Structure. The mythical structure brought us the experience of the dimension of time—in addition to the "point" dimension—making consciousness two dimensional. *Homo sapiens sapiens* (HSS) experienced living in a certain time. Mythic stories are temporal, for example, "once upon a time." While emotion is the source of the magical, imagination propels the mythical. We find the work of imagination already in Neanderthal men, who buried their dead in a birth position, imagining launching them into another world. But it was the Cro-Magnon man in whom mythical consciousness came to life in the fullest. Next to the imagination of the artist who painted those beautiful cave pictures, the Cro-Magnon carved feminine figures, indications of the symbolic worshiping of the feminine. "The full sweep of mythic imagination, and along with it the entire mythic structure of consciousness, did not break free of the older structures and come to its own, however, until the advent of the neolithic farming revolution" (Combs, 1996, p. 104), some time around 10,000 B.C. This mythic experience carried over into the ancient civilizations.

> The profound richness of the human imagination derives from the outward projection of imagery of the deep psyche, that is, from the human soul. In mythic consciousness humankind discovers the inner wealth of its own depths. Imagination, expressing itself through myth, renders the soul visible so that it may be visualized, represented, heard, and made audible. (p. 106)

In addition, rich mythical imagination became the "internal combustion engine" of the rapid acceleration of technology and artistic expression.

2.5.7.2.d The Mental Structure of Consciousness. The mental structure of consciousness came to full fruition in the classical Greek culture through the work of Parmenides, who said: "For thinking and being is one of the same" and then, through the ideas of Aristotle, Plato, and Socrates, who equated the soul and afterlife with pure thought. The roots of mental structure go back to earlier times as it overlapped with the preceding structure of mythic consciousness. During the late Roman empire, the manifestation of the mental structure of consciousness became clear in "ego" dominance. The "ego became established as the self-reflective center of the inner life" (Combs, 1996, p. 108). The writings of Plotinus represent the continuing ascent of self-reflective awareness and, later, St. Augustine's *Confessions* became a watershed in self-reflection. During the Renaissance, the emergence of the spatial perspective rounded out the three-dimensional world view. It became visible in Renaissance art as well as in Renaissance philosophy, mathematics, and literature.

Objectivity took on a new meaning as the ego distanced itself from the rest of the world. Objectivity was essential to analytical thought and it was the core idea of reductionist Newtonian science and the Cartesian world view. The mental structure of consciousness seeks efficient forms, reason, cause and effect, and "ratio" or rationality. It "places the sense of self in objective space somewhere in the head, as contrasted with the experience of the ancient Greek and Native Americans who would point to the heart. The heart is the source of the experienced soul, as the head is the source of ratio" (Combs, 1996, p. 110). "Ego" fixation led to isolation that was reflected in many forms: in the isolation of the individual, ideological monopolism, lack of extended relationships, lack of experiencing wholeness, and atomization of forms and experience. Given the effect of all these on the human condition: "Is it any wonder that Gebser saw hope for the future only in a large-scale shift to a new and more holistic structure of experience, the Integral Consciousness" (Combs, 1996, p. 111)?

2.5.7.2.e The Integral Structure of Consciousness. In our present era, the integral structure of consciousness is in the process of "becoming." It has various manifestations in the human experience, while the mental structure still dominates. Each structure of consciousness has its own kind of experience of time. In magical consciousness time is experienced vaguely as the present, in the mythical as temporacity. In the mental, time is an abstract quantity. In integral consciousness, the experience of time becomes a concrete, "living in the manifest world of the present," when various viewpoints are integrated and become a whole. There is spirituality in integral consciousness. Its transparent quality "is suffused with the light of the spirit, the animating force of the origin, and to a greater degree than any dominant structure since the archaic consciousness, but here with a solid clarity previously absent" (Combs, 1996, p. 113). The new human order would unfold as a work of the integral structure of consciousness. It comes as an ideological ferment, a threat to the conservative world of the mental consciousness structure.

Integral consciousness offers a "delightful play of creativity." This playful creativity becomes dominant in a range of endeavors, in new physics, in social and technological design, and in art and literature. As Combs (1996) says, the:

> emerging integral consciousness may become an important player in the world culture of the twenty-first century. Like all chaotic attractors its effects are difficult to foresee in detail. We can anticipate, however, that short of a major regressive trend, opinions and perspectives in all realms of culture will become more fluid and playful, while conservative perspectival factions resist these trends as if their very existence were threatened to its roots, as indeed it is. (p. 120)

I close this discussion of the structure of consciousness with Feuerstein's (1987) observation. The emerging world view of integral consciousness cannot be framed by the mythical world view; it cannot be grasped by the mental conception of consciousness. But it can be made understandable to us by direct participation in it. Such participation renders the self and world transparent so that the spiritual foundation becomes obvious.

2.5.8 Ervin Laszlo

I introduce below Laszlo's key ideas on the emergence of a new evolutionary paradigm and on sociocultural evolution.

2.5.8.1 The Rise of a New Evolutionary Paradigm

Laszlo (1996) sets the stage for the introduction of his general theory of evolution by reviewing the fate of scientific paradigms and the work of science in general. Scientific paradigms come and go. In its prime, each paradigm is a triumphant success; in its decay it is an obstructive nuisance. "Darwin's synthetic theory remained an obstruction in the way of the newer, more adequate theories of species evolution in the 1970s, and the positivistic conception of history remains an obstacle in the way of sustained attempts to inquire into patterns in the evolution of human societies to this day" (Laszlo, 1996, p. 11). He suggests that after a long and sustained debate on evolution, by now science has sufficiently advanced to demonstrate evolution's consistency in all realms. We can now set forth an evolutionary theory based on scientific concepts that are mutually consistent. It is now possible to establish the foundations for a transdisciplinary program by "assembling and analyzing the basic concepts of the contemporary sciences of evolution" (p. 21). This inquiry needs to pay special attention to recently emerged fields that are making significant contributions to evolutionary scholarship. These include the general theory of systems, cybernetics, nonequilibrium thermodynamics, autopoetic systems theory, a general theory of dynamic systems, and chaos theory. All these—collectively called the sciences of complexity—are considered by Laszlo as essential knowledge bases in formulating a new evolutionary paradigm.

Laszlo's basic proposition for this new paradigm challenges the notions of equilibrium and determinacy. The scientific laws conceptualized for the new paradigm are neither deterministic nor prescriptive; they posit "ensembles of possibilities within which evolutionary processes can enfold" (p. 23). Dynamic systems do not have determined individual evolutionary paths but bundles of possible paths. From identical initial conditions, different systems configurations can emerge within the limits of evolutionary laws and—within a specific setting—based on the skills and dispositions of the "players." "Evolution is always a possibility and never a destiny. Its course is logical and comprehensible, but it is not predetermined and thus not predictable" (p. 23). I close with a quotation from Laszlo (1996):

> The exploration of the paradigm and the creation, criticism and elaboration of progressively more refined general theories is a challenge awaiting the contemporary community of natural, human and social scientist. It is a challenge well worth accepting. The elaboration of a sound general evolutionary theory will surely rank among the greatest achievements of human intellectual history. (p. 50)

What follows is a continuation of Laszlo's propositions relevant to cultural and societal evolution. These propositions also reflect his core ideas about a new evolutionary paradigm.

2.5.8.2 Societal Evolution

Laszlo (1987) also discusses societal evolution, its direction, the role of technology, and the mastery of evolution.

2.5.8.2.a Contemplating Directionality. Laszlo suggests that it seems that there is no underlying pattern to societal evolution. Yet, if we look at a large enough cross-section of societal evolution, we can perceive basic patterns "that are consistent with the general direction and the dominant dynamics of the evolution of life in the biosphere" (p. 72). Stone Age tribes and modern nation-states stand for two temporal poles of societal evolution and "In the intervening epochs, despite temporary regressions and dead-ends, structural complexities have grown and the dynamism and autonomy of societies have increased" (p. 73).

2.5.8.2.b The Role of Technology. Laszlo holds that technology has been the engine of evolutionary change. Broadly conceived, technology is an instrumentality that is reflected in all human activities as it extends our physical and mental powers. Technological innovations stretch the imagination, and challenge people's values and practices. "The impact of technology on society is generally proportionate to the flexibility of the dominant modes of social organization, that is, to the ability of the society to apply internally or externally generated innovations" (1987, p. 73). Societal evolution is best conceptualized, says Laszlo, in terms of the transformation of the dominant technology and the institutional changes that are catalyzed by it. "Technological societies set forth the evolutionary progression toward more dynamic and autonomous systems by exploiting increasingly abundant, increasingly dense energy fluxes through correspondingly more complex societal structures" (p. 81).

2.5.8.2.c The Mastery of Societal Evolution. It seems that technological changes can release such quantities of energy that can inflict great harm to the human population and its life-sustaining environment. "Can *Homo* survive technological societies?" ask Laszlo. Can *Homo* "master the evolutionary process of high energy technological societies by purposeful action, based on sound knowledge of socio-developmental dynamics? Will *Homo* make use of his capacities for knowledge and action to steer the evolution of technological within the limits of survival?"

> The evolution of our societies, and therewith the future of our species, is in our hands. If we make use of our capacity to think rationally and act purposefully, we can defuse immediate treats and promote the creation of more mature, more autonomous, more dynamically stable societies. Only by becoming conscious of evolution can we make evolution conscious and only by conscious evolution can we have the assurance that we can survive the awesome combination of highly evolved order and complexity in our brain and immature and underdeveloped order and complexity in our societies. (p. 102)

Rephrasing Hamlet's question, Laszlo holds that the issue is not to evolve or not to evolve but "whether to evolve with distinction or devolve to extinction" (p. 126). Evolution does not let us stay fixed in our tracks, nor does it permit us to reverse our steps. It allows us to become the best we can. "Know thyself" was the challenge of the oracle at ancient Delphi. "Know how to evolve is the challenge we face today" (p. 122). This quote points to the thrust of this work: guided societal evolution.

Activity 4

There are two tasks in the final activity of this chapter.

Task 1. Review the statements of the ten evolutionary scholars and, author by author, identify and record their core ideas relevant to sociocultural evolution.

Task 2. Compare and contrast the recorded core ideas and see if there is internal consistency or contradiction among them. Record your findings. Your findings will help in addressing the issues in the chapters that follow.

Reflections

This chapter mapped out the historical landscape of the evolution of evolutionary ideas. We traced back the emergence of evolutionary ideas to the scholars of ancient Greece and continued to visit with scholars and philosophers through the next two millennia. We stopped at the nineteenth century, which was the watershed era of evolutionary science. It was during this century that the great pioneers of evolutionary scholarship lived and worked. They elevated the idea system of evolution to the level of true science. Then, we went on and witnessed the explosion of evolutionary science in the twentieth century.

In the course of working with this chapter, you were challenged to capture the core ideas of evolution and create your own understanding of general evolution and specifically cultural and societal evolution. Working with the text and the activities has prepared you to enter into Chapter 3, and join in reconstructing the story of human and societal evolution. The evolutionary ideas you encountered and interpreted in this chapter will help in developing that story.

3

The Story of Human and Societal Evolution: The Evolutionary Journey of Our Ancestors

In the previous chapter we traced the development of the evolutionary idea through the ages. Over a period of two millennia, starting with the Greek classical movement, we visited with philosophers and scientists who contemplated evolution in general and human and societal evolution in particular. We heard a great variety of ideas, most of them speculations but some providing rare insights. It was not until the nineteenth century that our journey led to a great discovery: evolutionary science. The lifelong work of Lamarck, Darwin, Wallace, and Spencer drastically changed the evolutionary idea landscape from one of hazy features to a clearly formed scholarly endeavor. From the work of these pioneers, the domain of evolutionary science exploded into a cross-disciplinary science, the literature of which takes up whole sections in libraries.

Probably no other domain of disciplined inquiry invites the interest of such a wide range of disciplines than does evolutionary scholarship. Today, evolutionary science is attended by biologists; paleontologists; sociobiologists; philosophers; paleoanthropologists; archaeologists; cultural anthropologists; social psychologists; systems scientists; chaos theorists; evolutionary psychologists; ecologists; and theorists of consciousness, art, and creativity. As we continue, we shall put the evolutionary ideas, introduced in Chapter 2, into the functional prehistorical and historical contexts of the evolutionary journey of the humanoid and human species. This journey is mapped in the section that follows.

3.1 *The Map of the Journey*

Our evolutionary journey is marked by several milestones, each representing a major evolutionary transformation. The first phase—reported in this chapter—has two legs, each marked by a milestone. The first leg covers several million years of the evolutionary saga of the "humanoid" species and marks the transformation from the origin—through the life span of three humanoid species—to the time of the migration of *Homo erectus* out of Africa to the Old World and the emergence of the Archaic *H. sapiens* and the Neanderthals. When we complete this part of our journey, we reach the first milestone which marks the "human revolution," as we witness a revolutionary transformation, the emergence of *Homo sapiens*—represented by the Cro-Magnons.

In the present chapter, traveling through the prehistoric period of human evolution, we tell the story of our ancestors. Their history was brought to light by paleontologists and archeologists, who used fossil evidence, analyzed remains on campsites, and examined climatic and vegetation evidences. Then, in Chapter 4, I give an account of the last three miles of the evolutionary journey as first we meet modern humans—the first generation of *Homo sapiens sapiens*, called Cro-Magnons—who travel with us and tell their story. As we leave the second milestone, we visit with agricultural *Homo sapiens* and people of the ancient civilizations. Then we meet the classical Greek philosophers, and walk through the Middle Ages. As we pass by the third milestone, we reach the Renaissance period, which marks the onset of the Scientific Revolution, where we meet "Homo Technologica," who leads us through the Industrial Revolution. Finally, we arrive at the post-industrial information/knowledge age: our current era. Here we stop for a while, and as we reflect upon the unmarked road of the next leg of our evolutionary journey, we ask: *Quo vadis hominem?* (Where do we go from here?) Answering this ancient question is the task of this work.

Figure 3.1 presents a map of the journey on which are marked the milestones that designate the various legs of the journey. The critical juncture of the journey is shown on the figure at the point where biological evolution is completed and cultural evolution becomes the organizing force in the evolutionary design space.

3.2 *The Story of Our Ancestors*

As we travel through the first mile of our journey, we are guided by scholars of human evolution who devoted their lives to searching for

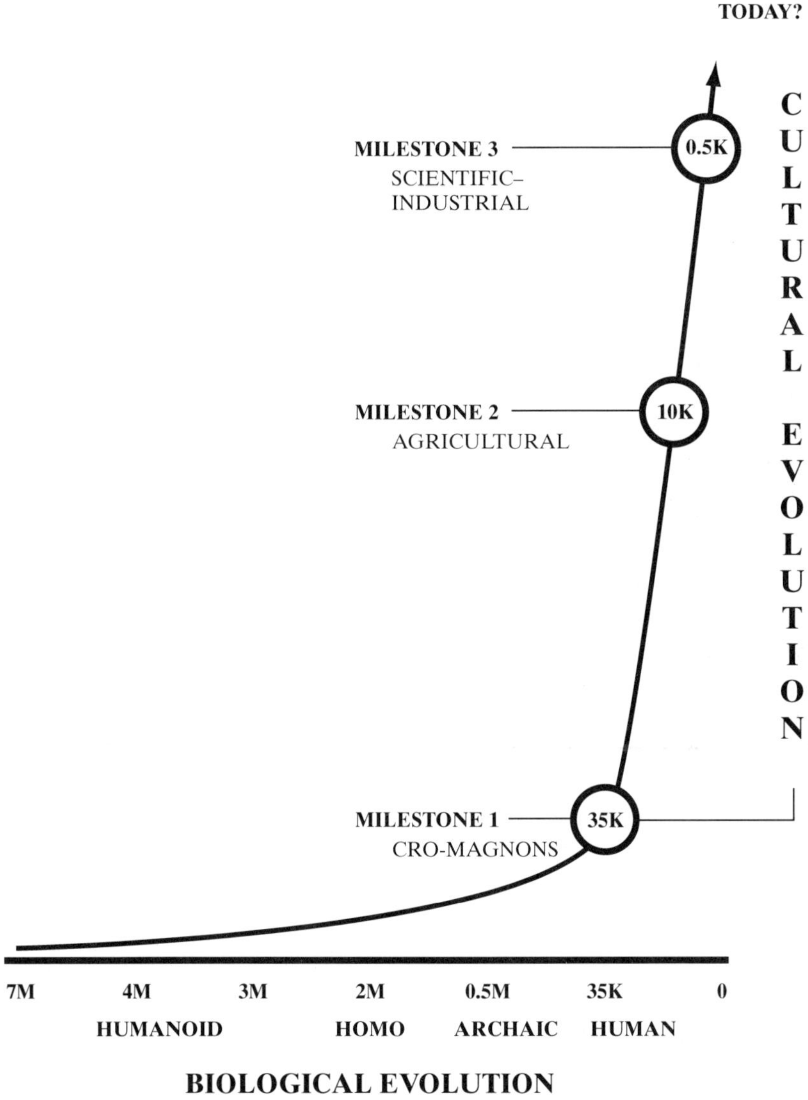

FIGURE 3.1 A Map of the Journey

evidence that tells the story of the what, when, where, and why of the evolution of our ancestors. The storytellers are palaeontologists, paleoanthropologists, biologists, archaeologists, and cultural anthropologists. They are discoverers and interpreters of early skeletal humanoid remains, tools, and home bases of our ancestors.

Commenting on the role of evolutionary scholars as "prehistory detectives," the Johansons (1994) say that after several "evolutionary experiments" of the humanoid species, in the course of the last five million years, we are the only humanoids alive today. "We live in the present but look constantly to the future and the past. We want to understand our beginnings, how we came to be and what we are. To this end, paleoanthropologists and archeologists are prehistory detectives piecing together a complicated puzzle" (p. 42). In search of understanding our ancient past, we ask questions such as: How and when did we depart from our primate ancestors and became humanoids, *Homos* species, and, then, humans? What kind of life did the earliest humanoids live? Where did they live and how did they survive? When, where, how, and why did *Homo sapiens* emerge?

We contemplate these questions by constructing a model of how to look at the evolutionary story of the first six plus million years of the humanoid species. We shall review the evolutionary transformation from primates to humanoids. Then, we listen to the story of Lucy and her family. Next, we listen to the story of "the handyman:" *Homo habilis*, the story of the Turkana Boy and his species: *Homo erectus*, and the stories of the Archaic *sapiens* and the Neanderthals.

3.2.1 The Model: Ladder or Tree?

The "ladder " and "tree" are metaphors for a fundamental issue of evolutionary theory. Was the evolution of the humanoid species a sequential ascension on the rungs of a ladder? Or can we model it as a tree in which branches emerge from a trunk? [By "ladder" Gould (1977) refers to the popular picture of evolution as a continuous sequence of ancestors and descendants.] Following an extended conversation, today paleontologists and archeologists converge on the "theory of the tree." It seems the more fossil remains are uncovered the more support is generated for the "tree model."

There are three evolutionary theories that elaborate the "tree model:" "speciation, "radiation," and "pruning." Gould (1977) explains the speciation theory of the tree model—which he calls the "bush" model:

> Evolution usually proceeds by "speciation"—the splitting of one lineage from a parental stock—not by slow and steady transformation of these large parental stocks. Repeated episodes of speciation produce a bush. Evolution-

> ary "sequences" are not rungs on a ladder, but the retrospective reconstructions of a circuitous path running like a labyrinth, branch to branch, from the base of the bush to a lineage now surviving at its top. (p. 61)

But how does this speciation occur? This issue is the topic of another ongoing conversation in the evolutionary scholarship community. By now, nearly all scholars agree on what is called "allopatric" theory, meaning "in another place," which was popularized by Mayr (1942). He held that new species emerge in small populations that become isolated from their parental group at the "periphery of the ancestral range" where very rapid evolutionary change—major genetic reorganization—occurs over hundreds of thousands of years—a microsecond on the evolutionary time scale. (Later when cultural change dominates, during the second, third, and fourth miles of the evolutionary journey, the time scale of major cultural reorganizations shrinks to tens of thousands, even hundreds, of years.)

The second theory of the tree model is called "adaptive radiation," which is defined as the norm in evolution (Leakey and Lewin, 1993). "A species evolves a novel adaptation—in the case of humanoids it was upright walking—and effectively becomes the founding member of a new evolutionary experiment. Very soon new species arise from this founding member, and a cluster of new descendant species appear through time. Evolutionary radiation enables the variability of living forms by abandoning existing patterns and opening up the evolutionary design space for the emergence of multiple changes"(p. 109).

The third theory is "pruning," which is a component of both speciation and adaptive radiation. It means that when some descendant variations of a species disappear, they are pruned off the tree or bush. That is precisely what happened several times in the evolutionary saga of our ancestors. The last pruning happened around 30,000 to 25,000 years ago, when the Neanderthals were pruned off, leaving modern man, *Homo sapiens sapiens*, the only lineage left of our species.

Today there is no consensus among paleontologists about exactly how the tree or bush looked through the evolutionary journey. Leakey and Lewin (1993) present what they call five possible hypotheses of our evolutionary tree in a set of diagrams of branching trees through time. Emerging from a common trunk of the accessorial tree, the pictures show various branching and pruning configurations of the humanoid and *Homo* species. Figure 3.2 shows an adaptation of one of the configurations.

3.2.2 The Story of the Original Branching

The most remarkable turn of evolution is the genesis of our species. The humanoid line diverged rather abruptly from the ape lineage into a new

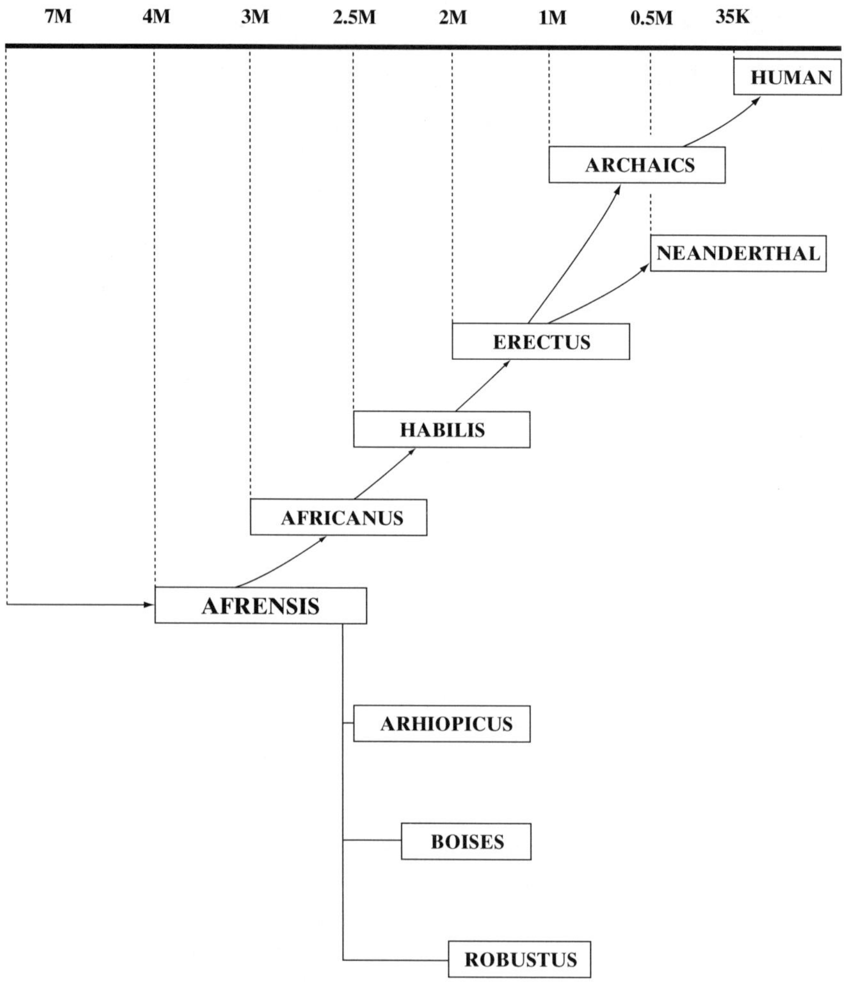

FIGURE 3.2 The Family Tree of Our Species

pattern of structural and behavioral evolution. The apes have not changed much since the branching, but the humanoid lineage has changed immensely.

After an extended debate, evolutionary scholars seem to agree that the original branching, called "evolutionary bifurcation," took place some five to seven million years ago. From a common line of primate ancestors—through "bifurcation"—two branches emerged, creating an evolutionary "Y." One fork of the "Y" left the chimpanzees with the primate community of the apes, while the other became the fork of origin of the humanoid species. Paleontologists label this ancient species of the human lineage *Australopithecus afarensis*. Perhaps the most visible manifestation of the appearance of this ancient ancestral species was an imagined event taking place one day when a group of our ancestors walked out from the forest to the grassland on two legs. Habitual two-legged movement, which is called "bipedalism," was the first distinguishing feature that separated humans from all other primates. It was the first crucial innovation in the evolutionary design space of the humanoids. "It is one of the oddest behaviors found in nature. Yet, two-legged walking must have provided some essential evolutionary advantage or it never would have developed or endured. Was it a key part of the first humanoid strategy paving the way for the appearance of modern humans a few million years down the road? "(Johanson and Johanson, 1994, p. 48).

Jared Diamond (1998) defined our original ancestor as the "third chimpanzee." He argues that the human species are actually most closely related to the chimpanzees than the chimpanzees are to other apes. Humanoids and chimpanzees occupy one branch and gorillas and other primates the other branch. We differ from the chimps by 1.6% of our DNA, diverging from a common ancestor some seven million years ago. The gorillas differ in their DNA from the chimps by 2.3%.

Potts (1997) suggests that terrestrial life is a pivotal doctrine of the investigation of human origin. The adoption of ground living is a durable symbol of humanness. It was accomplished by natural selection when trees became sparse after major climatic changes. The protection of the group shifted from "teeth to tools." Carrying and manipulation of tools, dietary changes, social life, and bipedalism were all tied together in what I call a closely knit system of evolutionary markers of a particular evolutionary entity of our species. These interdependent threads of humanness are all woven together, creating the conditions for one another as they reinforce and tighten the fabric of the tapestry of our existence. Later in evolutionary advancement, the "fine tuning of human adaptation largely involved the brain, which evolved under the pressure of more complex human life until the species *Homo sapiens* appeared" (Potts, p. 83).

Paleontologists seek fossil evidence, such as tools and campsite remains, on which they can base an interpretation of human evolution. Another source is molecular evidence, from which biologists tell us that it was some seven million years ago that an "evolutionary bifurcation" happened that created the humanoid species. But we do not yet really know what happened between the time of the original branching and the point in time about three million years ago. We lack fossil and other evidence that would enable us to tell the story of the first three million years of human evolution. But Leakey and Lewin (1993) tell us that:

> We can assume that soon afterwards (the branching) other humanoid species would have evolved, descendants of the founding species, variants on the theme of the bi-pedal ape. Humanoids, like other large terrestrial species, would have embarked on what biologists call an adaptive radiation. (p. 109)

Not having evidence of the earliest time of our evolutionary history, we shall start to tell the story at the point when a comprehensive set of fossil evidences was discovered some three million years ago. (Isolated single pieces of evidence were discovered even earlier.)

3.2.3 *Meet Lucy and Her Clan: The Australopithecus afarensis*

In 1973, Don Johanson and his colleagues in Ethiopia unearthed fossils that included a partial skeleton of the famous Lucy and a large collection of fossil fragments from fifteen other people. This collection of remains was called the "First Family." These fossils were said (Leakey and Lewin, 1993, p. 109) "to belong variously to a primitive *Homo* species and one, perhaps two, *Australophitecus* species. Again, the Y-shaped pattern seems to be present: *Australophitecus* on one hand, *Homo* on the other." The Johansons (1994, p. 78) suggest that, judging from the fossil remains of several males and females found at the First Family site, "*Australophitecus afrensis* was probably a social animal. Within a large group, males and females formed monogamous mating pairs and put equal energy into raising offsprings." Finds by paleontologist Meave Leakey in northen Kenya show that two-legged walking evolved by at least four million years ago. Her mother-in-law, Mary Leakey, discovered a trail of fossil footprints in Tanzania in open terrain, where three creatures walked erect on two legs 3.6 million years ago. The timing of the critical events, namely the walking on two legs and the adoption of ground living, are symbols of when humanness began. (Potts, 1997) suggests that:

> The selection of terrestriality over arboreality was accomplished by the process of natural selection as trees became sparse. This is the savanna hypothesis, an idea that has nurtured anthropologists for generations. Adaptation to open terrain was the spark that initiated the human lineage. Indeed, the fits and starts of the environmental decline and the rise of the savanna habitat during the late Cenozoic seem to confirm this perspective. (p. 82)

The move from the forest to open terrain of the savanna was the first event in our evolutionary story. It acted as a propellant in the process of becoming human. Wesson (1997) suggests that standing up made it easier to look over the grass, to carry food and infants during their first years, and to use tools. (The weaning period in humanoids became increasingly longer as they become more and more humanlike.) Although Lucy and her clan walked like humans, their brains were about the size of a chimpanzee's. However, they might have been more intelligent, because relative to their body size, their brains were larger compared to the chimpanzee's. (Brain growth occurred during the later part of the seven million years' separation from the ancestral stock. About two million years ago the enlargement of the brain advanced our species from humanoid to *Homo*.)

The key issue in the evolution of humanoids is not just their ability to adapt to grassland, but also their response to the increased variability of the topography as well as variations in the environment. "Versatility was the hallmark of the amazing risk of bipedal behavior" (Potts, 1997, p. 90). Seeking some other evidence of behaviors that followed two-legged walking, we find a package of: "stone-knapping," transportation of things, "trash-making," and use of campsites. Based on archeological evidence these behaviors first appeared between two and three million years ago. Around the same time, a monumental pruning and branching occurred in the humanoid family tree. Lucy and her clan were no longer around. The *A. afarensis* species had succumbed, passing on the humanoid lineage to the *Homo* species.

We can close our visit with the first lineage of the humanoid species by listening to a story told by Rick Potts (1997). The story relates our experiences as imaginary time travelers during a journey going back to three million years ago. We view from the air the African landscape where those fossils of our ancestors were unearthed. At times we descend and walk on the ancient savanna. (I have space here only for the following abbreviated adaptation of the story, described by Potts.)

> In our imaginary travel we rise high above the African plateau and view the landscape that has been shaken by unseen forces. Now we see a whole region consumed by flows of inexorable burning and by ash. In another region, we observe extraordinare bounty of vegetation and creatures. Among them we catch glimpses of two-legged creatures; first, one at a time, then groups. These bipedal beings are not always the same kind. They don't seem to be completely at ease in the open grass, neither are they attached to woodlands. While they inspire a sense of awe, we feel no immediate identity with them. As we journey through time, the grassland widens dramatically at certain intervals, but forests and woodlands still find refuge along the vast network of rivers and ponds. The landscape is continuously reforming itself. Here and there groups of bipeds enter and exit by no apparent design, settling in or moving on. They are pretty rare though amongst the constantly changing lines of large herds of mammals. We descend again on the terrain and notice a short, bulky biped, squatting and eating something in the grass. Tucked in his bent fingers is a stick blunted with wear. He digs with it for about two minutes and levers up

> a sizable tuber, which he chews for a long time, splitting out what is distasteful. Looking up he sees us, rises, gives an amazed hoot, utters a loud cry and runs towards a group of trees on two legs with an amazing speed.
>
> As we move on we see a short muscular female near a series of tall reeds working on an odorous antelope leg, scraping off meat pieces from it and eating the scraps. Then, she takes a rock and hits with it the bare end of a joint. She is after the marrow and licks it gingerly. Momentarily two others with short legs arrive, one carrying a rock. They see us. Without a sound all three flee to a grove of trees. Further to the north we see a much larger biped making tools in a dry river bed. His braincase is much larger than what we had seen before. It dawns on us that the stones the biped makes is an "out-of-body food-processor." He uses his stone tools for crashing and grinding: as a kind of teeth: As the first "denture." (pp. 101–104)

As we return from our time travel, we now realize that in our imaginary journey, we met with the humanoid species we described earlier in this section, but we also saw others that appeared to be more human than humanoid. Yes, in one of the sightings we met *Homo*, the "handyman." His story follows next.

3.2.4 *Meet the "Handyman": Homo habilis*

During the 1960s and 1970s, the paleontologist Leakey family helped us to recognize a new era of human evolution by the discovery of *Homo habilis*. In 1960, Jonathan Leakey found the first fragments of *H. habilis*. In 1964, his father Louis unearthed fossils that were definitely more human than any other humanoid species of that time. In 1973 Louis's son Richard discovered a nearly complete skull with a cranial capacity about twice that of previous discoveries, and in 1975, his mother, Mary, found fossils she classified as remains of the first genus *Homo*: *Homo habilis*, the "handyman." This was definitely a new species: a tool maker, with a large brain, and believed to be the ancestor of later humans. By now we know enough about the interpretation of various paleontology evidence to introduce a profile of *H. habilis*. We can describe what he looked like, his lifestyle, how he survived, his social nature, his culture, the technology he used, his communication ability, his level of consciousness, and his role in human evolution.

3.2.4.1 What He Looked Like

H. habilis is considered to be an evolutionary intermediate between the small *A. afarensis* and the large *Homo erectus*. From a distance, *H. habilis* may look like an early *H. erectus* (Leakey and Lewis, 1993). But at a closer vantage point, we can see the difference. *H. habilis* is smaller than *H.*

erectus. He is larger than his mate. From fossil evidence we also know that *H. habilis*, the toolmaker, was right-handed.

3.2.4.2 His Lifestyle and How He Survived

The Johansons (1994) call *H. habilis* the "Scavenger." In the debate over "hunter vs. scavenger," they suggest that *H. habilis* may not have hunted at all and that he survived as a scavenger. Compared to hunting, the food attainable from a carcass provided large payoffs for the energy and time expended. Carcasses had substantial nutritional windfalls as critical food sources for *H. habilis*, who still relied on plants and fruits in his diet. Rather than chasing down prey, it was brought for him by animals who chased down the prey. Next to scraping the shreds of meat from bones with his stone instruments, *H. habilis* was after the marrow by breaking the bones with his stone hammer. (He is also called the toolmaker.) It is estimated that marrow from the limb bones of an impala provided about 60% of *H. habilis* daily caloric requirements. Even today there is a heated debate about the time in his evolution when *Homo* began to hunt. On the hunter vs. scavenger debate, the Johansons (1994, p. 121) hold that "habitual hunting did not begin until late in human evolution, perhaps not until after the appearance of modern humans."

3.2.4.3 Sociocultural Characteristics

H. Habilis was an intensely social creature (Leakey and Lewis, 1993). Bonds between kin were strong, particularly between mothers, offsprings, and siblings. Strong friendships formed often as "political" alliances—in struggle for power and status. When reaching maturity, males did not leave their clan, as did other primates. They remained in the kin group cooperating with each other. Another difference is that the social structure was more tightly knit to enhance the nurturing and protection of offsprings. In prolonged childhood, there is much more opportunity to learn. The kinship group lived in campsite clusters of females and infants, children and maturing youngsters, and adult males. When they searched for food "the females are confident of returning laden with nutritious roots and fruit. When they are lucky, the males may bring back a bonanza of meat, which will be noisily shared by all members of the band" (p. 141).

3.2.4.4 His Technology

H. Habilis found and chose raw material for stone tools from great distances (Johansons, 1994). Basalt tools were carried on to their foraging for scavenger hunts. "Basalt tools came to the bones." In contrast, quartz tools were confined to central processing places: "Bones were brought to the tools." The early discoverers of *H. habilis* maintain that he was a designer of tools. Potts (1997) says that "By knapping stone against stone and removing slivers piece by piece, an implement preconceived in the mind could be fashioned from a lump of rock. There was a design for choppers, another for scrapers, another for polyhedrons, and so on. Each type of tool was considered a separate target produced by adept hands, learned and taught by generations of 'handymen'" (p. 187). (Recently, this handymen "designer" hypothesis has been questioned by some paleontologists.)

3.2.4.5 His Communication

Did he have language as a mode of communication? Looking for language ability in the fossil records, the high larynx in our ancestors would imply apelike language, a low larynx humanlike ability of speech. The shape of the bottom of the skull is related to the position of the larynx. In mammals it is flat and in humans it is distinctly arched. The first time we find a fully arched structure in fossils was about 100,000 years ago, in people we call Archaic *H. sapiens*. In *H. erectus* we found a structure that enabled rudimentary spoken language. How about *H. habilis*? Richard Leakey suggests that none of the crania fossil we have available is sufficiently intact. However, if we find an intact *H. habilis* cranium, Leakey predicts that we shall most likely see the beginnings of the flexion of the base and the lowering of the larynx, and the beginnings of languagelike speech.

3.2.4.6 Level of Consciousness

The lifestyle and technology of *H. habilis* and his cultural/social characteristics described earlier indicate that specific social skills had to be developed that enabled him to cooperate in small and large groups and live by certain social rules and standards. All of these imply some level of awareness. Commenting on *H. habilis'* degree of consciousness, Leakey and Lewin (1993) say that "we can only speculate that elements of consciousness—the sense of self, the tendency to attribute feelings to others, the facility to know the world better, and the raw emotion of compassion—all were enhanced through time as the evolutionary ratchet advanced" (p. 305).

3.2.4.7 Role in Evolution

Homo habilis is a connecting link in the evolution of our ancestry. He is at a boundary of connecting the humanoid lineage with the *Homo* species.

He is our great-grandparent. When in 1964 he was announced as a new species of hominid lineage, an extensive debate started. But as fossil and other evidence surfaced, "eventually the decibels of the debate died down, and *Homo habilis* became accepted as a valid species, the immediate precursor of *Homo erectus*" (Leakey and Lewis, 1993, p. 102).

3.2.4.8 Transition and Overlap

In the evolutionary history of the *Homo* lineage there have been changes in anatomy, changes in the ways of living and surviving, in social and cultural behavior, and in technology. These should be seen as a system of interacting changes, which shaped each other, realized evolutionary potential, and created new potentials that became manifested in further changes. We can now explore the three aforementioned changes. Were they gradual or were they evolutionary leaps? In the evolutionary trajectory of the humanoid and human lineages, can we speak of transition or sudden transformation not only between *H. habilis* and *H. erectus*, but also between the successive species of the human lineage?

Addressing these questions Lewin (1998) suggests that

> The tall stature—low-body-bulk-configuration so evident in *Homo erectus* almost certainly emerged with *habilis*. The transition from *habilis* to *erectus* involved a slight enlargement of the brain (to almost 900 ccs) and a change in the facial features, including the development of prominent ridges about the eyes, the brow-ridges. Once this configuration evolved, it persisted throughout the history of the *Homo* lineage—clearly a stable adaptation. (p. 30)

The overlap between *H. habilis* and *H. erectus* is quite clear. From the fossil findings we know that *H. habilis* first appeared some 2.5 million years ago and departed close to a million years ago. We can place *H. erectus'* appearance about 1.7 million years ago and estimate his departure somewhere around half million years ago. There is thus a definite overlap, which is a typical evolutionary occurrence. So, we travel back in time a million and half years and meet on the evolutionary trail *Homo erectus*.

3.2.5 Meet the Turkana Boy: Homo erectus

In 1984, Richard Leakey and Alan Walker's team excavated a well-preserved skeleton of a young boy—five feet, three inches tall, in today's terms about twelve years old. He became known as the "Turkana Boy:" a true representative of early *Homo erectus*. Richard Leakey characterizes *H. erectus* as follows (Leakey and Lewis, 1993):

> *Homo erectus* stands at a pivotal point in human evolutionary history, in a very real way it is the harbinger of humanity. *Homo erectus* was distinctly

> humanlike in behavior and form. The beginning of hunting and gathering was a life style that came with *Homo erectus*. Stone tools for the first time gave the impression of standardization, the imposition of a mental template, fire was harnessed for the first time, for the first time Hominids expended beyond the African continent. *Homo erectus* marks a clear move from apelike past to a humanlike future. If we are to understand the origin of humanity, we have to understands *Homo erectus*, its biology, its behavior. That's a tough challenge, given fragmentary fossils. But with a skeleton, a whole new anthropological world opens up. (pp. 46–47)

And that skeleton is the skeleton of our "Turkana Boy." He would help us to understand the extinct *H. erectus* almost as well as someone living today, says Leakey. In the Turkana Boy "the seeds of humanness we feel in us today were firmly planted" (p. 67). In *H. erectus* the brain size surged and the mental capacity was boosted. In him we see the beginning of compassion, morality, and awareness of awareness.

Here, we shall learn more about *H. erectus*. We will find out what he was like, how he lived, what he did, and where he lived. A characterization of the lifestyle of the *H. erectus* clan is described by Ornstein (1991) as follows.

> They cooked in pots, handled fairly advanced tools, and worked animal skins—all bespeaking of higher culture. They lived in an ice age and could construct shelters. They invented clothing, and helped prepare for the characteristics of human life: living in groups, being able to survive in new worlds, and adapting those worlds to them. Fire probably extended the day, giving more time to work and since fire is attractive, it is possible that groups of *H. erectus* huddled around the fire, were the first to use their larger brains to introduce language. (p. 35)

What follows is a set of specific descriptors that present *H. erectus*'s specific characteristics. These are derived from the works of Gould (1977), the Johansons (1994), Leakey and Lewin (1993), Lewin (1998), Potts (1997), and Shreeve (1996). These scholars developed their own interpretations from such evidence as fossils and other remains, from campsite study, from climate and earth formation analysis, from molecular and DNA studies, and by seeking and understanding the systemic interaction of these factors. A set of specific characterizations of the lifestyle of *H. erectus* follows.

3.2.5.1 Body and Brain

H. erectus was the first tall humanoid. He was about 20% larger and 14% heavier that *H. habilis*. The difference in size between the male and female *H. erectus* was significantly less than in case of the earlier humanoids. (*H. erectus* was taller compared to us.) He was adapted to long-distance travels on the savanna and was able to adapt to new environments. His features of height, long legs and long stride, strong bones, and massive muscles enabled him to withstand forceful and lengthy travels. These are abilities unknown by his predecessors and, curiously, unknown to us. He was more effective in walking and running than we are. These abilities increased his access to resources as well as his capability to carry large loads. An increase in his range of space as well as the increase of his social space demanded a more extensive capability of mental mapping and increase in brain complexity and size. (Compared to Lucy's 359 cc and *H. habilis'* 600 cc, *H. erectus'* brain reached 650 cc. The modern brain ranges from 1000 to 1400 cc.)

3.2.5.2 Mobility

On account of his superb capability of walking and endurance, *H. erectus* quickly dispersed and conquered new worlds. The migration of *H. erectus* populations out of Africa into various regions of the Old World is well documented. They expanded their range, spreading out into Europe, Southeast Asia, India, Indonesia, and China. There is evidence that by 300,000 years ago, *H. erectus* colonized the Northern Hemisphere, reaching as far as Britain. These expansions of his living space and his ability to conquer new and harsh habitats are eminent testimonials of *H. erectus'* advanced state.

3.2.5.3 Use of Fire

There is good evidence that *H. erectus* used fire a million years ago. He borrowed natural fire and kept it going at the home base. The availability of fire extended the time for his activities. It added to the accessible range of territory of finding food and gathering other resources. It improved the chance of avoiding predators. Most significantly, it enhanced and intensified the social interaction that bonded the *H. erectus* clans.

3.2.5.4 The Toolmaker

H. erectus established a tool industry. His heavy-duty hand axes, picks, and cleavers were sophisticated creations compared to the simple stone instruments used by *H. habilis*. Some hand axes weighed as much as twenty pounds, showing polish needed to carve both wood and meat. No instrument was uncovered that would indicate use for hunting. No arrowheads, darts, or spears were found at *H. erectus* sites. The same kinds of instruments were found at far-reaching *H. erectus* sites. But what is remarkable is that during a more than million year span of *H. erectus'* dominance the instruments show little change or improvement.

3.2.5.5 Communication

The anatomical conditions of speech and language use were described earlier. It is stipulated that it was not until some 300,000 to 400,000 years ago that those conditions began to emerge in Archaic *H. sapiens*. In *H. erectus*, the larynx was in an intermediate position between the ape configuration and that of modern man, enabling a partially developed speech capability. *H. erectus'* limited language capability might have been the reason for the unchanging nature of his stone tools. If members of this group had been able to talk with each other about making those tools one would expect the creation of increasingly more novel tools and implements over a million years.

It seems to me that the development of speech capability was enhanced for over a million years by: (1) the continuous enlargement and increasing complexity of the brain, (2) the enabling anatomical configuration described earlier, (3) the increasing of social interaction and cultural complexity, (4) the increasing space horizon that required mental mapping, and (5) the "inherent and expanding evolutionary potential," which is the "evolutionary seed" that, as it grows, reorganizes human existence and experience at ever higher levels of complexity. There is no single causation. It is the systemic interaction of these five components that created linguistic ability and enabled the use of speech. The ability to have linguistic communication is an eminent example of multiple, mutual, and recursive causation. It is a phenomenon exhibited throughout evolution.

3.2.5.6 Their Subsistence Strategy

In earlier times, a romantic picture presented *H. erectus* as a competent big-game hunter even responsible for the extinction of certain animal species. More recently, the accumulation and interpretation of archeological evidence paints a very different picture. There is little evidence to suggest that *H. erectus* exercised large-scale predatory skills beyond those of *H. habilis*. The mental and linguistic capacities needed for designing cooperative strategies and planning hunting events were not yet available to him. Furthermore, there is no evidence of *H. erectus* possessing hunting instruments, for example, arrows and spears. Next is a story of a day in the life of an *H. erectus* clan, which pictures their subsistence strategies. It is an abbreviated interpretation of the story of Leakey and Lewin (1993) describing finds at archeological site "50."

> We are in a position to observe a riverside camp of an *erectus* clan. At first light three men leave the camp to check the snares they constructed yesterday from strips of soft bark and sharpened sticks. They carry long sharpened sticks as much for self-defense as for spearing a fleeing prey. An hour later some of the mature women leave the camp, soft animal skin slung from their shoulder, serving as papoose and carrying bag. They also carry long sharp sticks for protection and shorter ones for digging. A few hours later they return laden with fruits, nuts, succulent tubers, and flamingo eggs. In the camp younger women keep an eye on children and engage in idle chatter with a man who stayed home today. Yesterday, during a hunting expedition for small game, he slipped down on a slope and gashed his leg on a jagged-piece of lava. One of his brothers found a clump sanseviera plant and squeezed it onto the open wound. Another brought some thorns from a nearby acacia tree and pierced the two sides of the gash together and held them in place with strips of bark.
>
> A woman, sitting on the edge of the camp, makes strips of bark for snares, with sharp lava flakes. Another, using stone flakes, is whittling wood, making digging sticks. Soon the hunting group can be heard returning with an antelope. It was wounded in one of the snares and the group spent most of the day tracking it down, finding it on the lake shore. While the group began dismembering the animal, one noticed vultures near the lake edge and went to investigate. He found a hippo carcass. Tomorrow they will go back to see if it's worth scavenging. Just before sunset, children are coming back to camp, some with catfish they speared. Darkness falls, and lightning can be seen in the distant hills. After some days at the camp they decide that it is time to move on as the rain is coming, flooding the river and their campsite. They move on, leaving at the camp site scattered broken bones, heads of catfish, and ends of tubers too bitter to eat. (pp. 178–179)

The story tells us how *H. erectus* lived for hundreds of thousand of years. They made a home base for several days, or a week or two. They foraged their surroundings or plant food and got meat in any way they could. As resources began to thin out, they moved on.

3.2.5.7 Consciousness

One can only speculate about the development of consciousness in *H. erectus*. Had *H. erectus* already attained such elements of consciousness as the sense of self, the attribution of feelings to others, the emotion of compassion? If yes, to what degree? If a certain degree of language development was already achieved by him, then it indicates the attainment of a certain level of consciousness. But this is a big "if." Language and consciousness coevolve in the context of the complexities of social and experiential life. They jointly enhance the ability to operate in that context and shape each other. It is clear that the consequences of this coevolutionary development go far beyond natural selection. To the extent that we can "read" from *H. erectus'* archeological evidence the level of his social development, we can speculate about some level of development of his consciousness.

3.2.5.8 The Place of H. erectus in Human Evolution

In our evolutionary history *H. erectus* stands at a pivotal point (Leakey and Lewin, 1993). Humanoids before him were apelike, after him humanlike. He was humanlike in body and behavior. He signals the onset of a hunting–gathering style of life. The standardization of his tools indicates a mental template. He used fire, began to build shelters, and spread out around the Old World. Most significantly, in him we see the birth of language, and coevolving with it, some movement toward the threshold of consciousness.

3.2.6 Summary: From Ape to Homo

During the first phase of the first mile of our journey of human and societal evolution, we listened to the story of our evolutionary ancestors. We started at a point in time where the humanoid species and the chimpanzee lineage departed from a common evolutionary lineage. Several evolutionary scholars, representing various disciplines, told us the story based on such evidence as fossil remains, tools, other finds at campsites, climatic conditions, terrain features, and plant and animal life existing then. It is truly remarkable how these evolutionary scholars can interpret from archeological evidence and paint for us pictures of the life of our ancestors. As we kept on with our journey from visiting Lucy and her family, then the "handyman," and last the Turkana Boy, the pictures became ever more clear. To moderate my enthusiasm, in closing, I quote Richard Leakey, "But how confident could we be that the calculations from life history theory were valid for extinct humanoids? How confident could we be that australopithecines were bi-pedal apes; that *Homo* was a humanized bi-pedal ape?" (Leakey and Lewin, 1993, p. 152.) Before we continue our journey toward the first milestone, an activity will connect Chapters 2 and 3. It will help to gain new perspectives and new insights in understanding human and societal evolution.

Activity 5

In the previous chapter, sets of evolutionary perspectives were introduced. Working with those, you were asked to select and note core ideas that represent the evolutionary thinking of several evolutionary scholars. In the present chapter, we visited with our ancestors. With these two experiences in mind, you are asked to accomplish two tasks.

Task 1. Review the description of the three humanoid/*Homo* species as captured by the stories of Lucy, the "handyman," and the Turkana Boy and identify and note in your activity workbook key evolutionary ideas you derive from the descriptions.

Task 2. You are asked to interface the evolutionary ideas you identified in Chapter 2 with those you identified in this chapter. In Chapter 2, several issues were defined and others left open, such as: inheritance of acquired characteristics, gradualism vs. sudden change, natural vs. cultural selection, evolutionary potential, continuity vs. discontinuity, the role of "memes," the manifestation of altruism and consciousness, etc. These are just examples. Select from the large range of issues/core ideas those that are most interesting to you and construct your understanding. Introduce your findings in your activity workbook.

3.3 *The Crucial Transition Phase toward Homo sapiens*

This phase of the journey represents the crucial transition from *Homo* toward *Homo sapiens*. In this phase we follow the evolution of the humanoid lineage as we leave *H. erectus* behind and listen to the evolutionary stories of Archaic *sapiens* and the Neanderthal man. It is estimated that this transition began some 500,000 years ago and ended about 35,000 years ago, with the emergence of the Cro-Magnons, the first generation of *Homo sapiens sapiens*. It marked the onset of the age of modern man. Before we enter this transition phase of our journey, we develop three perspectives for thinking about human evolution.

3.3.1 Three Perspectives on Describing the Two Species

First, I present the historical context of human and societal evolution of the transition era. Second, a system of cultural components are introduced by which we can view and understand the life experiences of the Archaic *sapiens* and the Neanderthal *sapiens*. As a third perspective, the design space is mapped out, within which the human and social evolution of our species has unfolded.

3.3.1.1 The Historical Context of the Transition Era

The historical context has three aspects. The first is the evolution of the *Homo* lineages in the course of the last half million years. The second is the story of *Homo*'s migration to the Old World, and the third is the replacement of Archaic species. These three perspectives are informed by the work of Lewin (1998).

3.3.1.1.a The Evolution of Two Sapiens Lineages. From a common ancestral lineage two species emerged in this transition period: The first more than 500,000 years ago as the Asian/European Neanderthal *sapiens* and the second, the African Archaic *sapiens*.

3.3.1.1.b The Story of Homo's Migrations. There are two models that represent two alternatives of the migration of *Homo* beyond Africa. The proponents of the two models agree that the first migration brought *H. erectus* into the Middle East, India, and China a million years ago and to Europe 0.7 million years ago. Here the agreement ends. The "multiregional evolutionary model" holds that in the new regions, *H. erectus* evolved toward Archaic populations and it was from those populations that *Homo sapiens* emerged.

The second, the "Out-of Africa" model, stipulates that anatomically modern humans arose somewhere in sub-Saharan Africa and starting about 100,000 years ago, they left Africa, reaching all of Europe, the Middle East, all of Asia, Australia, and North America, where modern human populations were established between 50,000 and 35,000 years ago. This second model is supported by the findings that there seems to be no genetic continuity between the Archaic populations of the first migration, including the Neanderthal, and the modern human species: *Homo sapiens*.

3.3.1.3.c Replacement of Archaic Populations. The second migration model is supported by recent DNA research. The findings of the extensive research of Allan Wilson and his colleagues established that all living human beings are descendants from a single type, namely, from an anatomically modern African population, which emerged some 200,000 years ago. We are descended from a single female who lived in Africa at that time. The research also indicated the absence of any archaic DNA in modern populations. The implications of this finding means that anatomically modern humans—migrating out from Africa—did not interbreed with established archaic populations in the Old World and eventually completely replaced them. The many humanoid experiments left only *one species*: *Homo sapiens sapiens*: Us. (One might wonder, What would life on earth be like if two or more *Homo* species were to exist today?)

3.3.1.4 The Pyramid of a System of Cultural Components

Potts (1997) discusses in detail the development of cultural behavior in the humanoid, *Homo*, and human species. He created a model of the evolution of cultural behavior by asking the question (p. 188): "What if we had to rebuild our concept of culture from the ground up?" He answers this question by creating a model of a "cultural monolith," which I prefer to call the Pyramid of Cultural Evolution. The model has seven interdependent components: transmission, memory, reiteration, invention, selection, symbolic coding, and institutions. These components are defined next.

3.3.1.4.a Transmission. Transmission implies observation and instruction, inasmuch "the transfer of information between individuals depends on the frequency and intensity of social encounters and the attention paid while observing one another" (p. 189).

3.3.1.4.b Memory. Memory is the capacity to retain information to which the person is exposed. "The brain's role in memory determines a great deal about the amount and kind of information that can be stored" (p. 189). (The storage potential is determined by the number of connections between neurons in the brain. It defines the complexity of the brain.)

3.3.1.4.c Reiteration. Reiteration is the third component, which Potts calls the "chemistry of social tradition, the tendency to reproduce or imitate stored behavior patterns or other transmitted information" (p. 189). The various neural pathways and the social influences on them enable the retained information to be translated into cultural behavior and mental functions.

3.3.1.4.d Innovation. Innovation is the capacity to alter transmitted information (or generate novelty) and as a result "new skills and variations on a behavioral scheme may occur within the group" (p. 190). However, not every innovation is retained, as we shall see next.

3.3.1.4.e Selection. Selection is the process by which the social group might filter out or reject certain innovations and retain and reiterate some others on a selective basis.

Cultural behavioral traditions arise from the interaction and merger of the five cultural components introduced in the preceding. They are passed on across generations. "There is little to suggest that hominids in the oldest periods had access to any but these basic elements of socially learned behavior" (Potts, 1997, p. 190). As we have already seen, the last two components—innovation and selection—had been quite constrained during the first few million years of human evolution. Even during *H. erectus'* life cycle we noted the scarcity of innovation. This scarcity was reflected in the lack of invention of new tools and the lack of change in lifestyles. This lack might be due to a "battle between invention and reiteration." It seems to be an ongoing characteristic of our species. Today, it is manifested in the resistance to change of our institutions, many of which are out-of-sync with the emerged new realities of our age. Another cause of the scarcity of innovation was the lack of developed language ability and language use. It was "only show and not to tell." Once language developed, it accelerated innovation. Below, I define Potts' last two markers.

3.3.1.4.f Symbolic Coding. Symbolic Coding is our ability to use language by which we can understand each other.

> Symbolic coding enables us to construct large edifices of social and personal significance out of the most inconsequential building blocks. Language is a form of social communication. Perhaps language evolved because of the value of stories in relying on information within the social group, or because it reinforced the bond within larger social groups. Symbolic coding also enables people to refer to the invisible, allowing information to be passed on in ways that transcend the need for direct observation. And symbols can be created for things never seen: a principle, an abstraction, a god." (Potts, 1997, pp. 190–91)

Beyond the internally generated necessity for language capability, there were external triggers that demanded its availability. Language has the power to express contingencies of time, place, and conditions. This power was needed when *H. sapiens* experienced dramatically changing environmental conditions, to which he had to adapt and learn to cope. This happened during evolution, when there were repeated shifts in available resources, and when drastic and recurring changes came about in the climate. Language was the primary means that helped *H. sapiens* to cope with these demands.

3.3.1.4.g Institutions. This cultural component raises us to the top of the Cultural Pyramid.

> "Institution" is the name we give to the highest order of complexity in human lives. It denotes the hyper-reaction between symbolic codes and social behavior. Because of this interplay, social relationships are catalyzed in a manner distinctive to human beings. The result is what anthropologists call cultural institutions, in which information and symbolic codes are organized into complex social order. (Potts, 1997, p. 195)

The cultural pyramid represents the ascent of *Homo* from the earliest ancestral times to our own time. We can create an image of the pyramid. It is constructed from seven stepping stones, representing the seven cultural components. Figure 3.3 depicts the image. In the course of

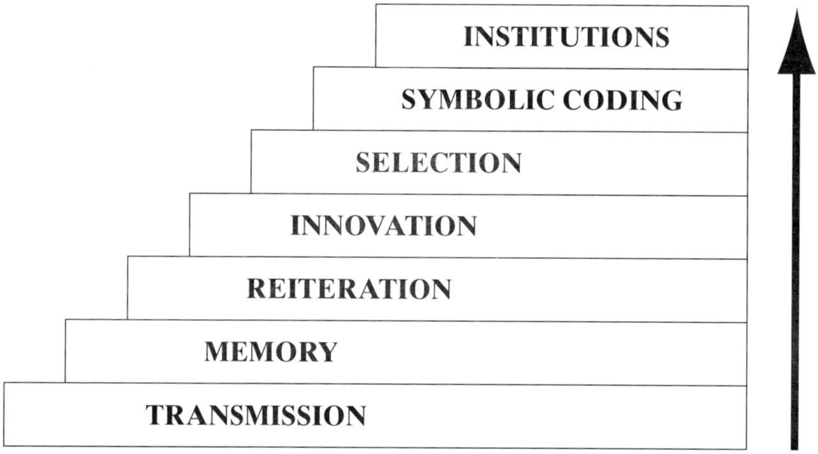

FIGURE 3.3 The Pyramid of Cultural Advancement

his evolution, our species ascended by taking steps up toward the top of the Pyramid. At any given time period we can gauge the advancement of cultural evolution by asking how far up we reached on the Pyramid and how long it took to climb up to the next step.

3.3.1.5 The Design Space of Evolution

In general, the design space of evolution is that space in which evolutionary experiments happen with life forms. The design space of human evolution is situated within the overall, general design space of evolution, and it is a specific design space in which experiments had been going on with humanoid and *Homo* species. Originating some seven million years ago, the design experimentation had produced numerous variations of Humanoid/*homo* species which—in the course of human evolution—died out, at the end leaving only one species: Us (*Homo sapiens sapiens*). In the general design space of life, the growth and death of the various branches obey the law of natural selection. But one branch, the human branch—over its course of evolution—created its own design space in which the law of natural selection gradually lost its force and the law of cultural selection took over the task of design. In the text that follows, I elaborate this special case of evolution—the evolution of the human branch of the Tree of Life—as I define the specific markers of the cultural design space. We shall see as we listen to the evolutionary story how the various design configurations and how the change of markers characterize and define a particular evolutionary stage.

Below, I introduce markers that I plan to use to explore and characterize the life of various evolutionary stages. In this section, the focus is on the description of the Archaic *sapiens* and the Neanderthals. We shall see that at some stages, for example, the stage of Archaic *sapiens*, we cannot say much about certain markers. On the other hand, as our journey progresses through the stages, we may have to nominate some new markers to accommodate a particular stage. I describe these markers in the following sections.

3.3.1.5.a Consciousness. Consciousness is the primary marker of the progression of evolution from here on, as it best defines the degree of "humanness." An evolutionary stage reaches its maturity when it is organized and animated by collective consciousness, shared by the evolutionary entity of a specific evolutionary stage. (At the current evolutionary stage, conscious evolution and finally evolutionary consciousness mark the advancement of evolution.) World view is derived from collective consciousness as it reflects how humans and human populations think and view themselves and the world around them.

3.3.1.5.b Communication. The mode of communication is another key marker as it evolved through our evolutionary history from signs to speech to writing, to print, to wired and wireless, electronic communication, and the Internet. It extends the boundaries of human experience in space and time.

3.3.1.5.c Social and Societal Organization and Way of Life. Social and societal organization and way of life are becoming ever more complex as mankind's evolutionary journey progresses from stage to stage.

3.3.1.5.d Subsistence. Subsistence marks the ways and means by which a population nourishes and sustains itself. During the evolutionary stages of our ancestors it was a dominant marker.

3.3.1.5.e Technology and Science. Technology and science are markers that change significantly as our journey progresses from stage to stage.

3.3.1.5.f Spirituality. Spirituality is a key expression of relationship with the Divine, and it coevolves with both consciousness and world view. It is an important evolutionary marker of any particular evolutionary stage.

3.3.1.5.g Ethics and Morality. Ethics and morality are markers that relate to consciousness, spirituality, and the prevailing world view.

3.3.1.5.h Aesthetics. Aesthetics and its various forms of expressions in the arts are distinctive markers of "humanness."

3.3.1.5.i Relationship with Nature. Relationship with nature was expressed in the life and behavior of our ancestors as "oneness with nature." However, from the time of the emergence of Cro-Magnons this relationship became one of increasing separation from and later even exploitation of nature.

The Cultural Pyramid and these markers create a dual framework within which the Archaic *sapiens* and the Neanderthals are characterized.

Activity 6
Go back to the section on *Homo erectus* and (1) define his evolutionary status in terms of the Cultural Pyramid; then (2) describe his design space, with the use of those evolutionary markers that seem to be relevant to the *H. erectus* population. Enter your descriptions in your workbook.

3.3.2 Meet the "Dawn Man:" The Story of Archaic sapiens

First, I place Archaic *sapiens* in the historical frame of human evolution, then develop his characterization by the use of the evolutionary markers described previously.

3.3.2.1 Time Frame

The findings of DNA testing, described earlier, indicated that all currently living human beings are descendants of a woman who lived in Africa some 200,000 years ago. Fossil evidence suggests that from her lineage anatomically modern human populations emerged some 100,000 years ago in Africa. These populations initiated another migration into the Old World and replaced the *H. erectus* populations, who migrated out of Africa over a million years ago. Then, Archaic *sapiens* were replaced by modern *Homo sapiens sapiens*, between 50,000 and 35,000 years ago. Lewin suggests (1998) that

> The use of the term Archaic *sapiens* indicates an uncertainty on the part of anthropologists as to the status of such individuals: they are neither *Homo erectus* nor *Homo sapiens*, but may be a transitional form between the two. The term Archaic *sapiens* is therefore more of a descriptive shorthand than a formal classification. (p. 29)

The Johansons (1994) hold that the fossil remains of Archaic *sapiens* share some features with *H. erectus*, but their brains are larger, and there are other differences in the size and proportion of the skull that characterizes *H. sapiens*. Furthermore, Africa is the only place where paleontologists found fossils more than 100,000 years old that "show the blend of primitive and incipient modern human features." Thus, the "Out-of-Africa" migration model, the DNA evidence, and fossil remains all support the place and time of the direct origin of *H. sapiens sapiens*, with his parent Archaic *H. sapiens*.

3.3.2.2 The Design Configuration of Archaic *H. sapiens*

In Section 3.3.1, the general design space of human evolution was defined and its markers described. The nine markers of the design space are expected to aid us in describing the specific design configuration that characterizes a particular population at a specific evolutionary stage. The number of evolutionary markers we use at a specific evolutionary stage depends upon the evolutionary advancement and the complexity of the specific population we wish to describe. As we shall see, in the present case of describing Archaic *sapiens*, several of the markers are subsumed by consciousness, which becomes the dominant marker at all later stages of human/societal evolution.

3.3.2.2.a Consciousness. We defined consciousness as the dominant marker of human and societal evolution. Therefore, we should explore its nature in some depth. In Chapter 2, discussing the evolutionary ideas of Gebser (1986), we briefly reviewed five successively emerging levels of consciousness: the archaic, the magical, the mythical, the mental/rational, and the integrative. The archaic represents a kind of consciousness that is primal, undifferentiated, dreamy, in which man did not know himself as separate from wholeness and separate from nature.

Feuerstein (1987) defines the features of this state of consciousness as: (a) egolessness; (b) identity with the group; (c) one-dimensional perception, meaning that space and time have not yet unfolded on consciousness; and (d) richly interwoven life with nature. "Dawn Man," says Wilber (1981, p. 22), "began his career immersed in the subconscious realm of nature and body, basically undifferentiated, embedded, fused and confused. His self was his naturic world; his naturic world was his self; neither was clearly demarcated." Elgin (1993) calls the consciousness of the archaic era: "contracted" inward by the gravitational pull of deep unconscious. He says:

> A pre-reflecting consciousness was characteristic of our earliest human ancestors, who were largely running on automatic—relying primarily on instinct and habit. Their way of life remained virtually unchanged over thousands of generations. Some degree of reflective consciousness must have begun to

> awaken, when *Homo erectus* migrated out of Africa and, to cope with harsh ice-age climate, learned to make warm clothing, construct shelters, and controlled the use of fire. Nonetheless it is only with the evidence of burials, dating from roughly 60,000 years ago, that we find a clear recognition of death and, presumably, conscious reflection on the 'self' that lives. (p. 30)

How far the consciousness of Archaic *sapiens* advanced from the state of consciousness described previously becomes a subject of speculation. Elgin's quote gives us some hint about it. A review of Archaic *Homo* emergence from two lineages places additional light on this. The first lineage merged from the continuous evolution advancement of *H. erectus*. Following his migration to the Old World a million years ago, *H. erectus* dispersed throughout Europe, the Middle East, and Asia. The second lineage gradually emerged during the migration of our direct ancestors, who left Africa some 100,000 years ago. As they arrived in the harsh climate of the Old World, their advancement toward reflecting consciousness was the result of the conditions described by Elgin in the preceding quotation. It is this second population from which *H. sapiens* emerged.

Some other markers of the design configuration of Archaic *sapiens*, such as spirituality, world view, morality, and relationship with nature, are all nested in archaic consciousness of Archaic *sapiens*. These markers will become gradually differentiated from consciousness and they will be eventually fully manifest in the behavior and culture of *H. sapiens sapiens*.

3.3.2.2.b Communication. The earliest forms of communication were not speechlike. (Deacon, 1997). They most likely included conventional/ritual gestures, activities/objects, and some form of vocalizations, each dependent on the other. Vocalizing played a minor role at the early stages. But by the time *H. erectus* was on the scene, some words were available and were used symbolically. "What drove this shift to speech was not just constraints on the ease of use of objects and performances and manual gestures, but also the reciprocal effect of cortical enlargement in response to the demands of symbol learning and the consequences for manual and vocal flexibility" (p. 497). By the time Archaic *sapiens* lived in the new environments of the Old World, his behavior was supported by a complex symbolic structure that enabled him to pass on information about how to survive in the new environments, and how to organize work, social interaction, and community life.

But what was his vocalization performance like? Based on our fossil evidence of the brain size and speech enabling anatomical structure, we can speculate on the speech performance of late Archaic *sapiens*. His speech most likely possessed some of the consonantal features of modern speech, but evidence of various human activities implies the absence of what we would recognize as human language.

Language and speech communication is in a coevolutionary relationship with brain development, the development of consciousness, speech-enabling anatomical structure, social complexity, environmental changes, and space and time contingencies. One enables and put demands on the others. Here, again, we can see the necessity of a system of interacting markers and conditions to be present to enable evolutionary advancement. It is consciousness-grounded in language and language-grounded in consciousness that make us truly human. Language differentiates us from all other forms of life. It has an infinite capacity to express contingencies of time and place (Potts, 1997).

> It has been long presumed that language, perhaps the most amazing and complex of all the higher faculties in humans, arose in increasingly open, arid, glacial environments during the Pleistocene. We know these were bracketed by incursions of warmer, moister, more vegetated climates, that the sequence was periodically interrupted by unpredictable events, and that kind of variability was by no means unique to northern settings. Knowing this environmental past, we can conceive how changing contingencies—repeated shifts in resources, periodic redefinition of the environment furnishes the vital spark to symbolic coding. The social and mental bases of symbolic coding increased as, over time, ecological settings varied. (p. 194)

These conditions became the experienced life for Archaic *sapiens* as he migrated out of Africa to the various regions of the Old World.

3.3.2.2.c Social Behavior. Social interaction is a transaction between partners and members of a group. Its is tied up with consciousness, which enables individuals to predict the behavior of others. Humphrey (1986) notes that once social skills become significant, cultural selection will further sharpen them, enhancing the individual's ability within the complexities of social life. Social thinking and interdependence among members of the group became especially important during changes of the habitat and fluctuating climate changes. Evoking social thinking and interdependence became all important in times of environmental uncertainty and in responding to novel circumstances.

The existence of home bases is evidence of intensive social interaction in the life of Archaic *sapiens*. The family, the kinship group, gathers around the fire (Potts, 1997). Resources of all kind obtained on the outside are brought back to the home base, which was the center of social life. It was home. "It is the place where infants are nurtured, children have the opportunity to play and the elderly might be provided for. The sick may become better by resting there, and adults play out the main emotional tracks of their life" (p. 177). Food was brought to the home base and was shared with others. This was uniquely human social behavior. It generated social cohesion among Archaic *sapiens*.

3.3.2.2.d Subsistence. Most evolutionary scholars hold that A. *sapiens* practiced some glimmering of hunting but they were basically opportunistic scavengers, exploiting carnivore kills, occasionally hunting small prey, but nothing that can be described as systematic hunting. They simply wandered around taking prey and carcasses they stumbled over. They lacked the competence and language to plan ahead, they did not have the economic organization that represented the lifestyle of hunters. But Leakey and Lewin (1993) hold a different view. They suggest that early *Homo* acquired the rudiments of hunting–gathering almost two million years ago. (For their view, see their story of the life of *H. erectus* around the campsite, discussed earlier.)

3.3.2.2.e Technology. The Middle Stone Age began some 250,000 years ago, during the reign of Archaic *sapiens* populations. Lewin (1998), discussing the technology of this age, suggests that it ushered in an innovation, namely, the preparation of raw materials from which artifacts were made. For example, such preparation involved removing flakes from a lump of rock to make a core with a flat top, from which many flakes were sliced, having similar thinness. Implements produced by the archaic population included side scrapers, backed knives, hand axes, denticulates, and points. The preparation of these required designlike thinking, which was a major shift, a "cognitive transition" from earlier stone technology.

3.3.2.3 The Cultural Pyramid: How Far Up?

Earlier, I suggested applying Potts' (1997) seven culture-building components to assess cultural advancement at the various stages of human evolution. In the context of the progressive steps of *transmission, memory, reiteration, innovation, selection, symbolic coding,* and *institutions,* it seems that Archaic *sapiens* advanced up toward the innovation and selection steps, and made limited progress toward symbolic coding (Figure 3.4).

Viewing this advancement from the larger view point of evolution, it becomes evident that in the life of Archaic *sapiens* we witness the beginning of cultural selection surpassing natural selection and taking over as the evolutionary force. Compared to natural selection, cultural selection is far more speedy, far more responsive, and far more powerful to generate behavioral adaptations that enhance and sustain life.

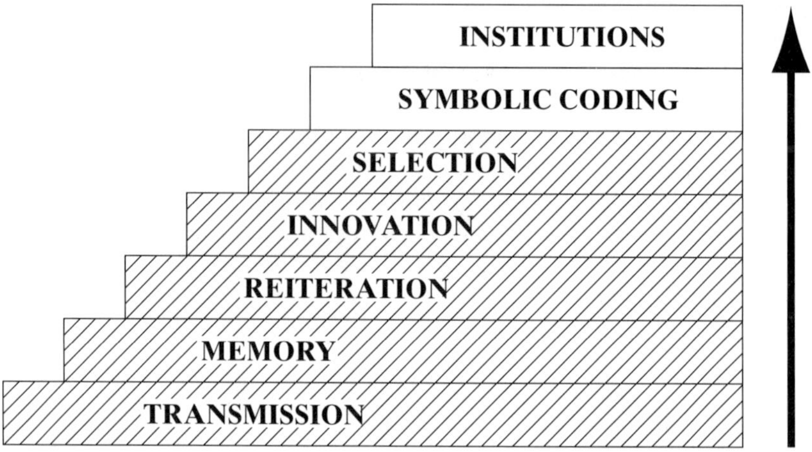

FIGURE 3.4 The Pyramid of the Archaic *sapiens*

Reflections

In the overall context of our evolutionary journey, we defined the last phase of the first mile journey as an evolutionary transition from the *Homo* state of our species to modern man. This transition happened during the life cycles of two species: Archaic *sapiens* and the Neanderthal. The story we just listened to, the story of the Archaic *sapiens*, was laden with uncertainties, paradoxes, and differences of opinions among evolutionary scholars. An obvious reason for this is that, while the earlier humanoid and *Homo* species lived in specific and restricted regions of Africa in similar environments, Archaic *sapiens*, during their life cycle, inhabited practically the whole world. The environmental and resource conditions of the regions they inhabited showed great variations, requiring a wide scope of behavioral adaptations. The role of Archaic *sapiens* in the "evolutionary transition phase" was therefore crucial as well as admirable. We now say goodbye to our evolutionary parent as we continue our journey and meet with the very special Neanderthals.

3.3.3 Meet the Neanderthal Man: The Evolutionary Enigma

> Neanderthal: The enigma group, the source of all our popular 'caveman' stereotypes of prehistoric humans has seized the modern imagination like no other early ancestor. There are, of course, good historical and cultural reasons behind the lengthy love/hate relationships both scientists and the general public have conducted with the Neanderthals. Theirs were the first human fossils found in Europe, where most of the early anthropologists worked and

> where these investigators located—mistakenly, it turned out—the birthplace of humankind. In the end, ironically, the much maligned Neanderthals would prove to be Europe's only original contribution to the human evolutionary record, and a side branch at that. (Johansons, 1994, pp. 254–255)

First, I locate Neanderthals in the time/place context of evolution and provide a general definition of him. Then, I describe him by using the markers of the evolutionary design space. In conclusion, the "step-up" components of the Cultural Pyramid will give us another picture of his evolutionary advancement.

3.3.3.1 Who Was This "Enigma Man?"

Homo sapiens Neanderthal is the scholarly designation of this species. His most likely ancestor—*Homo erectus*—came from Africa possibly earlier than 300,000 years ago. I say most likely, because this proposition is hard to identify from fossil remains. By the so-called Y bifurcation of evolutionary departure, he split from *H. erectus*, and entered the evolutionary scene some 200,000 to 120,000 years ago and departed soon after *Homo sapiens sapiens* showed up on the evolutionary scene. Based on some recent fossil evidence, Tattersall (1998) places another *Homo* species between *H. erectus* and the Neanderthals, called *Homo heidelbergensis*, who came to the evolutionary scene about half million years ago. The Neanderthals flourished through tens of thousands of years through the upcoming Ice Age and by about 70,000 years ago they spread all over Europe and the Middle East. We call him the "Enigma Man." Even though we know much about him in terms of fossil evidence, opinions about him vary widely. It is quite a task to characterize him. I attempt to walk a road between the two extreme positions taken by some of the experts.

3.3.3.2 First, What Was He Like?

Chris Stringer, the noted palaeontologist, said that "what made the Neanderthals was ice." His anatomy was suited well to the extreme cold of the Ice Age. The Johansons (1994) say that he lived in rock and open-air shelters and in caves. His bulbous protruding nose was his most striking feature, warming incoming air and radiating excess body heat to prevent overheating from exertion. His musculature can only be called awesome. His bones withstood extreme stress. He was extremely active physically. His brain size exceeded ours. Unfortunately, we do not know what he used his large brain for, since his tools—beautifully crafted flint tools—remained unchanged throughout his life cycle.

3.3.3.3 Mapping the Design Space of Neanderthals

Here, I characterize the Neanderthals by applying the set of markers operating in the evolutionary design space of our species.

3.3.3.3.a Consciousness. I repeatedly said that oneness with nature and oneness with wholeness was the original state of consciousness. In his *The Neanderthal Enigma*, Shreeve (1996) elegantly voices this "oneness" as he summarizes the state of consciousness of the Neanderthals:

> The purpose of knowledge to a Neanderthal would not be to gain control, but to increase intimacy, not just between individuals but between the individual mind and whatever it sees, touches, smells, and remembers. The plurality of 'selves' we invent to negotiate our guarded social encounters would be a waste of psychic energy for Neanderthals. Instead, let's give them a single but infinitely graded ego, an analog self, as opposed to our digitized identity. A Neanderthal thought would be much harder to abstract from the thing or circumstances that he thought about. The perception of a tree in Neanderthal mind feels like the tree. Neanderthal psyche floats on the moment, where the metaphor of consciousness as a moving stream is perfect, the motion is serene and unimpeded by countercurrents of re-think, counter think, and double-think. (pp. 340–341)

The presence of burial sites indicates some element of consciousness, namely death awareness. The story is told of the burial of an old man in the Neanderthal period, who was laid out on a bed of plant material, surrounded by many types of spring flowers, many with medicinal value, indicating that the old man was a medicine man or a shaman. (As a dramatization of differences between interpretations, one expert says that the flower petals were blown into the cave and covered the body.)

A young man's body was found in France, arranged in a flexed position, sprinkled with red ocher. The burial ceremony expressed the commitment of the group to a kin after his death. Commitment to the living is indicated by skeletal remains that are evidence for Neanderthals caring for the sick and wounded.

3.3.3.3.b Communication. There is an intensive debate about the language ability of the Neanderthals. Their anatomical feature—higher larynx position in the throat—indicates that they were limited in language use, and thus in cognitive ability. Fossil remains indicate that, on a scale of 1 to 10, the vocal track structure of early Archaic *sapiens* was 10—a fully human scale. But the Neanderthal's structure regressed to 5 to 8 (Leakey and Lewin, 1993). Their speech, if they had any, must have been strongly nasalized and extremely slow. They were unable to construct complex sentences.

3.3.3.3.c Social Structure and Style of Life. The Neanderthal was not organized in a nuclear family (Shreeve, 1996). There is no evidence for biparental provisioning for the young or division of labor or food sharing. They did not live in male-protected family arrangements. They did not transport food back to share with females and the young. Usually males were eating with their peers, living separately from females and children. They solved the vulnerability problem of the young by settling in spaces where food resources were available for females with young within day range. Such a profound difference from Archaic and *H. sapiens* bolsters the position that the Neanderthals and *H. sapiens* were separate species. Furthermore, Neanderthals could not build alliances with other groups through the exchange of mates—a custom of *H. sapiens*—because they were essentially a mateless society. Their strategy was to avoid contacts, stick together, and stay close to areas that provided resources. "Neanderthals seem to have passed their days in almost unrelieved isolation. The sparsity and the small size of Neanderthal sites suggest extremely low human population densities. In the whole Aquitaine area, for example, there may have been only a few hundred people living at one time" (p. 145).

3.3.3.3.d Subsistence. Their diet was a mixture of meat and plant food (Shreeve, 1996). They earth-baked their food in pits lined with fire-heated rocks and covered with soil. They scavenged large animals and hunted small and medium size prey. Their pattern of food acquisition was fishing and scavenging or hunting animals that were walking or flying nearby rather than organizing hunting parties and going after food or migrating to hunt particular seasonally available animals.

3.3.3.3.e Technology and Planning. As previously mentioned, the Neanderthals' tool technology was unchanged during their life cycle. Middle Stone Age flint flakes are associated with Neanderthal fossils. Their tool industry included backed knives, side scrapers, points, and hand axes, altogether some sixty tool types. There is some evidence that by the end of their life cycle as a species, they adopted the tool making technologies of *H. sapiens*. What is intriguing is that they usually left their tools behind when they abandoned their home base. This is one piece of evidence that they lacked the ability to plan ahead and to anticipate the future, including the future availability of food sources. They did not modify their environment for comfort in their caves; they slept taking advantage of the contours of the earth. There are strong indicators that they could start fires and often spread on the surface a thick layer of ash to make a warm bed for sleeping. But they did not innovate, it was anathema to them. Novelty and strangeness in all forms were avoided. They did not like newness or change. They had stability in technology and in lifestyle for some 100,000 years, another indication of a lack of language ability.

3.3.3.3.f Aesthetics. While we cannot point to an aesthetic culture of the Neanderthals in their remains, there is some evidence of the beginning of artistic creativity, at least the beginning of symbolic representations, for example, scratched lines on an ox rib, zigzag marking, etched ostrich shells, and use of some color. But they did not invent new technology.

3.3.3.3.g Ethics/Morality. For the Neanderthal there was "good" or "bad" but not ethical or moral good or bad (Dennett, 1996). They had good fires and bad ones, good hunting and bad, good shelters and bad ones, but no concept of goodness, no concept of a moral person or a moral act. "They had no concept of right or wrong because these are things that relate to man in Society not in solitude" (p. 454).

3.3.3.3.h Spirituality. The burial of kin in the caves of the Neanderthals indicates some spiritual connections with the dead. Their spirituality was also revealed in rituals (Shreeve, 1996), connected with ritualized cannibalism. Not far from Rome, an 80,000-year-old home site tells the story of a bear cult. They found carefully arranged bones of a brown bear, placed in a stone-lined pit, along with the skeleton of a Neanderthal youth, who might have been sacrificed. His skull was smashed at the base, presumably to extract the brain. It was surrounded by a circle of stones. "The rituals of Neanderthals may have been macabre, but they were also loaded with human meaning and import" (p. 52).

3.3.3.3.i Relationship with Nature. "Oneness with nature" characterized our original state. This state seems to have been lived by the Neanderthals. They perceived a single vital force that did not separate them from nature. "They had no need to distill life into representations, because its essences were already revealed to their senses. They had no drums or bone flutes, but they can listen to the booming rhythms of the wind, the earth, and each other's heartbeats, and be transported" (Shreeve, 1996, p. 341). In the Neanderthal epoch, there was no separation yet from wholeness and nature.

This observation leads me to think about our evolutionary future. Looking ahead toward evolution yet to come, I wonder if in an ultimate evolutionary state we will not have again oneness with wholeness and oneness with nature, but in this state we shall have *conscious oneness* and *conscious wholeness*, guided by shared ethics and morality.

3.3.3.4 Neanderthals and the Cultural Pyramid

On the advancing steps of the Cultural Pyramid of *transmission, memory, reiteration, innovation, selection, symbolic coding,* and *institutions,* the Neanderthals did not move beyond the first three steps. They rejected innovation and novelty. As we discussed earlier, they were conservative; their technology was unchanged for more than 100,000 years. Figure 3.5 depicts Neanderthal's advancement on the Cultural Pyramid.

3.3.3.5 Relationship with *Homo sapiens* and the Extinction of the Neanderthal

In the evolutionary scholarship community no one doubts that the Neanderthals were a species of their own. They were what is often called the last experimental variation in the design space of human evolution, which ended in extinction. *Homo sapiens* came on the scene some 35,000 to 50,000 years ago and by about 30,000 years, the Neanderthals disappeared. Had the two species encountered each other? Had they any influence on each other? There is no evidence of interbreeding. We know that there was a lower life expectancy for the Neanderthals and a lower rate of death in

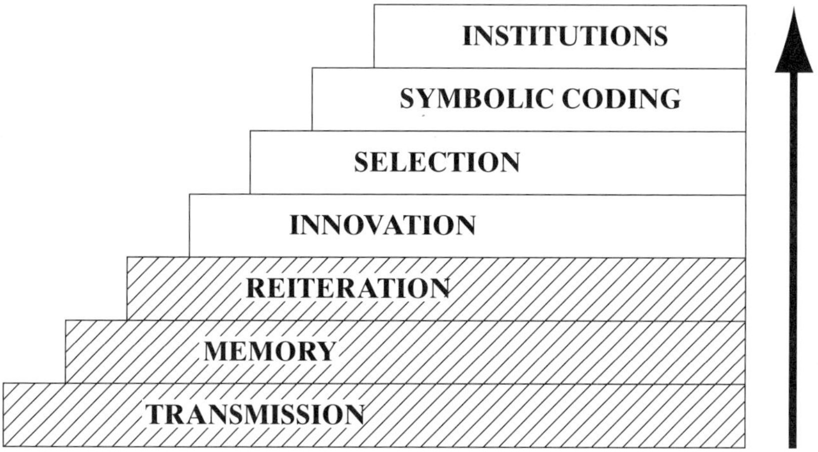

FIGURE 3.5 The Pyramid of the Neanderthals

H. sapiens. The difference in lower death rate was manifested primarily among the young. "It is easy to assume, that *Homo sapiens* did something right and the Neanderthals something wrong because we are here and they are not. We don't need to feel sorry for them. They had a respectable species life and left their mark" (Johansens, 1994, p. 289).

Reflections

I designated the last phase of the first mile of the evolutionary journey as the "transition" from Archaic *sapiens* to *Homo sapiens sapiens*. Our visits with the "Dawn Man"—the Archaic *sapiens*—and the "Enigma Man"—the Neanderthal—discovered that although these two species had reached the threshold of becoming modern man, they did not step over it. They disappeared. This tendency of "disappearing" is part of the human condition. Its source is *always* resistance to change and trying to keep things as they are—even at times of major transformations—when the choice is between evolution and extinction. We have much evidence of this phenomenon even today, when, in the face of massive societal transformations and emerging new realities, many of our social and societal institutions resist change and find themselves declining and eventually disappearing.

Now, we say goodbye to our "ancestors" as we reach the first milestone of our evolutionary journey and embark on the second leg of our travel, as we meet the First Generation of *Homo sapiens sapiens*: Modern Man, the Cro-Magnon. Before we meet him, you are asked to reflect on the evolutionary "transition" discussed in this chapter and take on an activity that helps you the create your own meaning of this transition.

Activity 7

Task 1. Review the transition phase and identify and note core evolutionary ideas of the story of the Archaic *sapiens* and the Neanderthals. Enter these ideas in your workbook.

Task 2. Review Figures 3.4 and 3.5. Compare the evolutionary advancement of the two species. Speculate about the meaning of the difference between the two species. What brought this difference about? Enter your findings in your workbook.

Summary

In the present chapter we followed humanity's descent from the time when by an evolutionary "Y," the humanoid race split from the chimpanzees and from a line of common ancestry some seven million years ago, and embarked on its own evolutionary journey that lead to the threshold of reflective consciousness and the revolutionary emergence of *Homo sapiens sapiens*. In the course of this evolutionary descent, we witnessed several "evolutionary experiments" of the humanoids, *Homo*, and Archaic *sapiens* species. During the first leg of our journey, we visited with Lucy and her family, met the "handyman," *Homo habilis*; and the Turkana Boy of the *Homo erectus* species. The second leg of the first mile of the journey represented a transition phase from *Homo* to *Homo sapiens*, as we described the evolutionary story of the Archaic *sapiens* and the Neanderthals.

The completion of transition marks the end of the primacy of biological evolution as *Homo* reached the threshold of becoming modern human. From here on, his evolution becomes solely cultural. It also represents a major evolutionary shift from the *descent* of the humanoid race to the *ascent* of modern man. The story of this ascent is described in Chapter 4 as a progression through the stages of several revolutions: the human revolution of the Cro-Magnons, the agricultural revolution and the emergence of ancient civilization, and the scientific and industrial revolutions, leading up to the information/knowledge revolution of our era.

4

The Story of the Evolution of Homo sapiens sapiens

As we meet *Homo sapiens sapiens* at the first milestone of our evolutionary journey, we are surprised to recognize him as one of us: an anatomically modern man. He can talk to us and shows us his beautiful cave art. The emergence of *Homo sapiens sapiens* completely changed the story of human and societal evolution. The phenomenal speed of his emergence actually changed human evolution into *human revolution,* which then became a turning point in the evolutionary saga of the human race. It completed the transformation from *Homo* to *Homo sapiens sapiens* to modern man.

It is remarkable how insightful was the writer of the book of Genesis. The depiction of a blissful life in Eden (for millions of years of life of the humanoid), the eating of the apple from the tree of knowledge (the emergence of consciousness), and, as a result, departing from Paradise (leaving dream time) remarkably parallels the evolutionary story of our species. The Genesis story is a true metaphor for the emergence of modern man. Once the first man and woman took the apple from the tree of knowledge, each became a *Homo sapiens sapiens,* "wise man." As they left the dream world of the "Garden of Eden," they became "knowing that we know" people. Out of Eden, they woke up from the dream state of millions of years and now had the privilege and the burden of self-reflection and making conscious choices. The price that they had to pay for this "knowing" was the loss of the security of the "oneness with wholeness" and the "oneness with nature." This "knowing" also represents a major shift from biological evolution to cultural evolution. Stepping over the threshold of consciousness, *Homo sapiens sapiens* (HSS) awakened to the sunlight of knowing that they know. They "self-reflected" as they asked: Who am I? Who are you? Who are we? and What is happening to us? Leaving the Garden of Eden, they became separated from nature. They now were left on their own and began to create their own world.

And what a wonderful world they did create! They immersed in self-reflecting consciousness, which led them to creative action. They experienced spirituality, had a conscious view of the world, and developed new modes of symbolic communication: speech. They created beautiful visual art and music and designed a new generation of tools and ornaments. They developed and formalized new social relationships, built trading networks, and generated oral traditions. *They created and lived culture. They were us.*

From the overall perspective of evolution, the emergence of HSS and the disappearance of other humanoid species mark the end of several millions of years of "experimentation" in the evolutionary design space of the humanoid and *Homo* species. When *Homo sapiens sapiens* emerged, he became the only of our species left. (The use of "*sapiens sapiens*" underlines the difference between *Homo sapiens sapiens* and the Archaic *H. sapiens* and the Neanderthal *sapiens*.)

In the present chapter, we continue our evolutionary journey in the course of which we observe several structures of consciousness emerging—the magical, the mythical, deep consciousness, and the ego-focused mental consciousness. We shall see the development of several modes of communication—speech, writing, print, wired and wireless, electronic, and now the Internet. We shall account for three major, revolutionary transformations. Following the first, the "human revolution," which brought forth the emergence of HSS, we witness the agricultural revolution and the emergence of ancient civilizations, and, next, the third transformation, the scientific/industrial revolutions. In Figure 4.1 these three revolutions mark three successive evolutionary stages, signaling the emergence of three successive generations of HSS. On the figure, the widening spiral depicts the ever expanding boundaries of time and life space as well as the increase in complexity of sociocultural evolution.

4.1 Meet the First Generation of Homo sapiens: The Magician Artist

In evolutionary scholarship, the first HSS became designated as the Cro-Magnon man. The source of this designation is related to the name of the geographic region in southern France where, in 1868, in a rock shelter, fossilized skeletons were discovered in a cave. They looked very much like the skeletons of people today. In describing the story of the Cro-Magnons, I first identify the time and place of the appearance of HSS on the evolutionary scene, followed by an exploration of the evolutionary design space

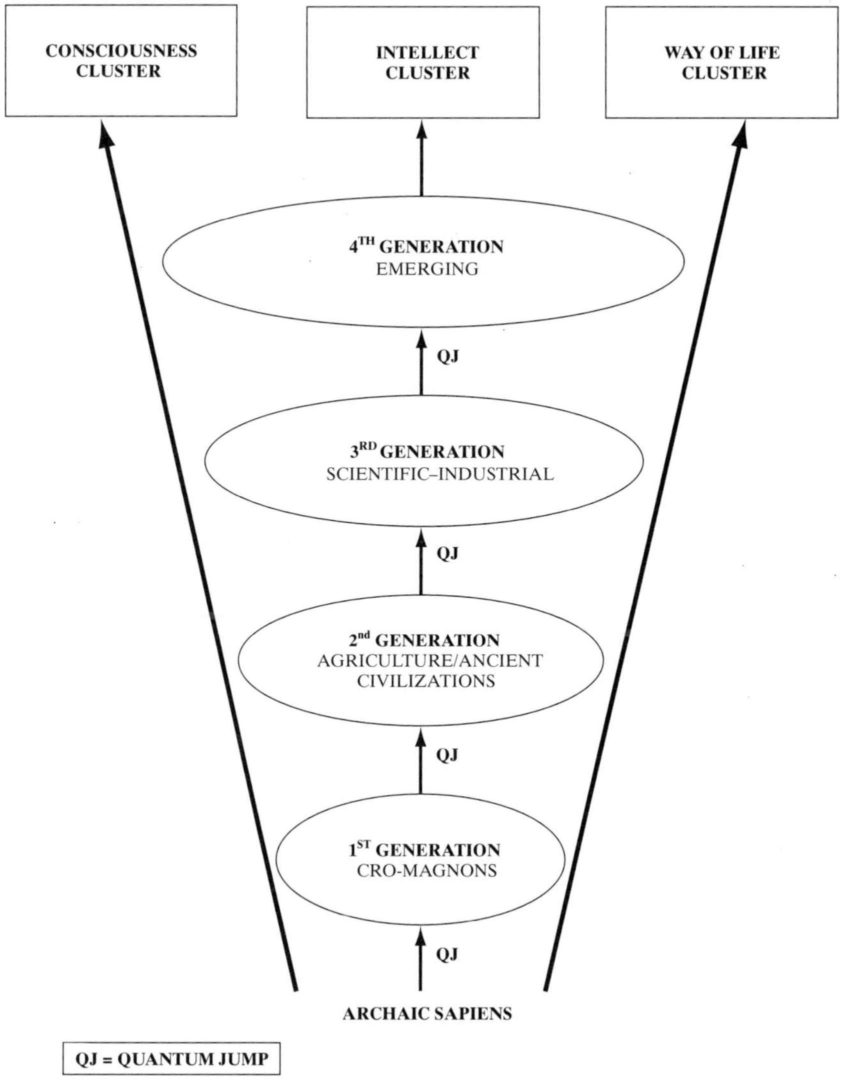

FIGURE 4.1 The Evolutionary Generations of HSS

of the First Generation HSS. Then I ask: How far did Cro-Magnons advance on the ascending steps of the evolutionary Pyramid?

4.1.1 When and Where

Some 40,000 to 35,000 years ago, at the beginning of the Upper Paleolithic era, a new and radically different human species appeared on the evolutionary scene: *Homo sapiens sapiens*. It was also the era of the last Ice Age, when much of the northern half of Europe was polar desert. But the climates in southwestern France and Spain and the river valleys of the Central Russian Plain were reasonably hospitable, with rivers full with salmon and valleys with abundant trees and plant life, offering "comparatively balmy 'refuge areas' where humans were able to survive on a permanent basis" (Potts, 1997, p. 315).

But what evidence do we have regarding the "when" and "where" of this emergence? To explore these questions we seek two major kinds of evidence: biological and paleoarcheological. For biological evidence we turn to Allan Wilson and his colleagues (1992), who in the early 1990s decoded the evolution of the modern human by an analysis of mitochondrial DNA. (Human mitochondria is a cellular "organelle" that is responsible for generating energy.) Their analysis determined that the transformation of Archaic *sapiens* to *Homo sapiens* occurred more than 100,000 years ago in southern Africa. All humans living today are descendants of that population. At the September 1998 meeting of the National Academy of Science, population genetics researchers at the University of Texas at Houston confirmed the findings of Wilson and his team. These studies confirmed the "Out-of-Africa" hypothesis. Modern humans evolved in southern Africa and migrated to Europe, Asia, and eventually to the Americas.

Paleoarcheological findings at the Qafzeh and Klasies River Mouth indicate that, about 100,000 years ago, *Homo* had acguired the identifying physical features of modern man, namely, the high rounded cranium, reduced teeth, prominent chin, and relatively gracile body. But our distinctive culture did not emerge until much later in our history as a species (Johansons, 1994, p. 301). Thus, the existence of anatomically modern individuals in southern Africa makes them contemporaries of the Neanderthals and Archaic *sapiens* in Europe and Asia. It was not the attainment of modern human anatomy but *the emergence of human culture* that propelled *Homo* some 35,000 years ago to become *Homo sapiens sapiens*. This dramatic change is manifested in the explosive spread of ideas across the Old World in the Middle East and in Australia. These changes can be recognized immediately through the work of artists of that period, who in human history have always played a key role in cultural and political revolutions. (From here on I will name the First Generation of HSS as Cro-Magnons, knowing well that they are only one of the HSS populations.)

4.1.2 The Cultural Evolutionary Model of the Cro-Magnons

In building a cultural model in the evolutionary design space of Cro-Magnons, I continue to use the design markers applied to *Archaic sapiens* and the Neanderthals, but with some modifications. Contemplating the use of those markers, I revisited them with the intent to see if it might be possible to move away from their earlier linear presentation by organizing them into a cluster of markers and into a system of clusters.

What emerged was a systemic image of a possible cultural evolutionary model having three interdependent and interacting clusters or subsystems of markers. Furthermore, within each of the subsystems, the component markers are also interdependent and interactive, creating an internally consistent whole: *The cultural system model of the Cro-Magnons.* In the following paragraphs, I introduce the component markers of the three clusters (subsystems) and explain their systemic interaction. My expectation is that this kind of systemic arrangement of markers might lead to the establishment of a "generic" systems model of cultural evolution, which might be adapted to portray the several evolutionary generations of HSS.

4.1.2.1 The Consciousness Cluster

Consciousness, world view, spirituality, and ethics/morality constitute one subsystem, one cluster of the Cultural Model. Consciousness is manifested in what we believe in, how we behave, and what our a view of the world is. It was also the source of the Cro-Magnons' spirituality as well as their art, ethics, and morality. From the interaction of these we gain insight into what the Cro-Magnons believed in, how they viewed the world, and how their ethics guided their behavior. Inasmuch as consciousness is the source of the other markers, I label this subsystem the *primary cluster of the model, the consciousness cluster.*

4.1.2.2 The Intellect Cluster

Communicating, thinking, and learning are interacting and interdependent markers. We have already observed the recursive and mutually influencing effect of these on each other and on the development of intellect and as a force to generate increasing complexity. Thus, these markers comprise another subsystem that tells us how the Cro-Magnons communicated through speech, how they thought and learned. This cluster created the identity of their "symbolic intellect." I label this the *intellect cluster.*

4.1.2.3 The Way of Life Cluster

Social behavior and organization, mode of subsistence, technology, and relationship with nature are interrelated markers that constitute another subsystem of the Cultural Model. This subsystem reveals what they do, how, with what, and how they organize their lives. Their social organization is the arrangement through which they relate to each, secure their means of subsistence, design and develop and use their instruments, and work with nature and use natural resources. The label for this subsystem might be *the way of life cluster.*

4.1.2.4 Cross-Boundary Clusters

The internal relationship of markers within each of the subsystems is only one kind of systemic interacting arrangement. Another type cuts across the boundaries of the three subsystems. For example, language, consciousness, and social behavior act and interact recursively. They mutually influence and reinforce each other to reach higher levels of complexity and attain increased evolutionary competence. Another cross-boundary cluster can be constructed around technology. Developing new technology is enhanced by language. We could continue this relating and influencing "coevolutionary" process of the markers. It would eventually lead to connecting every marker with all other markers. This speaks for the systems principle that *everything is connected with everything.*

4.1.2.5 The Core Marker: Creativity

At the very center of the Evolutionary Cultural Model is *creativity*, which is the core marker of the process of sociocultural evolution. It is the core marker in that evolution is continuous *creation.* It drives and organizes evolutionary experimentation. It is the guiding force of the process of emergence. For five to six million years in the life of our ancestors, in the design space of human evolution, it was natural/biological selection that played the creating role. When the transition point was reached from humanoid to *Homo*, cultural evolution began to share increasingly more the design role with the biological. By the time the Cro-Magnons emerged, the biological role of evolution had been phased out and cultural evolution took over.

Figure 4.2 depicts the relationships among the three subsystems or clusters of the evolutionary cultural model as well as the arrangement of the component markers within the clusters.

4.1.2.6 Four Design Roles

I suggest that in the evolutionary design space of the Cro-Magnons, there were four key central design roles that brought forth the transformation from Archaic *sapiens* to *Homo sapiens sapiens* and kept the system evolving: the priming, the enabling, the instrumental, and the driving roles.

Consciousness played the *priming* role. The Cro-Magnons, stepping over the threshold of self-reflecting consciousness and asking: Who am I? Who are you? What is out there? triggered the "human revolution" and primed the emergence of HSS.

The development of the brain and the intellect, speech as symbolization, new means of communication, and conscious learning *enabled* the activation of latent evolutionary potentials that became realized in all other markers of the Cultural Model. These played the *enabling* role.

Technology and its various manifestations played the *instrumental* role as it actualized and operationalized evolutionary advancements in the everyday life of Cro-Magnons.

Creativity played the *driving* role. Creativity is the *driving force* of evolution. It brings forth novelty after novelty and advances human evolution.

4.1.3 The Cultural Evolutionary Model of the Cro-Magnons: An Elaboration

The Cultural Model of the Cro-Magnons has three clusters of evolutionary markers that—in their interaction—create the model. These clusters are now elaborated, followed by examples of formulating cross-boundary clusters.

4.1.3.1 The Consciousness Cluster

This subsystem of the systems model of cultural evolution is clustered from the interaction of four evolutionary markers: consciousness and a view of the world, spirituality, aesthetics, and ethics/morality. Inasmuch as consciousness is the priming source of the rest of these markers, I labeled this cluster earlier the *consciousness cluster*.

4.1.3.1.a Consciousness. The journey of the modern human began in earnest with the emergence of the Cro-Magnons some 35,000 years ago

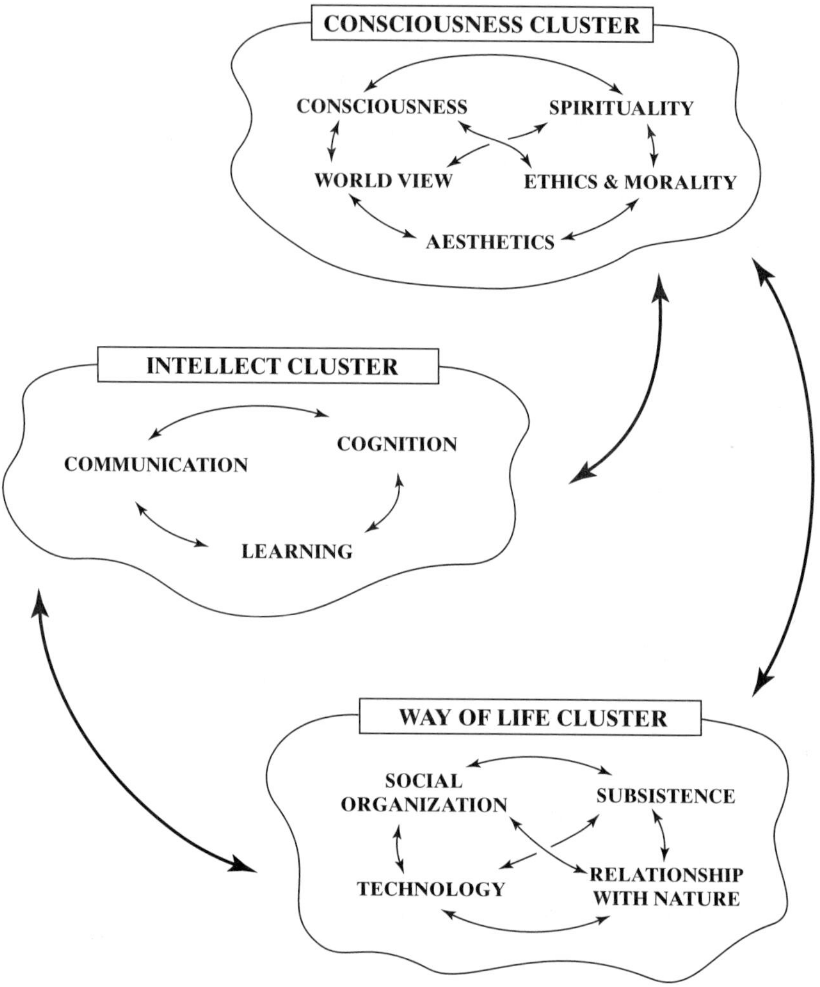

FIGURE 4.2 The Cultural Model of the Cro-Magnons

(Elgin, 1993). "At this time, the glacially slow development of the human culture and consciousness suddenly achieved a critical mass and begun its take-off into sustained development that leads directly to the modern era. Humanity finally had broken free from its "contracted within" consciousness and moved into the world with a dramatic burst of creative energy" (p. 45). This newly emerged consciousness enabled humans to step back and observe themselves. This self-reflective capability led Cro-Magnons to acknowledge the simple fact of their existence and created the capacity for "knowing that we know." Elgin says that this capacity brought them into a world of new feelings, such as mystery and awe, laughter and tears, and love and fear. Emotions such as these made consciousness a subjective experience.

In the consciousness of Cro-Magnons, the ultimate vicarious experience was the fear of death (Leakey and Levin, 1993). Archeological evidence attests to some deliberate burials before the Cro-Magnons, but their subjective consciousness made burials much more elaborate and magic laden and their emotional experience tied in with the ritual so much more keenly. "They experienced the world as immediate, a place with many unknown and uncontrollable forces, magical events, and strange happenings" (Elgin, 1993, p. 52). Frankfort (1946) remarked that while we see our environment as "it," the Cro-Magons saw it as "thou," as sacred and full of life. Sensations, thoughts, and feelings integrated into a direct experience. Furthermore, Leakey and Lewin (1993, p. 305) points out that: "We can only speculate that elements of consciousness—the sense of self, the tendency to attribute feelings to others, the facility to know the world better, and the emotion of compassion—all were enhanced through time as the evolutionary ratchet advanced." Once the "Cro-Magnons passed the threshold of self-awareness, there welled up in their mind the Big Question: *WHY? This question was a search for meaning* in the midst of uncertainty."

Reflecting consciousness was the "priming" force—the springboard—of creating consciousness which brought forth an explosive evolution—a revolution—in the arts and spirituality; in communication, cognition and learning; in social and institutional development; in new modes of securing subsistence; and in technology. The *interplay between reflecting and creating consciousness opened up a new world and created a new view of the world.*

4.1.3.1.b Spirituality. For the Cro-Magnons, nature and the supernatural were intimately connected (Elgin, 1993). Today we understand that a rainstorm is caused by changing currents and temperatures. The Cro-Magnons attributed the storm to one of their many gods. They had gods for all the important aspects of their life. It was the role of the shaman to intervene and gain favor from these gods. It was believed that shamans possessed special powers that enabled them to make contacts with the spirit world (Lewin, 1998). The shaman played a special role in both defining and maintaining a society's mythology. Among the San people in southern Africa their functions included self-induced trance during ritual ceremonies, followed by painting artistic images. And most of the visual art of other HSS people in this period consisted of expressions of spirituality. Shamanism can be defined as a group of practices (Krippner, 1997) by which shamans "enter the 'spirit world,' obtaining power and knowledge that is used to help and heal members of the social group that has given them shamanic status"(p. 1). Krippner (1998) further says that:

> For the shaman, the totality of inner and outer reality was fundamentally an immense signal system. Shamanic states of consciousness were the first step to deciphering this signal system. Shamanic technology yielded information from a database consisting of dreams, visions, intuitions as well as keen observation of the natural and social world. As language moved from a purely social function to include a general-purpose function, consciousness shifted from a means of predicting other individuals' behavior to a mode of managing mental databases of information relating to all domains of activity. (p. 2)

Every action, every place, and every situation was infused with a mysterious magical presence that worked in unseen ways as an omen for good or ill. Through magic, they substituted one object for another (a practice that survives to this day in the voodoo culture). They invented rituals and ceremonies to gain favor with the powers of nature. They did not understand how nature worked and saw it governed by unseen forces (Walsh, 1990). Therefore:

> It was vital to learn how to control these forces—and this brings in the role of the shaman. The shaman was the man or woman able to reach behind the vile of the visible world and engage the Life-force behind all things in order to serve the community—healing the sick, assuring good hunting, preserving the ceremonies and rituals of the group, and much more. (p. 15)

For every event and every aspect of their lives they called upon the assistance of the Life Force (Elgin, 1993). Rituals, chanting, and trance dancing gave them a direct experience of union with the Life Force and took them into the world of spirits. Furthermore, they sought the favor of the spirits by offering something that was precious to them, for example, an animal or even one of their own fingers. They expected something equally precious in return.

Archeological records dating back to well before 20,000 years ago (Combs, 1996) show in the art of Cro-Magnons numerous carvings of feminine figures. These figures strongly suggest the symbolic worship of the feminine principle and fertility. Campbell (1988) suggests that the figure of the Naked Goddess represented a central mythology of Cro-Magnons. This mythology was carried over to the next stage of human evolution to people of the agricultural villages. Furthermore, they worshiped "grave gods" at burial ceremonies.

The *Homo sapiens* Aborigines substituted mental complexity with layers of drama, magic, and spirituality (Johansons, 1994). "Aboriginal culture revolves around the concept of Dream-time, that is the ancestral past, when mythological beings created the land, gave its shape and life, and passed on a secret law" (p. 305). There was an indissoluble connection between the land and the spirit of their ancestors, which was the cornerstone of their spirituality and their culture.

4.1.3.1.c Their World View. A world view is a composite by which we conceive meaning and construct our own reality. The Cro-Magnons' consciousness and their spiritual and artistic state comprised such a composite, which represented a revolution in the evolution of our species. It was "a new perception of the world and the possibilities in it" (Johansons, 1994, p. 306). It was a radical shift far more explicit and powerful than the world view shifts at later evolutionary stages for example, the world view shifts of the agricultural and the scientific/industrial revolutions. In the archaic state, the view of the world was "cosmocentric," unconsciously embedded in the material cosmos (Wilber, 1981). But the Cro-Magnons gradually separated the self from the cosmos-centered consciousness. Within the self, there was an awakening of the personal "in here," which was separated from "the world-out-there" (p. 40). Even though there was this separation, the self was still intermingled through magic with the "out there." The emergence of the focused self was constrained by living only in the present with no perception of time, or no perception of the future. They lived in a moment-to-moment state, focusing on self-preservation and believing in magic as a defense against—and even denial of—death. Campbell says (1959): "Where there is magic there is no death."

4.1.3.1.d Their Ethics and Morality. Using their new reflective consciousness, the Cro-Magnons recognized the advantage of intensive social cooperation and created a truly egalitarian society. The social ethics of cooperation guided their mode of life in the home base. Cooperation was crucial in acquiring food by their hunting ventures and gathering other sources of subsistence. In the service of " the common good" they were to overcome the biologically determined selfishness as they created an altruistic culture. Food and other possessions were readily shared. The notion of private property was unknown. Particularly unknown was "owning land," which, at the next stage of evolution, during the agricultural stage, became an overriding objective. As we know too well, it resulted in envy, hostility, and even warfare between groups and societies. In human and societal evolution the era of the Cro-Magnons was indeed uniquely idyllic.

4.1.3.1.e Their Art. In the history of humankind, artists always played a central symbolic role. For the first time in the evolutionary saga of the human race it was the Cro-Magnon artists who gave us remarkable art that best expressed their spirit, soul, and thoughts. Art in itself was not the source of the emergence of modern humanity but it was the most remarkable expression of it. *It expresses what it means to be a modern human.* No other evolutionary stage was so centered on art and aesthetics as the life period of the Cro-Magnons. The Cro-Magnons, the First Generation modern humans, had a deep spiritual and aesthetic soul, which animated their quality of life. (The question that begs to be answered is: How can we regain that soul?) Commenting on the beauty and perfection of Cro-Magnon art, Lewin (1998) describes a scene in one of the sites as follows:

> Exquisitely carved horses: miniature, near perfect renderings of "horseness" in ivory: bold, colorful images of bison, horses, and mammoth painted on cave walls, dwarfing the awed observers, curious compositions (a seal, a salmon, a snake, a tiny flower, and other markings) etched into a piece of reindeer antler; a preponderance of carnivore (lion and fox) teeth as pendants; mysterious chimeras: part human part beast. By now more than 200 decorated caves have been discovered in Europe and more than 10,000 decorated objects. (pp. 138–139)

The Cro-Magnon artists who painted the caves are "us" says Gould (1996). "One look at the incredible richness and beauty of this work convinces us, in the most immediate and visceral way, that Picasso held no edge in mental sophistication over these ancestors with identical brains" (p. 220). In his most recent work, Gould (1998) says that in the contemporary evolutionary conversation "one subject—and only one—elicits absolute unanimity of judgement of ancient life, though for reasons more visceral than intellectual. Every last mother's son and daughter among us stands in reverent awe and amazement before the great cave paintings done by our ancestors in southern and central Europe between roughly thirty thousand and ten thousand years ago" (p. 162).

The sites and functions of the Cro-Magnons' art should be seen in the larger context of their life experience. The cave art at the easy to reach dwelling spaces had a decorative and life-representational function. But art that represented magical, mythical, ceremonial, and spiritual functions was located at difficult to reach, often the most remote recesses. Tattersall (1998) described the mythical art with great awe. At one cave in France, one has to paddle up an underground stream, and wriggle and crawl for hours before, a mile underground, he reaches a chamber at the end. Here we find a statue, sculpted out of the clay floor of the cave. It is a magnificent statue of a pair of bison, one male and one female. Looking at them we are overcome with deep emotion and admiration. It is if we are seeing the great work of a Renaissance artist.

Considering art in the context of consciousness and spirituality we can only say that the images had to be meaningful to the Cro-Magnons in a deeply mythological and spiritual sense. Cro-Magnon art was rich in its diversity. It was manifested in wall paintings, carved statues, body ornaments, and ritual dancing. Music was also very much a part of their culture. They made flutes from bird, reindeer, and bear bones (Lewin, 1998). Far away from the Cro-Magnons, music was also an important part of the life of the Australian Aborigines as well as the South African San people.

As we build an image of *Homo sapiens sapiens* from the interacting markers of self-reflecting consciousness, spirituality, and multiple modality of art, we must feel very close to them and deeply admire the quality of their humanity. This admiration will no doubt grow as we explore the other markers of their culture.

4.1.3.2 The Intellect Cluster

This cluster comprises the markers of communication, cognition, and learning art. These three markers continuously interact and depend on the state of each other. As stated earlier, the speech modality of communication, namely language, is a core marker of the First Generation HSS. It enabled cognition and conscious reflection—therefore thinking—and the two enhanced and expedited the potential for learning.

4.1.3.2.a Language. Earlier, we explored the conditions necessary for the development of language. In the Cro-Magnons, all the requisite conditions were present. Their anatomical features were ready to enable speech, and this capacity, coupled with the emergence of consciousness and the social pressures to communicate, brought forth the systemic requisites needed to create language. Consciousness, language, and social behavior created the critical coevolutionary interactions that brought forth the conditions necessary for modern humans to realize their evolutionary potential in general and make contributions to the realization of the potential in all the other markers of sociocultural evolution. As we discuss the other markers, we shall recognize the critical role that language has played in the evolutionary development of the Cro-Magnons.

"Speech symbols, as distinct from signals, served as proxies for direct experience" (Curtis, 1982, p. 91). Signals point to—and are bound up with—their referents, calling for action. Signals were the means of communication of the ancestors of *Homo sapiens*. Symbols do not point to, but represent. They can be separated from their referents and can be attached to many referents. Speech symbols represent our experiences and permit us to reflect on ourselves and our experiences. Speech symbols served to separate man from his habitat. The developmental advancement from signal to symbol was indeed a revolutionary change.

Language ability enabled the Cro-Magnons to develop an oral tradition. Stories told around the campfire became the vehicle for talking over the events of the day as well as passing on the stories of past events. The more time passed in the life journey of the Cro-Magnons, the richer and longer these stories became. Telling these stories placed more and more pressure on language development as their life events and social interactions became increasingly more complex. Furthermore, the increase of the quality and refinement of their language skills resulted from the need to build mental models that stood for their construction of their "reality" as they perceived it. The refinement of their speech symbols, which represented their spirituality, intermeshed with their artistic symbolisms and provided the bases for their magico–spiritual expression and experience, around which much of their knowledge was centered.

4.1.3.2.b Cognition. Thinking is the invisible source of language. Thought and language are not only linked, but also inseparable. Thinking is in coevolutionary and interacting relationship with language and consciousness. It was the attainment of a certain level of complexity and a requisite architectural arrangement of the brain that provided the anatomical potential for thought to develop in coevolution with language and consciousness.

The emergence of Cro-Magnons with the cognitive capability of modern man marked a major evolutionary transformation. The anatomically modern brain, which enabled ever greater cognitive capacity, can be equated with growing intelligence. The Johansons (1994) suggest that the cause of the evolutionary transformation of the brain that created an intelligent mind "may have been some kind of neurological leap forward, a biological re-wiring in the brain some 40,000 or 50,000 years ago that enabled humans to manipulate culture and the environment in a way—and to an extent—never possible before" (p. 302).

Potts (1997) suggests that "the brain works in the midst of other brains. Our mental organs operate in a social context, and human actions and thinking are manifested in the collective and interactive spheres of the human experience. The results far transcend the biological functions and products of any one person's brain. The driving force behind mental evolution is the complexity of social life. High intelligence means that organisms adapt their strategies to continually changing conditions" (pp. 206–207).

The more complex Cro-Magnon society became, the more creative and calculating the cognitive functions of its members must have become. The *complexity of their mental functions increased in proportion to the increase in complexity of their various behaviors*, such as their spiritual and artistic practices, their home-based social and subsistence functions, their hunting and gathering strategies, their collective tool making, and their trading practices. Furthermore, the distinctive character of their mental capacity enabled them to adapt to the complexities of environmental contingencies as well as make adaptations in their environment to meet their increasingly more complex needs.

4.1.3.2.c Learning. We can project and recognize various kinds of learning taking place in the course of evolution. Erich Jantsch, a leading evolutionary scholar, distinguished four basic learning processes working in human/societal evolution: virtual, functional, conscious, and superconscious. "Two highly significant characteristics apply to all four modes of learning: First, *learning is always open-ended* and never deterministic. It is geared to a process of open-ended experimentation, comparable to the strategic exploration of available options with subsequent vindication of the choice made. In other words, it is learning by doing in an informed way. Second, what is learned is not places or relations, but a sense of direction, thus evolutionary learning inherently refers to process" (Jantsch and Waddington, 1976, p. 50). Let us see what these learning types are and how they manifest the learning modes of Cro-Magnons.

Virtual learning is characteristic of nonreflective archaic consciousness, which was the mode of learning of our ancestors from the time of Lucy and her species to the era of Archaic *sapiens*. Earlier, we saw how *Homo habilis* and *Homo erectus* learned by observing and imitating the use of tools. They learned the modes and manners of making and using instruments, gathering and processing scavenged meat, and establishing home bases.

Functional learning was the baseline mode of learning of the Cro-Magnons. Jantsch and Waddington suggest (1976) that functional learning can be characterized as feedback–interaction between consciousness and environment, constituting a binary link. This kind of interaction is typical in primitive social processes, such as in the social life of Cro-Magnons. Functional learning accelerates evolution by establishing cybernetic links with the environment. In this mode of learning, the perception of variety increases.

Conscious learning is a product of self-reflecting consciousness. It may be viewed as multiple/recursive feedback–interaction. It was the normal mode of learning of the Cro-Magnons. Conscious learning developed and transformed communication from its archaic form of signals and limited speech to the full richness of symbolic human language, and by so doing it further accelerated evolution, enabling and enhancing advancement in all evolutionary markers.

Superconscious learning is transpersonal. It leads to social consciousness. Through superconscious learning we develop our culture. The Cro-Magnons developed their rich sharing which was manifested in internally coherent social, spiritual, and artistic institutions. In general, superconscious learning is the vehicle of social and societal evolution, even the evolution of humanity as a whole. Through superconscious learning we can prefigure the emerging order of the human community, thereby guiding history. "Super-conscious learning provides a sense of direction for cultural and mankind processes by 'illuminating' the process from the far end in terms of guiding images" (Jantsch and Waddington, 1976, p. 42).

4.1.3.3 The Way of Life Cluster

The way of life cluster is a systemic arrangement of four markers of the Cultural Evolutionary Model: social organization and manner of social life, subsistence and modes of its acquisition, technology and its use, and relationship with nature.

4.1.3.3.a Social Organization. The Upper Paleolithic era of the Cro-Magnons represents the most fundamental change in human and social behavior and social organization since the emergence of the humanoid species. Whatever triggered this change, it brought forth new social/behavioral relationships within kinship and tribal groups as well as among groups. With the emergence of HHS—in contrast to the Archaic *sapiens*—a totally new and different social order and intra- and intergroup relationships were established, characterized by a creative balance between competition and cooperation. These newly emerged relationships turned the social organization and life of previous humanoid species into human society. After hundreds of thousands of years of hardly noticeable change the Cro-Magnons created the First Generation world of modern humans—in a very short period of time. In that world they had, essentially, the first manifestation of everything we have today.

In their social life, the Cro-Magnons demonstrated that human groups can help each other without being threatened by other groups; in fact they depend on each other. One manifestation of this was the development of extensive trade relationships. After millions of years of stasis in humanoid history, with the emergence of Cro-Magnons, ever extending relationships were developed between groups and regions, coupled with exploding creative cultures everywhere.

4.1.3.3.b Subsistence. "No one doubts that early humans were accomplished hunters, pursuing the ecologically powerful mixed economy of hunting meat and gathering plant foods" (Lewin, 1998, p. 130). Remains at the home bases of *Homo sapiens sapiens* reveal a wide broadening and variation of their diet as compared to that of Archaic *sapiens*. The ability to use language greatly facilitated skillful and well-organized hunting at larger distances as well as more sophisticated techniques of gathering plant foods. Typically, the Cro-Magnons lived in a two tiered social organization: "in small mobile bands of about 25 individuals, which were members of a larger group of some 500 individuals. In the course of the life-span of Cro-Magnons, their social organization was developed in ever larger settlements. Economic activity was often divided, with males principally responsible for hunting and females the principal gatherers of plant foods (typically the bulk of the diet). The hunting and gathering lifestyle, given its broad geographical distribution, was clearly a successful adaptation" (Lewin, 1998, p. 34).

4.1.3.3.c Technology. Shreeve (1996) noted that: "If human evolution were an epic, the Upper Paleolithic would be the chapter where the hero comes of age" (p. 266). After millennia of hardly noticeable progress, that didn't seem to be progress at all, human culture erupted into a creating explosion. Sophisticated tools, completely different from those used by Archaic *sapiens* proliferated. Now we see regional variations where before there was 'global monotony.' Elegant implements are carved from bones and ivory. In the creation of technology, change replaces stasis. Like fashion today—new industries dominate one day and disappear tomorrow. From Spain to the Urals, site lists begin to read like Sears catalogs, sewing needles, barbed projectile point, fish-hooks, rope, meat-drying racks, stone lamps, temperature controlled hearths, complex dwelling structures" (p. 266).

In Central Europe at the famous Dolni Vestonice site a cluster of five dwellings was discovered (Potts, 1997). The largest measures fifty by thirty feet. This village was identified as a kin aggregation site, "a promise of society, an opportunity for relayed bands to exchange information, find mates, and generally to massage relationships" (p. 279). A few hundred feet from this village, another structure was discovered, dating back 26,000 years. It was a circular structure about twenty-nine feet in diameter, with its roof supported by posts. Inside was a kiln; on its floor more than two thousand fragments of fired clay pieces were found. Close by another kiln was found loaded with ceramic pieces. These sites represent the earliest application of ceramic technology.

4.1.3.3.d Relationship with Nature. The life of the Cro-Magnons was closely and consciously connected and interactive with nature. As a result, human qualities were assigned to the various aspects of nature. "The migratory herd had to be treated with respect, otherwise it would refuse to return next season. Appropriate gifts had to be made to the sun, otherwise it would become angry and not rise. The spring had to be constantly blessed, otherwise it would choose to flow elsewhere" (Leakey and Lewin, 1993, p. 307). Respect for nature was also manifested in the art of the Cro-Magnons. As expressions and images of the spirit world, often animal forms were combined with human forms. One of the best examples of this art is the so-called "sorcerer" painted in an inner section of a cave in the French Pyrenees. It is believed that the image represents a kind of human/animal chimera of spiritual/magical shamanistic experience.

4.1.3.4 The Cro-Magnons' Cultural Pyramid

After millions of years of climbing, the last vestige of our species, the Cro-Magnons, reached the top of the Cultural Pyramid. Climbing upward on the stepping stones of the "Cultural Pyramid" of transmission, memory, reiteration, innovation, selection, symbolic coding, and institution, the Cro-Magnons ascended to the top by mastering symbolic coding and establishing social and spiritual institutions.

Activity 8
In this activity you are asked to complete two tasks.

Task 1. Speculate on what caused the sudden leap from Archaic *sapiens* to HSS. Enter your thoughts in your workbook.

Task 2. Reflect on the Cultural Model presented in the evolutionary design space of the Cro-Magnons. The key organizing principle of the model is that human culture emerged—as manifested in the story of the Cro-Magnons—from the interactions both among the specific cultural markers within their subsystem clusters and among the subsystems of clusters. Examples of these interactions have been provided. Your task is now to describe all possible configurations of enhancing/reinforcing relationships within the markers and among the clusters. Enter these ideas into your workbook.

Reflections

The life-cycle period of the Cro-Magnons was just a second in the evolutionary time scale of our species. But that magical second was time enough for the Cro-Magnons to bring forth the human revolution: the creative explosion of human culture, the unconstrained emergence of Spirit in our species. It was a revolution that superseded the often static and slow process of the mostly biological evolution of our ancestors and replaced it with the swift emergence of modern man and his culture. Spurred by reflective consciousness, innovation and creation were the dynamic forces that motivated and inspired this revolution. These forces brought forth conditions that were well balanced and internally consistent and ensured a harmonious way of life for the Cro-Magnons. The Cro-Magnon period was a time of unprecedented change and fermentation. Tattersall (1998) notes that we still cannot see where this ferment will end, but it is amply evident that with the arrival on Earth of modern *Homo sapiens*, a truly new kind of entity had emerged, one whose potentials we are still exploring and enlarging today. It is the task of this work to search for ways by which we can give direction to the exploration of this potential.

The Cro-Magnons were inspiring and fascinating companions during the second mile of our evolutionary journey. They showed us so much about how the Spirit manifests itself in us, how humans can reflect upon their destiny, create beauty in all forms, love, and create a compassionate society, and live in harmony with their fellow creatures and with nature.

As we now have arrived at the second milestone of our journey, we reluctantly say goodbye to our newly discovered brothers and sisters, the First Generation of *Homo sapiens sapiens*, the Cro-Magnons. Before we depart, a quotation from Stephen Jay Gould (1998) is an appropriate farewell message to them.

> We have loved the rainbow for thirty thousand unbroken years and more. We have struggled to depict the beauty and power of nature across all these ages. The art of Chauvet—and Lascaux, and Altamira, and a hundred other sites—makes our hearts leap because we see our own beginnings on these walls, and know that we were, even then, worthy of greatness. (p. 178)

We are now ready to meet with the next revolutionaries, members of the Second Generation of *Homo sapiens sapiens*, the farmers of the agricultural villages.

4.2 *The Agricultural Revolution and the Emergence of Agriculture-Based Civilizations*

We now meet the first representative of the Second Generation of *Homo sapiens*: the "Farmers," who established the agricultural villages some 10,000 years ago. As we travel toward the third milestone, we shall meet other representatives of this generation: the engineer, the priest, the scholar, the tradesman, the philosopher, the teacher, and the ruler. They lived in the large cities of the ancient civilizations. Later, we visit with the Greek philosophers, the early Christians, and people of the Byzantine and Roman empires.

Leakey and Lewin (1993) see a transitional period from the last 10,000 years of the Cro-Magnon era into the agricultural era. During this period, people began to organize their lives so that they were able to exploit food sources to such an extent that it required less mobility, which thereby enabled them to establish more permanent home bases. More permanent settlements led them to begin to explore food production rather than just hunting and gathering. Thus, development slowly begun to bring to an end the nomadic lifestyle of our ancestors that lasted several million years.

The beginning of the era of the Second Generation of *Homo sapiens* can be placed around 10,000 to 12,000 years ago. We can define the evolutionary design space of this era by the three major clusters of evolutionary markers: consciousness, intellect, and way of life. Different evolutionary scholars give primacy to different evolutionary markers. Some define the period of the Second Generation as the era of "depth or mythical consciousness." Others say it was the emergence of writing as a new mode of communication that best defines the driving force of this era. Many designate the style of life as primary, and label this period the Agricultural Revolution.

A systems view leads us to shy away from an either/or single designation and suggest that it is from the mutually influencing and shaping interaction of the emerging evolutionary markers (in the evolutionary design space) that the sociocultural model of the Second Generation of *Homo sapiens* can be constructed. That is why I elected to designate this evolutionary stage the era of the Second Generation of HSS.

Within the span of life of the Second Generation, we can distinguish distinctive substages. During the first substage, "Farmer *Homo sapiens*" settled in small-scale, low-density villages. They established subsistence agriculture, producing food enough for the people of the village. Occasional hunting enriched their diet. In addition, they slowly initiated the domestication of animals, some for farming and transportation, while others provided subsidence.

The second substage, lasting from 3500 B.C. to the end of the Middle Ages, saw several civilizations emerging, living in high-density settlements. The establishment of these civilizations was enabled by a new level of consciousness, by a new mode of communication, by surplus agricultural production, and by new technologies, all of which made it possible to create large cities in which large numbers of people pursued productive lives in different occupations. What follows is a portrayal of the design space of these two stages in which the cultural models of the Second Generation HSS comes to life.

4.2.1 The Cultural Model of the Agricultural Village

In the evolutionary design space of the Second Generation HSS, we can see a major transformation, a revolutionary advancement of all of the markers of the cultural model of HSS. The cultural model is constructed from the three clusters of the evolutionary markers: consciousness, intellect, and way of life. First, I portray the style of life cluster, followed by the consciousness cluster and the intellect cluster.

4.2.1.1 The Way of Life Cluster of the Farmer

This cluster grew out of an integration of cooperative social organization, the mode of securing subsistence, and the developing technologies of planting, growing, harvesting, and storing.

The transformation of the hunting–gathering lifestyle of the First Generation HSS (the Cro-Magnons) to the farming life of the Second Generation was gradual. With the warming climate that came at the slow end of the Ice Age, wild wheat and barley offered ample supply of subsistence, once the technology developed to harvest this food source. To test how the first farmers could have harvested grain. Elgin (1993, p. 71) describes how an archaeologist used a nine-thousand-year-old flint sickle to simulate their harvest. The results were surprising. He found that a family could probably harvest enough wild wheat and barley in the space of only three or four weeks to feed themselves for a year.

The small village-based agriculture produced an egalitarian social structure. It provided great stability, a period of several thousand years of stasis. But slowly, over this period, people developed ever more efficient farming technology. They learned how to weed wild fields, plant seeds, increase the yield, protect their fields from animals, harvest with better tools, and store food. From such modest beginnings came one of the most fundamental transformations the world has ever known. The gradually increasing efficiency of farming produced increasingly more surplus food, which made possible the rise of large-scale urban civilizations.

In addition to the life space territory of the Cro-Magnons in Europe and the Middle East, the Agricultural Revolution emerged around 11,000 B.C. in southern Asia. This development included the domestication of bananas, yams, chickens, and pigs, followed a couple of thousands of years later by the farming of grains and the herding of sheep, goats, and cattle. In Central and South America farming was established somewhere around 7000 B.C. It should be noted that agricultural production developed independently in each of these regions. The life span of the Second Generation in the agrarian-based villages lasted at least 5000 years. These five millennia of stability were important for establishing a way of life that, in turn, provided the foundation for the eventual rise of the world civilizations.

4.2.1.2 The Consciousness Cluster

The consciousness cluster integrates consciousness, spirituality, world view, and moral/ethical behavior.

4.2.1.2.a Consciousness. During the first phase of the life cycle of the Second Generation, their collective consciousness was transformed from the magical to the mythical. The most significant manifestation of this transformation was the emergence of future consciousness. The world of the hunter–gatherers was focused on the present. It was a moment-to-moment survival, striving for instant gratification. Immortality for a hunter was to live until tomorrow. But the world view of the farmer became redefined in terms of a world of extended time. The focus of activity was making present preparations for future harvests, of being able to gear the actions of the present toward significant future goals and rewards.

4.2.1.2.b Spirituality. Spirituality was a reflection of deep consciousness that was "Earth centered." The many gods of the Cro-Magnons were replaced by Mother Earth or the Great Mother "who represented all nature, matter, instincts, body, crops, earth fertility, sexuality, emotions, desire, magic, and the beginning of myth" (Wilber, 1981, p. 122). The Great Mother had to be appeased by ritualistic sacrifices to evoke her help to ensure fertility, increase the yield of crops, and bring about rain. Sacrifices were offered also to prevent the Great Mother's wrathful vengeance. The ritual was most often blood sacrifice which took various forms. Initially it was almost always human beings, but later animals were substituted.

4.2.1.2.c Their World View. How they viewed the world was greatly influenced by the emergence of future consciousness, which became essential for planning food production. Another step forward was their observation of interrelationships and interdependence among the various aspects of farming activities: planting, cultivating, protecting crops, harvesting, storing, etc. This view is a dramatic manifestation of systems thinking.

4.2.1.2.d Ethics/Morality. The farming way of life in the small agricultural villages developed rules of behavior and expectations that reflected the need for, and the requirements of, cooperation, mutual respect, and carrying workload that served the collective. In the life of the farming village the concept of the *common good* came to life.

4.2.1.3 The Intellect Cluster

The intellect cluster integrates communication, cognition, learning, and aesthetics. I move aesthetics into this cluster because the artistic creations of the farmers were far less motivated by their spirituality than was the case in the life of the Cro-Magnons.

4.2.1.3.a Language. It was the full-fledged emergence of language that enabled the Second Generation HSS to make the quantum jump from the present focus to the temporal consciousness of the future. The farmers of the agricultural revolution needed to learn to picture the future and talk about it, and this extended their mental capacity. As Hall (1973) remarked, it is language that enables us to deal with the nonpresent world. We already remarked how language and consciousness coevolved in the behavior of the Cro-Magnons and how one without the others would not have developed. "With language, the verbal mind could differentiate itself out of the previous body-self, as it could rise above the prison of the immediate and conceive and sustain long-range goals and tasks" Wilber (1981, p. 93).

4.2.1.3.b Cognition. With the emergence of the Second Generation, cultural evolution came forth with full force. Verbal communication became the vehicle of building and passing on culture. Culture building is transbiological and transorganic; it moves us toward a higher cognitive plane, which Kohler called superorganic. Superorganic capacity enabled the first farmers to create mental symbols and concepts. The superorganic (Wilber, 1981) operates with those mental symbols directly without depending on the presence of their referents. The mind—because it transcends the physical world as it leaps to a higher plane—possesses the power to represent the world in symbols. Symbols represent a higher level of reality: the mental level.

4.2.1.3.c Learning. With future-focused language and symbolization, superconscious learning emerged. Superconscious learning is the vehicle of social and societal evolution. It is the vehicle of the evolution of humanity as a whole. Through superconscious learning we can prefigure the emerging order of the human community, thereby guiding history. Superconscious learning develops modes of sharing in the mind-at-large. Superconscious learning provides a sense of direction for cultural and human processes by illuminating the process from the far end in terms of guiding images (Jantsch and Waddington, 1976, p. 42). But what does this mean to societal evolution? In the present work, the notion of the evolutionary guidance system is proposed as an approach by which we can take charge of our future. The attainment of superconscious learning enables us to take the notion of guiding images a large step further. Guiding images are the seeds of the purposeful design of Evolutionary Guidance Systems. It is by such purposeful design that we can bring the image to life and thus create our future.

4.2.1.3.d Aesthetics. In the Cro-Magnons' cultural model this cluster included art. In the Cro-Magnons' life, art was an expression of their consciousness and spirituality. In the agricultural villages there was no art comparable to that of the Cro-Magnons. Their aesthetics were expressed in a different type of art: pottery, decorated with circular compositions of geometrical designs. "The abstract appearance of these well composed, abstract geometric designs in the centuries just preceding the development of large-scale civilizations suggests that a new kind of aesthetic consciousness—as a new objectivity in human perception was emerging rapidly" (Elgin, 1993, p. 88).

Activity 9

Task 1. Revisit this section and identify core ideas that characterize cultural evolution in the first phase of the Agricultural Revolution. Reflect upon these core ideas and—assuming you are a reporter—write a couple of pages on the story of the "Farmer."

Task 2. Review the three clusters of the Cultural Model of the Farmer and search for relational, interacting, coevolutionary characteristics between and among the clusters, as well as relational, interaction enhancing, coevolutionary characteristics between and among markers within and across the cluster boundaries. Record your findings in your workbook.

Reflections

In the preceding section, we developed a Cultural Model for the first phase of the life of the Second Generation of *Homo sapiens*. During this-phase—beginning some 12,000 years ago—we saw a major transformation in cultural evolution, called the Agricultural Revolution. This "second revolution"—following the first "human revolution" of the Cro-

Magnons—-changed all the evolutionary markers of the hunting–gathering Cro-Magnons into the emerging evolutionary markers of the Farmers who settled in small agricultural villages. These village communities flourished. Life in the village was a peaceful life of egalitarian cooperation. Everyone was working hard for the benefit of the entire village. For them, there was clearly something that we call today "the common good." These farmers were the first planners and the first social designers. They were the first future-oriented humans. As we repeatedly mentioned, the driving force of cultural evolution is *creativity*. The Farmers were exceptionally creative. They were planning for next year's crop; designing/ building safe settlements; and designing more effective tools for planting, cultivating, harvesting, storing, and transporting. They made excursions as they were searching for trading partners. They were designing rituals to ask for the help of the Great (Earth) Mother; pleading for fertility, for rain, and for a good yield of the crop, and trying to please her by making offerings to her. They established the first full-fledged society, governed by ethical and moral precepts. Life in the agricultural villages was most likely harmonious and well balanced, as was the life of the Cro-Magnons. The first 7000 years of the agricultural period of the village communities was a peaceful period—"a lull before the storm." The storm came with the emergence of the second phase in the life of the Second Generation HSS. The harmonious life was drastically changed when the large cities of the first agriculture-based civilizations emerged, relying upon, even exploiting, the surplus production of agricultural villages.

4.2.2 The Cultural Model of the Second Phase of the Second Generation HSS

The Second Phase is defined as the development of *Agriculture-Based Civilizations*. The emergence of this phase stands for a transformation from small agricultural villages to three sequential waves of civilizations: (1) the large cities of the first ancient civilizations, (2) the city-states of the Hellenic era, and (3) military empires, such as the Roman and the Byzantine empires.

In attempting to describe the evolutionary design model of these civilizations, we are faced with evolutionary markers that are very different from those of the small village agricultural community. The Cultural Models created in the evolutionary design space of the ancient civilizations, the city-states, and the empires of the Second Generation HSS are very different from the Cultural Model of the "Farmer's" small village culture.

It should be noted, however, that during the existence of these civilizations, the vast majority of the population still lived in farming communities, providing subsistence to the people who lived in the large settlements. In fact, it was the success of the small village in providing surplus food that made possible the establishment of large cities. Thus, the Second Generation HSS maintained two parallel cultural models. Even more, the cultural model of the agricultural communities is still with us in many regions of the globe. It can be said to be the most persistent and most resistant cultural model of HSS.

4.2.2.1 The General Characterization of Emerged Large-Scale Civilizations

People in both the small villages and the large cities were guided by what we called mythical, deep consciousness in contrast with the magical consciousness of the Cro-Magnons. From such a common base, the "pastoralist" culture of the small agricultural village was an Earth-focused and feminine-oriented culture worshiping the Earth Mother. The city culture was more aggressive, male-oriented, and sky-worshiping. The key and pragmatic connection between the two cultures was that the labors of the small village communities supported the "building of celestially based civilizations that were governed by a male-dominated priestly class (Elgin, 1993, p. 73).

4.2.2.2 When, Where, and What Kind?

4.2.2.2.a When and Where? The first "high civilizations" of the Second Generation HSS emerged independently in five regions of the Earth between 4000 B.C. and 1200 B.C. The first was the Sumer in Mesopotamia, then Egypt at 2800 B.C. and India around 2500 B.C., in China along the Yellow River in 1500 B.C., and the last in Mexico around 1200 B.C. There is some evidence of the existence of an ancient civilization in South America as early as 3000 B.C.

4.2.2.2.b What Kind? What is most fascinating is that isolated from each other, all these civilizations developed identical systems of activities, including writing, science—with a focus on astronomy and mathematics—then commerce, architecture, professional priesthood, various classes of occupations, and strong centralized government. This development suggests that "When a critical threshold is reached in the co-evolution of culture and consciousness, civilization will inevitably arise" (Elgin, 1993, p. 88). Elgin also presents a set of common features that characterize these early civilizations. Each civilization:

- Had several interconnected cities as its vital centers of the culture, religion, governance, trade, and culture.
- Developed some form of writing, philosophy, engineering, mathematics, and astronomy.
- Created its mythology around Sky God spirituality. With a powerful priestly class a patriarchal religion was allied with strong central government.
- Organized itself in distinct occupational classes, for example, merchants, engineers, scientists, traders, the priestly class, warriors, artisans, rulers.
- Developed armies, able to wage large-scale warfare.
- Established communication networks to support trade and governance.

These characteristics of ancient civilizations represent major changes, even radical transformations in the evolutionary markers of the cultural model of the Second Generation HSS.

4.2.2.3 The Cultural Model of the People of Ancient Civilizations

The Cultural Model of the Second Generation HSS—living in the ancient civilizations, in the city-states, and later in the empires—was markedly different from those living in the agricultural communities. There were major differences in all three clusters of the Cultural Model: in the way of life cluster, the consciousness cluster, and the intellect cluster.

4.2.2.3.a The Way of Life Cluster. For the first time in societal evolution, two new cultural markers must be recognized, markers that will play dominant roles in the life of successive evolutionary generations of HSS. These are *science* and *economics*. The surplus production of food and the material technology of producing clay containers for storing and transporting food enabled masses of the population to concentrate in large settlements and establish the "ancient civilizations" and later the city-states and the large empires. The second "enabler" was the invention of money in the form of metal coins, which became the currency with which to acquire food and goods. The creation of these "enablers" could not have been possible without the power of "symbolic consciousness" and the mind's ability to manifest this consciousness in "symbols that stand for." These two enabling conditions created what we can call "economy." Wilber (1981) provides a summary of the impact of this emerging economy of the *Social Organization* of the societies of the Second Generation.

For the first time in the saga of societal evolution, the availability of surplus food and other produced goods and money freed people to engage in such specialized fields of study as architecture, mathematics, and astronomy, collectively called *science.* Science-based cognition created the calendar and enabled the invention of writing. These truly mental productions were all crucial to the creation of civilizations. At the same time specialized classes emerged, such as administrators, priests, healers, educators, scientists, artists, various craftsmen, and tradesmen. The organized interactive and systemic arrangements of these occupations created what we earlier called the large settlement city-life of the emerged civilizations.

The *third cultural marker* of the style of life cluster is *technology.* It is a good place to tell the evolutionary story of and the meaning of technology. Technology has a range of meanings. (1) It stands for both "how" and "what" is produced by the "how." (2) In another sense, we can talk about "intellectual technology" that produces a mental creation, such as a process of governing, healing, planning, designing; and "material (substance) technology" which produces material products from raw materials. (3) In material technology we can differentiate between ornamental, for example, jewelry and functional products (vessels for grain). For the first time in the evolution of our species, the Second Generation HSS engaged in and produced all these types.

The main "how" technology of the Farmers was of course farming; the "what" technology was the instruments of farming, for example, the plow. They used clay to shape and fire large containers, decorate them, and use them for storing grain. Here art and material technology became integrated. Their "intellectual technology" focused on the future—as farmers must. They planned planting, cultivating, harvesting, storing, and transporting. They invented designs for governing the village.

Technology in all its forms and in all its meaning exploded in the city-states. Sass (1998), presenting a comprehensive review of the evolution of material and intellectual technology, tells us how—in Mesopotamia, the first ancient civilization—farming technology was married with engineering and technology, creating agricultural systems of irrigation such as the world had never seen. The Sumerian temples were architectural marvels and had a central role in both spiritual practices as well as secular affairs, including the planning and upkeep of the irrigation systems and the planning and governing of the affairs of the city.

The age of ancient city-states was the age of metallurgy technology. Technological specialization flourished. Gifted craftsmen now were free to develop their energy to experiment with and innovate all types of products. Metal was used to produce both ornamental and useful products of great variety. Just after 4000 B.C., in Mesopotamia, metal workers smelted scarcely available copper for fashioning jewelry. Later, when they imported copper they made vases, utensils, mirrors, hoes, plows, and all kinds of ornaments of gold and silver as prized possessions. These technological products heralded in the "age of greed." Whole civilizations were destroyed for gold.

Iron entered when—first—meteorites were discovered and were called "metal from heaven" and prized higher than gold. Then, when metallurgists discovered how to smelt steel from iron ore, the Iron Age began. It initiated a new revolution in the use of material substance. In addition to making possible the crafting of pickaxes, hammers, and plowshares, iron technology served "the military's insatiable appetite for weapons" (Sass, 1998, p. 94).

It was the period of ancient civilizations when—for the first time in societal evolution—humans *used technology for destruction as well as for the common good.* This trend continued throughout the ages and escalated to the point where we have now the technology to destroy the whole planet.

4.2.2.3.b The Consciousness Cluster. The era of the Second Generation HSS was guided by mythical/deep consciousness. Its emergence was a major shift from the surface consciousness of the Cro-Magnons, who had virtually no distance between them and the spiritual forces. They felt the immediate presence of these magical forces. In contrast, deep consciousness created a strong ego—both self and social. It brought about another division—a separation—between heaven and earth. This separation had to be bridged by new myth. In this myth, "The Sky God represented the awesome and remote powers of heaven, and the priest was the intermediary between the individual and the divine" (Elgin, 1993, p. 86).

Their world view represented a "dynamic stasis." Life was moving around in cycles, as seasons move in circular motion. This view was rooted in both "agricultural" and "sky" consciousness. The consciousness of the Second Generation HSS of the ancient civilization lacked historical perspective. The Sumerians did not write history. They did not have the intellectual tools of generalization so essential to gaining and imparting a historical perspective. Neither did they have the evolutionary consciousness needed to engage in historical interpretation.

4.2.2.3.c The Intellect Cluster. A most crucial development in the age of the ancient civilizations was the development of a new mode of communication, namely *writing*. For thousands of years, *Homo sapiens* drew pictures as references. These were visual marks for visual references. But gradually, these became more and more abbreviated marks as an alphabet was created. The oldest "discursive records" date back to the Sumarians after

4000 B.C. "By substituting visual marks for spoken utterances, man removed himself even further from directly experiencing his habitat, further severing the psychological umbilical cord tying him to that habitat" (Curtis, 1982, p. 102).

The availability of writing was crucial in the ancient city-states. Records based on human memory were inadequate in a complex society, such as the ancient cities of the Sumarians. Writing was an essential requirement for agriculture, for the economy in all its forms, for administration, for mathematics, astronomy, engineering, architecture, and the priesthood. Furthermore, through the ability to write, the "hard-won knowledge gained by one generation could be passed onto the next. With the ability to record, retrieve, and transport accurate information, a centralized government could now control distant colonies" (Sass, 1998, p. 33). Archeologists discovered vast libraries of clay tablets, recording the activities of life in Mesopotamia. A law—instituted around 2000 B.C.—required the recording of all transactions.

Curtis (1982) sums up the evolutionary and cultural significance of the development of writing. Visual symbols enabled the manipulation of ideas and the references. Writing extended the boundaries of consciousness within the transmitting distance of writing and made it possible to record collective experience through time. It extended man's consciousness through time. He could converse with his predecessors as well as his successors. As written symbols replaced oral utterances, they helped to develop the left hemisphere of the brain.

What became necessary for HSS became reality, says Curtis. Writing attests to this notion. The establishment and life in the city-states created an up to now unheard demand on the further development and enrichment of language capacity and capability. The explosive evolution of consciousness and the various domains of the experiential dimensions of the Second Generation HSS substantively advanced and enriched human cognition. The future consciousness of the Farmer and the multiplicity of human experience in science and engineering and in the various occupational domains of the city-state all demanded new and expanding language potential. Language and cognition coevolve and the evolution and enrichment of language through the development of writing—as described here—significantly extended the cognitive and intellectual power of the Second Generation. Furthermore, written language greatly enhanced the establishment of formal education.

In Sections 4.2.1 and 4.2.2, I presented the cultural models of the "Farmer" and the mega-city dwellers of ancient civilizations. But the evolutionary story of the Second Generation does not end here. We continue the evolutionary story as we visit the city-states of the classical Hellenic culture, the Byzantine and Roman empires, and witness the decline of the Second Generation HSS in the darkness of the Middle Ages.

Activity 10

Task 1. Review Sections 4.2.1 and 4.2.2, which presented two cultural models of the Second Generation HSS. Identify and describe core evolutionary ideas and use them to create a brief summary description of the two periods of the life of the Second Generation. Enter your findings in your workbook.

Task 2. Study the evolutionary clusters and markers within the clusters of the two stages. Seek and identify relational interactions among the markers. Describe how they influenced each other. Enter your findings in your workbook.

4.2.3 *The Decline of Ancient Civilizations and the Emergence of New Civilizations*

The ancient agriculture-based civilizations developed in a time span of 3000 years. Elgin (1993) suggests that their decline around 1000 B.C. was due to: (1) the acceleration of the pace of change such as the invention of Iron Age technology and new weapons of war, including the horse-drawn chariot, (2) the internal decay and corruption in the city-states, and (3) a series of barbarian invasions in Mesopotamia and the Nile Valley. Following the decline of these civilizations, after a period of transition, the world of the Second Generation HSS entered into a new phase, as new civilizations emerged in Greece, Rome, Persia, India, and China.

In telling the evolutionary story of the Second Generation HSS, there is no *intent to provide* a *historical account of this Generation.* We have whole libraries describing the history of this period. In this work, I focus on the unfolding of evolution, as I try to capture and share with you core *evolutionary ideas,* impart a gaining of evolutionary consciousness, and—eventually—put all these in the service of engaging in conscious and guided societal evolution thinking.

4.2.3.1 A General Characterization of the Classical Period of the Second Generation HSS: When, Where, and What?

We now enter the *Classical Age,* during which, from the ruins of the ancient civilizations, new civilizations emerged and were flourishing in Greece, Rome, Persia, India, the Americas, and China. Each of these civilizations made crucial contributions to science, philosophy, art, architecture, education, and government. A number of the great religions emerged in this period. Some of these classical civilizations lasted longer than others. For example, while the western section of the Roman empire collapsed around 500 A.D. into the "Dark Age," the eastern Byzantine era continued to flourish another 1000 years, as did the Moslem cultures, and cultures in China, India, and in the Mayan empire.

Of these various civilizations, I focus here on the evolutionary story of *Hellenism.* "Hellenism was a civilization which came into existence towards the end of the second millennium B.C. It preserved its identity from then onwards, until the seventh century of the Christian era" (Toynbee, 1959, p. 3). It was a classical cultural model that started in Greece and reached into much of Europe, the Middle East, North Africa, and as far as China.

I shall develop in the evolutionary design space of this period another Cultural Model that embraces the three component cultural clusters: consciousness, intellect, and way of life with their various cultural markers. It will be your task later to explore and describe synergistic relationships among the clusters and relationships of the evolutionary markers within the clusters.

4.2.3.2 The Cultural Model of the Hellenic Civilization

The emerging force of the *Mythical/Deep Consciousness* in the last 2000 years in the life cycle of the Second Generation HSS was *ego-consciousness.* It marked the beginning of the "mental-ego" state, which later became a dominant form of consciousness of the Third Generation HSS. "The ego brought introspection and self-analysis, penetrating science and philosophy. Most significantly, it marked the final emergence from the subconsciousness realm, which meant that the self can now become super-conscious in ways and to a degree never before quite possible" (Wilber, 1981, pp. 180–181).

Historical consciousness coevolved with ego-consciousness. Prior to the emergence of ego-consciousness history did not exist. It was Herodotus in the fifth century B.C. who is considered to be the "father of history." He wrote of the history of the antecedents and the outcome of the Persian–Hellenic war. "Historical consciousness is today's reflection on yesterday, the paradigm of reflective thinking in general. Historical consciousness is the epitome of polis–praxis" (Wilber, 1981, p. 203).

Ego-consciousness became manifested in *Hero Myth.* The Hero is the most visible image of the emerged ego-consciousness. For the Greeks, Zeus's victory over Typhon made Zeus the hero of the patriarchal gods of Mount Olympus, and he became the subject of worship in the early heroic ego period. The principle of the victory of the heroic ego with its moral corollary of individual responsibility are core characteristics of the Occidental myth and stand for the archetype of the uniqueness of the individual as well as a protest against Earth worship.

As stated earlier, the all-embracing designation of the Cultural Model of this period of societal evolution is called "Hellenism." Hellenism, says Toynbee (1959, 1981), is both cultural and social. It is a distinct way of life, embodied in cultural institutions of the city-states of Greece and others in the Medieval Western states, such as Venice, Milan, Florence, and towns of Northern Germany, as well as Barcelona and Marseilles. Hellenism made use of the cultural institutions in the city-states to give an expression of the world view of *humanism*. Humanism is manifested in the worship of what is the best, the most perfect in man. No other civilization committed itself so completely to humanism than Hellenism. Humanism had to attain supremacy over nature and the individual's dominance over the family. It was finally the law of the city-state that liberated the individual from the bondage of the family at the price of servitude to the city-state.

"Aristotle, in his treatise of the theory of the City-State, declares that the City-State comes into existence in order to make life possible, and remains in existence in order to make life worth living" (Toynbee, 1981, p. 61). For Aristotle, it was the "Agora" that was the essential place of the city-state. It was the Agora where people assembled to have lively social conversations. It was the place where a politician might address his fellow citizens or a philosopher might have a conversation, a "searching together"—called "sizitsis"—with his fellow inquirers about the nature of the Universe and Man. But, above all, it was the place to conduct the business of the city, to live democracy by making collective decisions. Self-governance was the objective in every Hellenic city-state. In Athens from 462 to 217 B.C. "every male citizen, rural as well as urban, had a vote and served on juries and could stand for elective office or win non-elective offices by lot" (Toynbee, 1959, p. 63).

The Hellenic city-states were fertile seed-beds of creativity and excellence in many fields, such as science, philosophy, poetry, architecture, and all forms of art. In Hellenic pottery the potter's three-dimensional art was married with the painter's two-dimensional art, creating harmony between the two. In Hellenic sculpture, harmony became unity as an art form, in which the Hellenic artist achieved true success. "In this field, the Hellenic Greeks' creative faculty lasted as long as the Hellenic civilization itself" (Toynbee, 1959, p. 56). In the course of centuries of Hellenic creativity, works of art were produced and scientific and philosophical theories developed that have yet to be surpassed. Adherents of the Hellenic schools of philosophy were found throughout the

Hellenic world of Europe, the Middle East, North Africa, India, and even in China. There were four major Hellenic schools: Plato's Academy, Aristotle's Peripatos, Zeno's Stona, and Epicurru's Garden. They drew to Athens students from above, and Athens became the university city of the Hellenic World as well as its political and commercial capital (Toynbee, 1981).

Together with athletics, the study of *literae humaniores* provided the materials for *paideia: the learning society*. The effectiveness of *the paideia* was immense. Its instruments were the products of the high point of the Hellenic genius. Hellenism—very specifically *paideia*—captivated the Roman conquerors of the Hellenes. From the close of the third century B.C. until well into the Christian era every educated Roman was bilingual and Latin literature and science and art were greatly influenced by the Hellenic culture.

4.2.3.3 The Gifts of the Classical Age

We might consider the classical age to be the most creative period of humankind, offering to future generations of *Homo sapiens* three gifts: Polis and Praxis, Democracy, and Philosophy.

4.2.3.3.a Polis and Praxis. One of the greatest achievements of the Classical Age and one of its greatest gifts to us is the twin ideas of *Polis* and *Praxis*. Wilber (1981) suggests that it was the emergence of intersubjective communication that led to the understanding that constituted what the Greeks called *polis*. Socrates said that I can know myself—I can understand who I am—only by having a conversation with my fellow men. "Polis was the first arena of truly human relationship, the relationship found nowhere else in nature, the relationship that *specifically* defined the new species of *Homo sapiens*" (Wilber, p. 161). Polis creates a *shared human community* that engages in unconstrained conversation in the arena of membership in the community.

Praxis is the activity of the polis. For Aristotle praxis is a purposive, enlightened, and moral behavior pursued in the company of polis. "I can only become truly human in polis, or in symbolic community, and I can only exercise my humanity in praxis, or social engagement and sharing with fellow communicators. The arena of that sharing is polis; the exercise of that sharing is praxis" (Wilber, 1981, p. 162). Polis–praxis was the way of life in the Hellenic city-states. As we contemplate in this work the evolution of the Fourth Generation of HSS, we cannot do any better than proposing polis and praxis—as lived by the classical Greeks—as key markers of guided evolution.

4.2.3.3.b Democracy. Democracy as a form of collective governance is grounded in the "polis–praxis" idea and the practice developed by the ancient Greeks, as described in the preceding. It is best expressed by three words (Banathy, 1996), described next.

> These are "democracy," "sizitsis," and "demosophia." *Democracy* means "the power of the people." *Sizitsis* stands for "searching together." *Demosophia* is the "wisdom of the people. If we integrate these three, we get a clear idea of what democracy meant for the people of Athene. The true meaning of democracy is that people have the power to make decisions about issues affecting their lives. These decisions are made by searching together, by disciplined and focused conversations, that is manifested in the wisdom of the people. (p. 338) (Socrates said that the only way to arrive at understanding and the truth, and attain wisdom is by searching together.)

In view of the above definition, it seems that today—mostly—we have given away the power of making decisions about issues that affect our lives and we do not know how to search together to attain wisdom that would enable us to exercise our power of collective decision making and future building.

4.2.3.3.c Philosophy: The Three King-Philosophers: Socrates, Plato, and Aristotle. In the cultural history of mankind no other period produced such intellectual giants in such a short time. Plato was Socrates' student and Aristotle was Plato's student. Each had his own philosophical position but contributed to a common theme: the ethical life, virtue, the truth, and ways of attaining wisdom. It is often said that all the philosophy that came after these three is just a footnote to their scholarship. Three vignettes of their work are offered here.

- We inherited *Socratic* (moral) *philosophy* and the Socratic method. Socrates held that: Knowing that I do not know is the true measure of wisdom. The highest good is caring for wisdom, truth, and virtue. Knowing the good is to do the good. And finally, the unexamined life is not worth living. The Socratic method, the *Method of Dialectic,* is an approach of searching together to seek knowledge and attain wisdom by a question-and-answer approach.
- *Plato* suggested that a just society will not be possible until it is ruled by a Philosopher King or a King Philosopher. Plato explored the meaning of good and evil, right and wrong, and duty and obligation. He held that eternal truth, goodness, and beauty are not subject to changing interpretations. He believed that through the power of reasoning and the love of truth we can attain justice and serve the common good. He claimed that there are absolutely

true and eternal forms of justice and the good. However, Plato rejected democracy and individualism and advocated government by an elite group.

- *Aristotle's* philosophical stance was different from Plato's. It was grounded in a different vision of the world. A Platonist stands for the abstract and the perfect truth of logic. An Aristotelian favors the concrete and deals with ongoing change. Aristotle's philosophy is a process philosophy, concerned with realizable ideals. Attaining happiness is the highest good. Virtue, justice, courage, temperance, and self-respect are *moral virtues*, to be established by practice. The contemplation of truth is an *intellectual virtue*.

4.2.3.4 An Astonishing Spiritual Transformation

The Cro-Magnons had spiritual connections with many gods through magic. The Farmers designed rituals and ritual sacrifices to please the Great (Earth) Mother and her lesser gods and to plead with her for rain, fertility, and a good yield of the crop. People in the cities of ancient civilizations worshiped the Sky Gods and asked their priests to intervene for them. In the Hellenic period, the Greeks worshiped the hero-gods on Mount Olympus, and the Egyptians worshiped their king-gods.

During the later life period of the Second Generation HHS human and social evolution experienced an astonishing and transforming spiritual emergence of the seven faiths that brought for humanity revolutionarily different "faces of the divine." These faiths invited humanity to the realization of what is best in us, what is most creative, what is the most sacred, what is the most noble, and what is highest in consciousness in us. Elgin (1993) says that the life forces of each of these faiths are inviting us to experience a different face of the divine:

> **Christianity** inviting us to bring love, forgiveness, and a concern for social justice into our community's life; **Judaism** inviting us to appreciate the divine force that is not aloof from this world but lives in and through human history and society; **Islam** inviting us to recognize the great value of the awakened individual whose life is surrendered to God; **Buddhism** inviting us to discipline our thinking minds and to develop the capacity for penetrating insight that reveals the core nature of our Being as an unbroken flow of pure awareness; **Hinduism** inviting us to experience the absolute reality behind all appearances and the moment-to-moment dance of creation and destruction of the cosmos; **Taoism** inviting us to appreciate the subtle, unceasing flow of reality and the wisdom to yielding to that flow; **American Indians** and other **native people** throughout the world inviting us to experience the living presence infusing all of nature's expressions—plants, animals, earth, wind, fire, and water. (p. 172)

Together, these spiritual faiths reveal to us the essentials of ethics and morality for living our lives. As I contemplate the contribution of these spiritual traditions to the evolution of humanity, a question arises: Could we synthesize the teachings of these faiths into an all-encompassing belief system, a spiritual force that would guide us toward creating our spiritual future, the spiritual force of the Fourth Generation of HSS?

4.2.3.5 The Last Thousand Years of the Evolutionary Story of the Second Generation

We bring the evolutionary story of the Second Generation to its conclusion by first exploring how the Hellenic culture melded into the Byzantine civilization and then following the evolutionary trail up to the time of the emergence of the Third Generation of HSS.

In the course of the first five hundred years A.D. there was an overlap between the Cultural Model of Hellenic Greeks and the Byzantine civilization. The Hellenic culture greatly influenced the emerging Byzantine culture. Still, the Byzantines rejected several aspects of Hellenism. The self-governing city-states could not serve as governing structures of the Byzantine empire. The Hellenic form of worship was also rejected as Christianity emerged as the spiritual domain of human experience in the Byzantine way of life. Furthermore, art, science, and philosophy were transformed under the influence of the ideology, practice, and mythology of Christianity.

The Byzantine and Hellenic artistic expressions of the human body were worlds apart (Toynbee, 1981). It reflected a difference in spirit and aim.The Hellenic expression aimed to give a naturalistic portrayal "on the assumption that this was the best, and indeed the only possible way of revealing human nature. On the other hand, Byzantine *eikins* were attempts to represent in visual form the invisible soul, on the assumption that the soul is Man's essence" (pp. 78–79).

An observation should be made here. Christianity embraced the concept of a perfect state of the human being, but without the idea of progress. Contrary to the idea of progress, it was expected that the return of Jesus will usher in a new Heaven and a new Earth, but it will be preceded by misery and anguish. The early Christians entertained no idea of progress. It was not until the time of the Renaissance and the scientific-industrial age of the Third Generation HSS that the notion of progress emerged.

4.2.3.5.a Strengths and Weaknesses. The Byzantines were strongest in religion, where the Hellenes were the weakest (Toynbee, 1981). Their greatest achievement was the establishment of the Eastern Orthodox version of Christianity. In their liturgy they created a poetical musical expression of art that satisfies the spiritual need for congregational worship, "But this collective expression of religious feeling has not inhibited the individual expression of it" (p. 87). In contrast with the more rigid liturgy of Roman Catholicism—even up to the most recent times—the Orthodox Church promoted both the individual and collective approaches to God.

In economics, for the first seven centuries the free peasantry of the Byzantine civilization was its mainstay. Evidence of the strong economy was the maintenance of the East Roman gold coin for more than seven centuries. Later, however, economic activity was passed on to foreign hands, and free peasantry—the backbone of the society at earlier times—lost its ground to large landowners.

4.3.2.5.b The Decline of Empires. Around the end of the seventh century the Byzantine Empire slowly dissolved into anarchy. The countryside was overrun by barbaric invaders and the cities became engaged in devastating civil wars. As in earlier times, hunger for land and power-grabbing through war destroyed entire cultures and civilizations. The rise of the Ottoman Empire—and its military ventures into southern and central Europe—caused a major setback to positive creative evolutionary developments in these regions. The rest of Europe was devastated for hundreds of years by warring empires. For almost a thousand years, cultural evolution was stifled. The fire of the creative force—which is the source of evolutionary advancement—turned into ashes. It was only the rise of the Third Generation of *Homo sapiens sapiens* that revitalized and brought another revolutionary advancement in the evolutionary journey of the human race.

Activity 11

Imagine that you are a reporter of a large news organization. You have a time machine at your disposal. Your assignment is to write vignettes of life during the agriculture-based civilizations. First you go back to the beginning of the Agricultural Revolution and spend a day—and observe life—in a small farming village. Your next stop is the largest city of the ancient civilization of the Sumerians. Then, you fly to Athene and spend a day at the Augora. Your final stop is the Byzantine Constantinople. Now you are back home and file your report. Enter your stories in your workbook. In writing your report, you might consult books in your local library that pertain to sites that you visited.

Reflections

In Section 4.2 we saw the unfolding of societal evolution, which comprised the longest period of time in recorded history. During the ten to twelve millennia of this period, we witnessed the trials and tribulations of modern man. Through its emerged mythical deep consciousness, the Second Generation of Homo *sapiens sapiens* discovered the depth and wealth of their inner selves, discovered their own ego and soul, and made that consciousness manifest through humanity's creative genius in many forms. They created agriculture through cooperative egalitarian working communities, and designed and built the large cities of the ancient civilizations with professional specializations. They established laws, advanced science in all its forms, and understood life and the human being through philosophy that is still with us. They created the collective self-governance of "polis and praxis," established democracy, crafted beauty in many forms of art, wrote poetry that was never surpassed, invented history, and established bureaucracies that ruled whole empires.

At the same time they brought forth social upheavals, crafted ever more efficient means of destruction, and launched devastating wars that ruined cultures and civilizations. For centuries a creating evolution was not only set back but reversed into devolution. Evolutionary balance, harmony, internal consistency—the creative forces of human advancement and betterment—were all sacrificed on the altar of greed and hate.

If we have learned anything from the evolutionary story of the Second Generation of HSS it must be the senselessness and evilness of war. So we yearn for a state of the human condition in which we attain a higher level of consciousness and live by ethics and morality that are shared by all humanity and promote and nourish peace from the family on to the global systems of humanity.

4.3 *The Sea Change of the Third Generation of Homo sapiens sapiens*

Out of the "dark age" of the last centuries of agrarian-based civilization, some five hundred years ago, a remarkable sea change—a total evolutionary transformation—marked the emergence of the Third Generation HSS. This new stage of societal evolution is often called the *Scientific–Industrial Era—the Era of Enlightenment—the Era of Reason—the Renaissance Era.*

It was truly a revolutionary transformation by which in the evolutionary design space of human evolution all cultural evolutionary markers had changed. One can only say that there was a total break—total discontinuity—between the Cultural Models of the Second Generation and the Third. An observer of the life of the Second Generation HSS could not have predicted the evolutionary advancement of the Third Generation, the same way as one could not have predicted the evolution of Cro-Magnons from the evolutionary state of Archaic *sapiens*. Elgin (1993) summed up these changes as follows:

> A number of powerful revolutions blossomed in Europe that continued to reverberate around the world: a scientific revolution challenged the belief in the supernatural and the authority of the church; a religious reformation questioned the role and function of the church; a renaissance brought a new perspective into the arts; an industrial revolution brought unprecedented material progress; an urban revolution brought masses of people together in new ways, breaking the feudal pattern of living; and a democratic revolution fostered a new level of citizen empowerment and involvement. These powerful revolutions were expressions of the opening of a new perceptual paradigm, and they mark a dramatic break with the era of agrarian-based civilization. (p. 92)

In telling the story of the emergence and evolution of the Third Generation of HSS, I first specify the "when" and "where" of the emergence of the Third Generation. Then, I define and organize the evolutionary markers in the evolutionary design space of the Third Generation and show how these markers influenced each other and created an internally consistent Cultural Model. In conclusion, I reflect upon this third stage of the evolutionary story of modern man. (Today we witness the last fading out period of the Third Generation. Its last vestiges are now struggling to keep alive as the Fourth Generation HSS begins to enter the evolutionary scene and tries to define itself.)

4.3.1 *When and Where?*

We can place the emergence of the Third Generation of HHS somewhere between 1300 and 1500 A.D. Its emergence marked a major transformation from the era of the agrarian-based civilizations. For the next few hundred years it continued to develop until it attained full maturity in the nineteenth century. Then, quite quickly, it became increasingly dysfunctional and begun to collapse by the middle of this century. Initially, the primary space of the evolution of the Third Generation had been limited to Europe but soon extended into North America. It also gradually influenced and is still influencing evolutionary developments in other parts of the world.

4.3.2 *The Evolutionary/Cultural Model of the Third Generation HSS*

It is in the evolutionary design space that the Cultural Model of a specific societal evolutionary stage is formed from the interaction of the various evolutionary markers. Evolution from one stage to the next happens by reorganizing life at a higher level of complexity, thus demonstrating a key characteristic of evolution, namely, increasing complexity. Increasing complexity is accommodated by an increase in the number of evolutionary markers and by an increase in the complexity of their relationships. This observation becomes quite evident as one aims to describe and characterize the evolutionary cultural model of the Third Generation.

First, the number of evolutionary markers at this stage increased, resulting in increased complexity. Markers representing this stage included: consciousness, world view, spirituality, science, economy/subsistence, mode of communication, cognition and learning, the social/societal organization of life, technology, aesthetics, economy, relationship with nature, and polity/government. Second, the complexity of the markers themselves increased also. Third, as the markers increased in number and complexity, as they interacted with each other and created powerful synergy, the complexity of their relationship increased significantly, leading to the emergence of a Cultural Model, which was different from the Cultural Model of the Second Generation of HSS.

The markers of the Third Generation Cultural Model are arranged in the three clusters introduced earlier: Consciousness, Intellect, and Way of Life. The driving force of the emergence and development of the markers is *creativity*. Interaction within the clusters and between the clusters animated the dynamics of the life of the Third Generation.

4.3.2.1 The Consciousness Cluster

The cluster emerges from a new structure of consciousness, which brings fourth a new world view and a new form of spirituality and morality. Thus, consciousness, world view, spirituality, and morality comprise this cluster.

4.3.2.1.a Consciousness. Different scholars label human consciousness at this stage of evolution differently, such as: *mental structure of consciousness* (Gebser, 1949; Combs, 1996), *ego-consciousness* (Wilber, 1981), and *four-dimensional dynamic consciousness* (Elgin, 1993). Their descriptions integrate well into a comprehensive characterization of the state of consciousness at this evolutionary stage.

New patterns of consciousness do not emerge with one grand swoop. They evolve over time. The initial pattern of mental structure of consciousness first supplemented the deep mythical structure of consciousness during the classical period of the Second Generation and then became the dominant paradigm of consciousness the middle of the second millennium. Parmenides, already in 489 B.C., said that "thinking and being is one and the same." Socrates equated the soul and the afterlife with pure thought. Later, during the Renaissance, we heard Descartes' famous statement: "I think, therefore I am." (How many times did I hear this statement from my philosophy teacher in my high school years?)

"The mental structure underwent further elaboration during the late Roman empire, when the ego first became fully established as a self-established center of the inner life" (Combs, 1996, p. 108). These were "forerunners" of mental-centered consciousness. When the Hellenic culture disappeared into the culture of Byzantium and, when around 500 A.D., the Roman empire collapsed into the "Dark Ages," mental focused consciousness had to wait another thousand years to become the source of emergence of a new state of consciousness, which then became the dominant consciousness of the Third Generation.

The best designation that I can think of that integrates the various labeling of consciousness of the Third Generation is *"the four-dimensional spatial perspective of consciousness"* (Elgin, 1993). Representing a major transformation of consciousness, it subsumes ego-consciousness and mental consciousness. In essence, it gave the Third Generation HSS the ability to accommodate the three-dimensional perspective within a four-dimensional perceptual context.

At this point, it is useful to review and clarify the role of "dimensions" in the evolution of human consciousness through the various sequential evolutionary stages (Elgin, 1993). In the course of human evolution, a one-dimensional perspective manifested itself in "contracted consciousness." Consciousness was contracted into a single point "within," in the dreamlike life of our earlier ancestors. The Archaic *sapiens* had to wait for the emergence of *Homo sapiens sapiens*, for the appearance of the first modern man, the Cro-Magnons, to wake up from their dream state and step out into the light and experience a two-dimensional "surface consciousness." During the next stage, the agriculture-based ancient civilizations—while focusing on the sky and finding deep consciousness within—the Second Generation extended its consciousness into the third dimension.

Elgin (1993) suggests that each dimension was designed to bring forth new potentials from within. "Each dimension provides a 'unique opportunity space' or learning context for people and their societies to fill out with their action. Each dimension embodies an enlarged frame of reference within which are nested all previous dimensions" (p. 26).

Dimensions have their own life cycles. They are born from an evolutionary potential within us as it is realized by our creating forces. They inspire the coevolution and development of new substances and forms of the evolutionary markers. They come to full fruition and create civilizations. Eventually they reach a point when they become both irrelevant to human dreams and aspirations, they become dysfunctional, and they produce increasingly more "waste" and surface "systems of problems" that cannot be addressed from the old consciousness. (Einstein remarked that you cannot address a problem from the same consciousness that created it. You have to learn to think anew.)

In the evolutionary design space, new evolutionary potentials surface, new consciousness emerges that organizes a new Cultural Model, and a new civilization is born. The new pattern of the four-dimensional perspectives was first expressed in art and later in the other evolutionary markers, for example, in the spiritual, in social organization, in economics, in science, in polity, and so forth. The Florentine artist Giotto was the first to paint three-dimensional scenes on a flat canvas. But remember, one of the Cro-Magnon cave paintings was also three-dimensional, although no other painting exhibited this form until Giotto's work. It was Leonardo da Vinci who brought dimensional perspective to full fruition in art. Then, soon, Descartes created the special mathematics of analytic geometry. Elgin (1993) explains the dynamics of the four-dimension perspective.

> To see how we see requires that we stand back from immersion in the process of seeing and look at both the scene and the seer simultaneously so as to put them in accurate relationship. Three dimensional depth perspectives, then, require a new step back in consciousness. A person must move, in their consciousness, from inside the three-dimensional reference frame to outside of it in order to see the relationship between the observer and the scene. *We began to see things in accurate three-dimensional perspective only when our experienced context for looking at the world began to expand into four dimensions.* (p. 97)

The fourth dimension fosters an existential sense of self, within which there is no special status of positions or places. Everything is relative. The point of view of everyone is equally legitimate. They are not subservient to the authorities of the state or the church. They see themselves as being able to act in their own self-interest socially, spiritually, economically, or politically. No previous civilization pursued material progress so relentlessly as did the Third Generation. People of the Third Generation were driven by competitive and aggressive urges. The materialistic orientation of this stage "is a double-edged sword" says Elgin (1993). "While it promoted aggressive economic growth, its rational orientation negated the validity of many subjective or spiritual experiences. Although this era is often called the Age of Enlightenment, ironically it has diminished our access to enlightening experiences" (p. 108).

The time consciousness of the agricultural man was circular. Seasons went and came back with regularity. The sky moved with precision, and time was defined by that precision. They planted the seed and harvested at the same time over and over again. For the scientific–industrial man of the Third Generation time moved out of the circle and became a spiral that moved ahead. The idea of progress through time became conscious in us and its consciousness guided life. The mode of progress was primarily material growth. The aim in life was to get more and more possessions and to get ahead. Elgin (1993) comments on the effect of time consciousness by saying that:

> Indeed the drive to material development is the most basic expression of the time experience of the fourth dimension. An expectation of perpetual novelty and innovation became widespread in popular consciousness. Along with the ability to imagine material progress came a new facility for practical invention and projecting plans into the future. No longer was progress something to be achieved by the few at the expense of the many: now it was possible for everyone to advance. (p. 99)

I conclude a review of the state of consciousness of the Third Generation HSS by suggesting that the characterization developed here still represents the state of consciousness of most people today even though we embarked on a new stage of the evolution, that of the Fourth Generation of HSS. Living and acting upon the consciousness of an earlier generation has opened up a consciousness gap. The existence of this gap has created increasingly more serious problems for us personally, for the systems in which we live and work, and for our society and for the global community.

4.3.2.1.b The World View of the Third Generation. It is interesting that Webster (1979) defines world view with the German term *Weltanschauung*, which the dictionary defines as "a comprehensive conception or apprehension of the world." Period.

Each one of us has our own view of the world, our own cognitive map of how we see the world, how we see our place and role in it, and how the world works and what it means for us. We construct our own view of the world. We confirm it, change it, and may even reject it. Our world view is in constant coevolutionary relationship with our consciousness. A view of the world shared by a community, a culture, or a society is often called a "paradigm." The source of a world view and its paradigm is the prevailing consciousness with which our world view and the shared world view of an evolutionary stage coevolves.

The world view of the four-dimensional perspective allowed people to step back and take a new look at their institutions that set for them specific and often absolute norms, such as the institutions of the state and the church. In the four-dimensional perspective everything is relative. This relativistic world view facilitated and nurtured a new level of individualism in science, religion, intellectual pursuits, the arts, as well as in economics.

In science there was freedom from dogma and superstition, which enhanced the ability of scientists and scholars to explore new ventures and take a new look at the nature of the universe. The church no longer had an exclusive and favored position. Reformation brought forth greater freedom in practicing worship and having one's own personal belief. In the economic domain, "while free-market economy liberated people from centuries of oppression by a landed aristocracy, it also transformed the primary purpose of life into the pursuit of material gains" (Elgin, 1993, p. 104). The attainment of an "autonomous sense of self," fueled by a materialistic world view, was paid for by some of the most antisocial behavior we have ever seen, such as unbridled greed, unrestrained competition, and lack of compassion for the less fortunate, for the "losers."

4.3.2.1.c Spirituality. Among the various revolutions that characterize the evolution of the Third Generation HSS, the "spiritual revolution" brought forth the Reformation. Initiated by Huss, Luther, and Calvin and their followers, the question was raised whether there is a need to have an intermediary, namely the church, between the individual and God. Reformation created a secular rationale for government as it challenged the divine rights of rulers. The success of spiritual enlightenment offered by the reformation was much enhanced by the communication revolution that printing brought about. The Gutenberg Bible was the first product of this revolution. Before the Reformation, only the clergy had access to the Bible. Printing and Reformation made it available to all. Prayer and the study of the Bible created a direct link between the individual and God.

In opposition to this spiritual revolution, the mechanistic world view of the scientific revolution held that science was able to describe the laws by which nature worked. Thus, "people no longer need to fear that some capricious and wrathful God will intervene arbitrarily in their lives. We can discover the laws of the universe and become masters of our fate" (Elgin, 1993, p. 100).

As in every other facet of life of the Third Generation, spirituality was also dominated by individualism, secularism, and expectations of material progress. Cambell (1959) summarized this state of affairs as follows: "In the broadest view of the history of world mythology, the chief creative development in the waning Middle Ages and approaching Reformation was the rise of the principle of individual conscience over ecclesiastical authority" (p. 146). The way to know God was by knowing the laws of nature and laws of the cosmos created by God. This form of spirituality became known as "Scientific Humanism." Mythology and mystery, the mainstay of the spirituality of the Second Generation HSS, were discarded in favor of Scientific Humanism, according to which science is the main vehicle of any inquiry

4.3.2.1.d Morality. The morality of the Third Generation reflects its consciousness and its view of the world. Morality became the morality of material progress, the ethics of the acquisition of more and more possessions, allowing the exploitation of nature and of the working class. As said earlier, the *attainment* of an "autonomous sense of self" fueled a materialistic world view, and was paid for by some of the most antisocial behavior we have ever seen, such as unbridled greed, unrestrained competition and capitalism, and a lack of compassion for the less fortunate—who were considered to be the "losers."

4.3.2.2 The Cluster of the Intellect

This cluster is formed from science, a new mode of communication: print; cognitive development; and the development of intellect, learning, and the arts.

4.3.2.2.a The Scientific Revolution. The last 1000 years of the Second Generation were devoid of any scientific advancement. The achievements of the great scientists and philosophers of Athene were all but forgotten. The collapse of the Roman Empire brought forth the Dark Ages in which some pseudo-science prevailed but any novel scientific thought and discovery was not only stifled but denied and punished by church authorities.

The scientific revolution brought forth the world of Galileo, Newton, Descartes, and their fellow scientists. The Cartesian, mechanistic clockwork view of the world was a world of reason in which the universe was governed by natural laws. It was believed that these laws can be discovered by proving or disproving scientific hypotheses.

> The cosmos of the fourth dimension bleeds the mystery from nature and replaces it with rational laws and machine like processes. Because of the "grand machine" of the universe can be understood by the analyzing intellect, there is the assumption that, in time, all questions about the cosmos will vanish before the dissecting power of science. (Elgin, 1993, p. 107)

Inspired by the Newtonian–Cartesian scientific view, the disciplined inquiry of the Third Generation scientist sought understanding by taking things apart, seeking the "ultimate part," and groping to reconstruct the whole by understanding its parts. This reductionist orientation—with analysis as its core method—was the hallmark of science for the next four hundred years, and is still very much alive today (Banathy, 1992). Another key marker of what we now call traditional science was the attainment of certainty of the outcomes of scientific investigation with its power of prediction and its faith in determinism. In the Cartesian view there is always a one-directional and predictable effect of a given cause. This reductionist, single-causality orientation was unwilling—unable—to consider *purpose* and *meaning*.

Classical science saw systems to be basically closed, having only limited and controlled interaction with their environment. Systems operate like a heating system. The thermostat is set and correction happens only when there is a deviation from the setting mark. Objectivity governs disciplined inquiry. The scientist (the observer) is detached from what is observed. It should be noted that this "detached" position was oddly out of step with the "perspectival" orientation of the four-dimensional geometry, which was so clearly appreciated and practiced by the artists of the Renaissance. Elgin (1993) observed that "it would take physics nearly four hundred years to catch up with and begin to explain the transformation of perceptual experience portrayed by artists in the East and West" (p. 98). One of the most significant scientific achievements of the Third Generation HSS was the emergence of the science of evolution through the scholarship of Darwin and his peers, discussed in Chapter 2.

4.3.2.2.b The Communication Revolution. Change in human communication was an unmistakable and powerful marker of each stage of human evolution. I consider *language* the greatest human creation. Speech with self-reflective consciousness brought forth what we called "the human revolution," the emergence of "us" modern humans. The emergence of agriculture-based ancient civilizations could not have happened without the invention of writing. Writing significantly extended the space and time boundaries of the human and social experience. The existence of the Third Generation HSS cannot be imagined without print, the third modality of human communication. It enabled the rapid extension of information and knowledge, the flourishing of science and technology, the spread of the Reformation with the aid of the printed Bible, the establishment of nation-states, the further development of cognitive and intellectual development, and the further extension of the space and time boundaries of the human experience.

In his *Evolution or Extinction* (1982), Curtis directly associates change in consciousness at the various stages of human evolution with change in the mode of communication. Speech made possible the human revolution. The invention of writing made possible the establishment of ancient civilizations. He holds that the emergence of the dynamic or four-dimensional consciousness was enabled by the development of movable type, which:

> tended to prescribe the spacial boundaries of consciousness to those within transmitting distance, e.g. the extended city or nation state. Printed symbols also tended to prescribe the time boundaries of that consciousness, opening up the first time in human history to the idea of progress. Literacy, until this time the prerogative of the privileged few, spread with the availability of manuscripts. One of the most profound changes induced by the growth of literacy

> was the birth of public education which, in turn, made possible for the first time collective decision making in the form of democratic government. Increased literacy also gave rise to an incremental increase in knowledge that, in turn, served as the catalyst for change in the Renaissance. Gradually, scientific law replaced the logico-philosophical universe as the all-embracing paradigm around which all knowledge was centered." (p. 127)

As indicated previously, the development of print brought forth a massive growth in literacy and had a massive effect on cognitive development and the development of the intellect. Here we have another manifestation of one evolutionary marker interacting with and influencing the other markers.

4.3.2.2.c Cognitive Development, the Development of the Intellect, and the Rise of Public Education. In addition to the effect of the printed book on cognitive development and the development of intellect, the impact of mental consciousness and the emergence of the fourth dimension were fostering freedom for disciplined inquiry. Freedom to pursue intellectual development was best aided by the availability of the print mode of communication. One of the most profound changes induced by the availability of printed text, and consequently the spread of literacy, was the birth of *public education.* The first compulsory public education system was established in Massachusetts by an act of law in 1646, a decade after the first printing press was set up (Curtis, 1982). Since that time, public education has been regarded as the birthright of all children.

4.3.2.2.d Aesthetics: The Creating Force of the Renaissance. We have seen repeatedly in the course of our evolutionary journey that it has been the "creating force" that fueled evolution and brought forth revolutionary advancements. "The Renaissance, in which our present civilization has its roots, may be considered as the creative surge that was party inspired by the creative surge of the ancient Greeks" (Bohm and Peat, 1987, p. 105). But after the Hellenic period, the creating force was dormant for a thousand years. It was not until the artist and the scientist of the Renaissance became part of the human experience that the creating force became again manifested.

As mentioned earlier, it was Giotto, the Florentine artist, who fully conceived and represented three dimensions in his paintings. Later, it was Leonardo da Vinci who brought perspective forcefully to life, as a source and manifestation of four-dimensional dynamic consciousness. In a broader sense:

> In the Renaissance, there appeared a whole new spirit of the age, which could be characterized as "the new secular order." In the previous eternal order, the

> highest value was that which never changed. But in the Renaissance, the way things changed—and the power to change them—became the highest value. For example, Galileo argued that it is precisely the transformations of nature that are the most interesting. Francis Bacon carried this further, opening up the prospect that through the understanding of such processes humanity could dominate nature for the general good. In sculpture, Michelangelo explored the dynamic transformation of human form that could be contained within a single block of marble, and Leonardo, in his notebooks, was occupied with energy, change, the Deluge and the general expression of power in nature. Throughout the whole culture, there spread an infectious exhilaration. This released the energy necessary for a vast surge of creative activity. (Bohm and Peat, 1987, pp. 207–208)

We have seen in the course of our evolutionary journey that each of the new stages of societal evolution emerged from an unexpected creative surge. In the Third Generation of HSS this creative surge was expressed most strongly by the artist of the Renaissance and the scientist of the era of reason.

4.3.2.3 The Way of Life Cluster

This third cluster of the Cultural Model of the Third Generation is comprised of five markers: technology, economics, social organization, polity/government, and relationship with nature.

4.3.2.3.a The Technological Revolution of the Industrial Era. Material progress was the overwhelming demand of the Third Generation, and the explosion of technology was the key to the tremendous material progress at this evolutionary stage. Such progress was dependent upon the replacement of human and animal energy by artificial energy. Replacement sources were extracted from coal, steam power, natural gas, oil, and electricity.

A brief review of the evolution of power sources is in order here. Initially, water and wind power were the energy sources of larger scale mechanization. This was followed by large-scale mining and smelting projects. By the sixteenth century, the demand for iron was driven by the appetite of the military (Sass, 1998). By the eighteenth century, coal became the primary source of energy, and cast iron supplanted wrought iron. Next, the steam engine was invented. It was used to drain mines, but it was inherently wasteful. Watt subsequently designed an economical steam engine. This chain of events led to the high achievements of the industrial era: the birth of railroads, machine-production-based factories that operated assembly lines, and the mass production of goods.

Furthermore, the industrial–technological revolution brought forth the second agricultural revolution. The use of improved plows, the mechanical tractor and reaper, and other agricultural technologies created huge food surpluses, freeing large populations to engage in industrial production (Elgin, 1993). Another result of the successes of technological invention was that it generated encouragement in engaging in social inventions.

But what of the relationship of science to technology? At the early stages of the Industrial Revolution, leading scientists such as Copernicus, Galileo, Kepler, and Newton aimed at understanding fundamental phenomena of how the world works (Sass, 1998) and discovering the laws of the universe, rather then solving practical problems. By the middle of the Industrial Revolution, however, many scientists became interested in improving technology. These scientists began to call themselves "chemists." By the end of the eighteenth century, the scientific method of controlled experiments, followed by an analysis of results, led to many technical advancements.

Finally, let us explore the relationship of technology and communication. We already saw that print technology played a key role in the emergence of the Third Generation HSS. It was during the last phase of this scientific/industrial era that electronic communication emerged. Electronics is a special case in the evolutionary advancement of the Third Generation and relates to the explosion of information. Moving at the speed of light, information transmission became instantaneous. For all practical purposes, electronic communication created the global village.

4.3.2.3.b Economics: The Second Agricultural Revolution and the Rise of Capitalism. The evolution of the subsistence economy of the Third Generation HSS marks the beginning of a new era of agricultural development. For most people the earlier agrarian era was associated with poverty, disease, and drudgery. But people of the Third Generation experienced just the opposite. They expected that with practical inventions and hard work, their life could be improved dramatically. They believed that "With ingenuity and entrepreneurship great material advances were possible in the here-and-now" (Elgin, 1993, p. 93).

The exploding technology created new and highly improved methods of agricultural production. New machines of planting, cultivating, and mass harvesting of crops were developed. Furthermore, advanced techniques were invented for transporting and storing food. The result of this new versatile agricultural technology was twofold. First, it resulted in a quantum jump in productivity, which amounted to nothing less than another agricultural revolution. Second, increased productivity led to a mass exodus of people from agricultural work and a mass increase of populations living in cities and working in factories and in the service occupations. The economy responded to an increasing obsession of people to possess more and more, to attain material progress, and get ahead in the material world. Free markets replaced state and feudal control. A new financial world was built to respond to and fuel the new world of material progress. New and diverse institutions were created as instruments of managing financial capital and advancing business ventures. As a result, capitalism emerged, often in rampant, exploitative, and merciless forms.

4.3.2.3.c Societal Organization: The Urban Revolution and the Creation of an Urban–Industrial Society. The application of the evolutionary markers described previously created a radical reorganization of the society. As already mentioned, the second agricultural revolution led to the relocation of masses of people in large cities of great complexity, coupled with the emergence of a large middle class. Whereas in the seventeenth century in the United States some 90% of people lived in farming communities by the end of the nineteenth century more than 90% lived in large urban centers.

The "all is relative" world view, the legitimization of each person's point of view, and the emancipation from church and state authority empowered people to act in their self-interest in all aspects of life. While in the Greek city-states the state was omnipotent, in the Third Generation the individual became omnipotent in pursuing his or her material gains. But Elgin (1993) suggests that:

> Material affluence was gained at considerable social cost. The urbanizing and industrializing process tore individuals from their traditional communities and extended families and made them autonomous workers—cogs in the great machines of industry. Instead of small-scale, emotionally bounded communities nurtured by many generations of families and relationships, people moved into impersonal cities with few ties to an extended family. Cities were not organized to serve a feeling of community and a sense of beauty, rather they were oriented to serve economic efficiency. (p. 104)

In the urban environment, the autonomous individual became the social unit of existence. The individual had the authority to pursue his economic and material goals. Still, living among the many meant living in isolation, nearly alone. For most people, the notions of being an integral part of a community and serving the common good evaporated in the increasing heat that was generated by the industrial machine age.

4.3.2.3.d The Polity of the Third Generation: The Democratic Revolution. In the arena of polity, the expression of individual autonomy and the increase in individual freedom have led to the emergence of self-determination. It was the revival of the Hellenic democracy even if in a different form. While Hellenic democracy was participatory, the democratic polity of the Third Generation was representative. People invested the power of political decision making and the responsibility of making laws in their elected representatives. New forms of government were invented, such as the governing practice in the United States, which was a complex arrangement of checks and balances.

Democracy was a faithful expression of the consciousness of the Third Generation HSS. Democracy mirrored well the dynamic consciousness and the perceptual paradigm of the fourth dimension. There were several particular ways that democracy reflected the new perceptual paradigm (Elgin, 1993). The relativistic view that emerged claimed that the position of each individual is to be respected and has equal validity with every other position. The scientific paradigm (discussed earlier) challenged supernatural claims, the authority of rules, and the church. The expression of the time concept of the perceptual paradigm opened up the "wheel of circularity" into a forward moving "spiral" of ever possible progress. Finally, the great success of technological inventions of engineering signaled the possibility of social engineering; the creation of new forms of polity: new forms of who govern and how.

4.3.2.3.e Relationship with Nature. "The oneness with nature"—which was the way of life of *Homo* through several millions of years—was still the dominant world view of the fist modern man, the Cro-Magnons. They revered nature and through the magic of the spirit were in constant communication with the gods of nature. Even for the Second Generation, oneness with nature was still the guiding consciousness of the Earth-centered and Earth-worshiping people of the agricultural villages. But as the Third Generation emerged there came about a major paradigm shift in all the evolutionary markers. This paradigm shift, then, became coupled with the resource demands of new energy-intensive technologies, leading not only to a total separation from nature but also to a domination over it. Nature was subjugated to material gains, technological progress, and often greed. By the second half of the nineteenth and during the twentieth century, this trend reached the systematic exploitation and large-scale destruction of nature. Nature, which had been a revered friend of man for several millions of years, now quickly became his slave. We no longer carry on loving conversations with nature, or even let it talk to us.

Reflections

In the preceding text I described the evolutionary markers that emerged in the design space of the Third Generation HSS. The markers were organized within the three clusters of the General Cultural Model of societal evolution. If we stop here for a moment, we can reflect and see how these markers not only interact with and influence each other, but also how they *define* each other. This recursiveness and "coevolution" are clear manifestations of the systemic and synergistic nature of societal evolution and indicate the dynamics from which the wholeness of an evolutionary stage emerges. For an evolutionary stage to be sustained and advance, its markers most show internal consistency and synergy, attained by the integrating creative force of the new collective consciousness. The following activity will help to create your own understanding of the systemic and mutually influencing nature of evolutionary markers.

Activity 12

I stated that the markers, characterized in this section of Chapter 4, constitute an internally consistent and mutually influencing and supportive system of markers. Your task now is to test this statement and describe how the markers interact, how the characteristics and specifics of each of the markers influence each other. You are asked to complete two tasks.

Task 1. Describe the systemic interaction and coevolution of markers within each of the three clusters.

Task 2. Describe the systemic interaction and coevolution among the three clusters within the Cultural Model of the Third Generation. Enter your findings in your workbook.

Reflections

The evolutionary journey of our ancestors took several millions of years. The First Generation of modern man, the Cro-Magnons, traveled for some thirty thousand years. The time span of the evolutionary journey of the Second Generation was more than ten thousand years. The Third Generation, emerging around the end of the fifteenth century, traveled on for five hundred years before it now gives way to the next Generation of HSS. The life span of the Third Generation was minuscule in the total context of the evolutionary saga. But this minuscule period created an evolutionary epoch of massive transformative changes in the Evolutionary Cultural Model. Elgin (1993) summarizing the achievements as well as the failures of the Third Generation, says that the scientific–industrial era generated:

> unprecedented social dynamism, moral relativism, intellectual absolutism, nation-state egotism, and technological giantism. Both humans and the rest of the life on the planet paid very high prize for the learning realized in this era. Although people in industrial societies are more intellectually sophisticated and psychologically differentiated, they are also more isolated; feeling separated from the Divine Life-force, from nature, and other persons. Feelings of companionship and community have been stripped away as many live nearly alone in vast urban regions of alienating scale and complexity. Unprecedented economic and political freedom have been won, but at great cost, when life seems to have little meaning or sense beyond ever more consumption. The

> perceptual paradigm of the scientific–industrial era has immense drive but virtually no sense of direction beyond sheer accumulation. It is a dynamic with no moral anchor or guiding ethic—an economic engine with no idea where it is going beyond the drive to acquire material things and material power. (p. 113)

The quote gives us a comprehensive reading of how far evolution advanced as it reached the twentieth century. It also confronts us with issues that we, the Fourth Generation HSS, need to address at the threshold of the new millennium.

An Overview of Part I

We commenced our journey of human and societal evolution in Chapter 3. In the course of traveling the first mile of the journey, we listened to the evolutionary story of our ancestors, revealed to us by paleonarchaeologists, paleontologists, cultural anthropologists, and evolutionary philosophers. As we entered the second mile of the journey, we met the first modern men, the Cro-Magnons, who revealed to us the inspiring story of their lives. We called them the First Generation of *Homo sapiens sapiens*: The double wise man. They knew that they knew. Next we visited the "Farmer" in the small agricultural villages. He was the first representative of the Second Generation. His advance in agricultural production enabled the establishment of agriculture-based ancient civilizations, and later the emergence of the Hellenic, the Roman, and the Byzantine cultures. The last thousand years of the Second Generation saw a gradual decline and loss of cultural identity. Out of the ashes of the Second Generation rose the Third Generation of the science-based industrial age, the Age of Reason and Enlightenment, the era of the Renaissance. This generation lived through several major revolutions that reached high points, but soon declined under the burden of the many problems that these revolutions brought about. Our journey—described in Part I—ends in the present, as we seek to embark on the evolutionary emergence of Fourth Generation of HSS.

In Part I, we traveled the evolutionary landscape of the biological and cultural evolution of our species. As we traveled the landscape, we saw its features and developed an understanding of how evolution works. Now, we are ready to ask questions: What does it all mean? What have we learned during our journey? And most importantly: What does it tell us to do and how, as we prepare ourselves to take responsibility for conscious and guided evolution?

II

The Journey from Evolutionary Consciousness to Conscious Evolution

Part I presented the story of human and societal evolution from the point in time when the humanoid species split from its primate ancestors to the point when modern man arrived to the present time. The story was presented as a journey that was marked by several milestones, each representing a major evolutionary transformation. During the first mile of the journey our species had a long walk, lasting more than six million years. At the end of the fist mile we met with *Homo sapiens sapiens* (HSS). They were the people of the *Human Revolution*. We named them the Cro-Magnons, the First Generation of modern man. We walked with them to the second milestone of the evolutionary journey, where we met with the first Farmers, the Second Generation of HSS. They started the *Agricultural Revolution* some 10,000 years ago and their surplus food production made possible the establishment of ancient civilizations and later the classical civilization. At the third milestone, some five hundred years ago, our evolutionary story turned into the *Scientific–Industrial Revolution* of the Third Generation of HSS. Now, at the beginning of the twenty-first century, we are faced with an evolutionary predicament, marked by the accelerating decay and decomposition of the industrial machine age. At the fourth milestone, we are at the threshold of the emergence of the Fourth Generation of HSS.

We do not know yet how we might call this new era, so we speak of it as the post-industrial age, the era of post-business, the post-modern, or even the post-post-modern age. We know that we are leaving the industrial age and we have to decide for ourselves what next. We are in search of a direction. Can we, and—are we willing to, guide it? These are the questions we attempt to answer in the rest of this work.

Part II is developed in three chapters. In Chapter 5, we jointly explore what we have learned about human and societal evolution from working with the evolutionary story in Part I. This exploration aims to lead us to achieve a state of evolutionary consciousness. In Chapter 6, the current evolutionary predicament is explored. In Chapter 7, we listen to those who are calling for conscious evolution. We explore how an understanding of evolutionary consciousness and an appreciation of our current evolutionary predicament may lead us to respond to the call for engaging in conscious, self-guided evolution.

5

What Have We Learned from the Evolutionary Story?

In this chapter we ask the questions: What have we learned from the evolutionary story? How does a reflection upon what we have learned help us to attain evolutionary consciousness? And: How may evolutionary consciousness lead us to engage in conscious evolution? To answer these questions, first, perspectives and propositions are introduced that build upon the core ideas we identified in Part I. Next, informed by these perspectives and propositions, three images are introduced that provide a summary characterization of the evolution of our species. In closing, a set of organizing principles is developed by a synthesis of the perspectives, propositions, and images. These principles, then, will be used to organize our thinking about the current state of societal evolution in Chapter 6.

5.1 Perspectives and Propositions on Sociocultural Evolution

Sociocultural evolution is an overarching term that denotes the sum of core ideas, concepts, and principles of sociocultural evolution. We derived these while working with the text and the activities in Part I. In the present chapter, the interpretation of these core ideas, concepts, and principles becomes the source for attaining evolutionary consciousness.

5.1.1 Reflecting upon the Evolutionary Story

A reflection upon the evolutionary story helps us to gain two sets of perspectives. The first set relates to the development of a systems view of evolution and thinking about evolution as a systemic process. The second set offers perspectives that pertain to the creation of our species in the design space of evolution. I shall also consider the function of evolutionary potential in this creating process as well as the limitations and constraints that act upon evolution.

5.1.1.1 The Systemic Nature of Evolution

The systemic nature of evolution became evident as I described the emergence of the various generations of *Homo sapiens sapiens*. They emerged from the creative systemic interaction and interpenetration of the evolutionary markers of a particular generation of HSS. Here, I begin to explore this systemic nature in a focused way. Taking a systems view of sociocultural evolution enables us to explore and describe the following:

5.1.1.1.a The Ongoing Coevolutionary Relationships and Interactions, and the information and energy exchanges between the evolving entity and its environment. Such a systems-environment view (Banathy, 1992) helps us to understand how a coevolutionary relationship between an emerging species and its environment shapes the purposes and functions of the species and how the species shapes and affects its environment.

5.1.1.1.b The Dynamics of Interaction, Interdependence, and Interpenetration of the evolutionary markers; and the *patterns that connect the markers and integrate* them into the whole of the evolving sociocultural system. We have already seen how these dynamics operate in the evolutionary design space of the various generations of *Homo sapiens sapiens*.

5.1.1.1.c The Emerging Wholeness. The whole organizes the parts. In evolution, the whole—the "evolutionary design concept" of the evolving entity—generates, organizes, and integrates the various "evolutionary markers." As a result of this generating organization and integration, systems properties emerge at the level of the whole that are *not* properties of the various evolutionary markers.

5.1.1.1.d The Dynamics of Change that operate through time in the evolving system, and the way by which the dynamics of change affect the evolutionary markers and the evolving system. The evolutionary play of these dynamics shows how a new evolutionary entity emerges from a creative surge, how it develops and maintains itself for a while, and how—eventually—it begins to decay.

5.1.1.2 The Systemic Perspective of Generative Order in Evolution

Bohm and Peat (1987) introduced the systemic perspective of "generative order" in evolution. Exploring their ideas, first, the notion of generative order is defined, followed by a description of how it works in general, how it works in evolution, how it works in society, and—eventually—how and why it breaks down.

5.1.1.2.a The Systemic Perspective of the Generative Order in Evolution. The systems view, which helps us to capture the systemic nature of the process of evolution, is explained in the work of Bohm and Peat (1987). They apply the concept of *generative order* to the evolution of life forms. Their *generative order* is not concerned with the outward side of development and evolution as a sequence of succession, but "with the deeper and more inward order, out of which the manifest order of things can emerge creatively" (p. 151). The generative order is manifested in many aspects of life. It can be seen in the work of the artist, who—making a portrait—first attempts to capture the overall form (of the sitter) with an initial sketch. Then, within the overall form the image is made more detailed as the work progresses. The artist is always working from the generative source of the initial big idea. It is the generative source that *enfolds* the idea. It is, then, the (enfolded) idea from which the picture is generated. It is by the process of painting that the work *unfolds* into ever more definite form.

We can transform the painting metaphor of generative order into the context of the evolution of a human species by an example. In the course of human and societal evolution, in the evolutionary design space, the generative source of human evolution enfolded the idea of the Cro-Magnons. The painter is evolution itself. It painted an overall sketch, in which the evolutionary core idea of the Cro-Magnons was *enfolded as an evolutionary potential.* It is from this enfolded idea that the painting process, namely, the process of evolutionary emergence, *unfolded* into the First Generation of modern humans. This unfolding produced an ever more delineated life form. This enfolded–unfolding process represents the true *mystery of evolution.*

5.1.1.2.b The Workings of the Generative Order in Evolution. The Darwinian and Neo-Darwinian approaches treat the notions of the "explicate" and "sequential order" as basic. "According to these notions, life arises as a "fortuitous" chance combination of molecules which leads, in a more or less mechanically determined way to further developments, which produce ever higher and more complex forms" (Bohm and Peat, 1987, p. 200). In contrast, the authors suggest that in light of the workings of the generative order, the deeper meaning of the evolution of life can be understood by exploring how it reveals the inward enfolded generative order of the whole, from which the new life form unfolds. Small variations cannot account for the appearance of new species. Bohm and Peat (1987) present a most compelling example of the systemic nature of the evolution of a species.

> It is not sufficient, for example, to give a bird wings, if it is to fly. In addition, its bones must be made lighter while, at the same time, maintaining their strength, feathers must be aerodynamically adapted, the center of gravity must shift, the breastbone and musculature must develop, and changes in metabolism must be acquired to provide sufficient energy for flight. If those changes do not all occur together and in a coordinated fashion, then they may be disadvantaged to survival. It is difficult to understand how many of the highly coordinated chances, demanded by evolution, could have come about by fortuitous chance combinations of small random mutations. (p. 202)

The generative order may enfold *totalities* of structures of the evolving entity that can emerge as a whole in a systemic interactive way. Bohm and Peat call this potential of the generative order of evolution "*proto intelligence.*" Proto intelligence guides the active generation of new order of totalities "from levels that are at the present 'hidden' to science" (p. 202). Proto intelligence is not a more or less mechanical adaptation to the existing order. The essential feature of proto intelligence is that totalities (wholes) are not formed in a random fashion, but emerge as integrated entities. The reason for the sudden formation of new totalities is not yet known to us. Bohm and Peat offer intensive *free play* as an explanation. In the course of free play, as discussed earlier, whole new totalities of structure may be put forth (in the evolutionary design space) for a test of their viability. For example, several forms of the humanoid species were put forth in the evolutionary design space and were rejected and "pruned off" from the evolutionary tree.

Bohm and Peat (1987) conclude their discussion on generative order in evolution by saying that generative order "goes on ultimately to include all life, all matter, the whole known universe, and what may lie beyond" (p. 204). Generative order enfolds evolutionary potential that emerges into the wholeness of a species in the evolutionary design space.

5.1.1.2.c The Generative Order in Society. Toynbee (1947) suggests that new civilizations emerge from creative surges in evolution. They flower and develop, but eventually decay. We have witnessed this process in the evolutionary story: (1) in the emergence and disappearance of the Archaic *sapiens* and the Neanderthals, (2) in the rise and decay of ancient civilizations and the Hellenic civilization, and (3) now we experience it as the decay of the scientific–industrial civilization. "This suggests that the generative order of societies has two basic sides, that of growth and decay" (Bohm and Peat, 1987, p. 205). They continue to say that:

A society that has gone beyond its first creative surge hangs on to the habitual orders that are contained in its customs, taboos, laws, and rules and that are held in its unconscious infrastructure. Because these orders are fixed and limited, they will be bound eventually to become inadequate, in the face of ever-changing reality. As a civilization grows more complex in its structure, what is most significant is the activities of the society itself. These may lead eventually to decay, more or less independently of the institutions, the will, and desire of the people who make up the society. (p. 206)

I suggested (Banathy, 1996) that if a society is to continue to evolve, *its evolution has to be manifested in a total transformative change. Its existing state has to be transcendent. A new creative surge has to generate a new image of the future in the evolutionary design space. And from this image new order has to unfold through the creative process of emergence.* The idea presented here is a core idea of the process of guided evolution. But, as a rule, societies resist this process. They want to hold on to the existing order. They put much energy into defending "the way it is," and create the illusion that their current way of life and their institutions can go on indefinitely. They fight against the emerging new realities that call for change.

Bohm and Peat (1987) hold that "not admitting to decay" and holding on to the illusion of viability becomes a destructive force that breaks down the generative order of the society. This process of resistence blocks the "free play" that is needed to bring about a creative surge that could bring about a new generative order.

To the extent to which a society is no longer creative in its basic generative order, it becomes destructive to itself and everything it touches. Thus, both with regard to human beings and society as a whole, the basic distinction in the generative order is not between growth and decay, but between creativity and destructiveness. Becoming caught up in the false process of a destructive generative order has been the ultimate fate of all known civilizations. But is this inevitable? Need this cataclysmic failure of the human necessarily take place? Can it be avoided by a creative approach to all areas of life? What is needed is not simply a new creative surge, but a new order of creative surge, one that extends into science, culture, social organization, and consciousness itself. (pp. 209–210)

The purpose of this work is to answer the questions as stated in the foregoing quote, by proposing a "new and consciously renewed generative order" that can guide the creative surge of conscious evolution.

5.1.1.2.d Systems Breakdown in Societies. In our evolutionary story we saw the systemic emergence of the various generations of *Homo sapiens sapiens*. It was by—and from—the systemic interaction and coevolutionary cooperation of the systems of evolutionary markers that a harmonious, internally consistent, and balanced new evolutionary form and order was created.

In the life cycle of an emerged society, following the initial creative surge, the culture of a society (of a new Generation of HSS) fully develops and reaches its high peak of maturity. Then, a point is reached where confidence in its own viability and the viability of its evolved knowledge, its institutions, and ways of life become so strong that they become rigid and inflexible. Any notion of change is rejected and the society becomes out-of-sync with its environment. Even if there are calls for change from within, these calls are dismissed. It is at this point that decay sets in. One of the best examples of this is the life story of the Neanderthals, who rejected change in any form and disappeared.

Decay folds into destruction when components of the cultural system run away and offset the balance and harmony of the system. An example of this is the decay and destructiveness of the scientific–industrial era which brought forth the decline and decay of the Third Generation of HSS. We reflected upon this in our evolutionary story. In its declining phase, according to Elgin (1993), the scientific–industrial era generated intellectual absolutism, moral relativism, national egotism, materialism, and technological "gigantism." People became isolated from nature and from each other in the industrialized mega-cities. The price paid for an unprecedented political freedom and unconstrained individualism was loss of direction and life having little meaning beyond ever more consumption. Life lost a moral direction or a guiding ethic. The economic engine was running the collective life of the society, but without having any idea of where to go—beyond the drive to achieve more and more material progress and material power. It is this state of affairs that confronts us as we stand at the threshold of the Fourth Generation of HSS, as we contemplate our evolutionary future. We can begin this contemplation as we take a systems view of designing human and social evolution in the evolutionary design space.

5.1.1.3 A Systems View of Creation in the Evolutionary Design Space

In our evolutionary story, we explored the general model of the process of design in the evolutionary design space, the role played by the evolutionary potential, and the constraints that limit evolution. We bring these together to see how they interact and constitute an operating evolution generating system. But first, I revisit the notion of the evolutionary design space.

5.1.1.3.a The Design Space of Evolution. Design is a creating activity that brings forth a potential-driven intended novel entity. Design activity takes place in the design space where alternative design ideas are proposed and tested for their viability and for their "goodness of fit" with their enfolded potential and the environment, which becomes their life space. In the general evolutionary design space of biological evolution, the process of testing life forms is the process of natural selection. The design space of human and societal evolution is situated within the overall evolutionary design space. But it is a very specific design space, unlike any other. Design experimentation has been taking place with humanoids, *Homo* species, and *Homo sapiens*. These experiments have been going on during the last seven million years. For more than six million years, the experimentation was primarily biological. But since the emergence of Cro-Magnons, the evolutionary process of natural/biological selection transformed into cultural selection.

5.1.1.3.b The General Model of Design Experimentation. In Chapter 3 we explored this model and called it the "tree" model. From the "trunk" of a parental stock a new species is split off. It is a split-off rather than evolution by a slow, steady transition. Repeated splitting produces the branches of the tree. The branches, representing some new variations, are tested in the design space. When a specific descended variation of a species disappears, it means that it failed the test of the evolutionary experimentation. It is "pruned off" of the evolutionary tree. After the humanoid race was split off from the ancestral tree some seven million years ago, repeated splitting happened as new branches appeared on the humanoid tree. As time progressed, more and more became pruned off. The last pruning took place when the Neanderthals were pruned off, leaving us modern humans the only lineage left.

5.1.1.3.c Evolutionary Potential. We can regard the splitting-off points where a new variation came into being, where a new branch appeared on the tree, as a threshold. A new variation, a new life form of the humanoid race, can burst forth across such a threshold only if it has in it the "enfolded" potential of a new design idea (not yet the design) of a life form that it can offer for evolutionary experimentation. The design idea can become "unfolded" into the design of the new offering by the creative burst that brings it to life. When a new offering failed the experiment, the failed species was pruned off. The pruning-off happened because the new variation was unable to realize its enfolded potential.

Evolutionary potential, even though it was enfolded in the original primordial, is *not* fixed. It evolves through time. New potential builds upon unfolded potential. Potential " can learn" even in its enfolded state. In the evolution of the *Homo* lineage there have been ever unfolding changes that became realized and unfolded from enfolded evolutionary potential. The unfolded potential creates new (enfolded) potential, which at the next stage of evolution emerges as a new, more advanced, and more complex entity. This new entity then can create more advanced potential. Evolution always offers new possibilities. It is neither predetermined nor predictable. From ensembles of possibilities of identical initial conditions, different variations/alternatives may emerge and are tested in the evolutionary design space. Understanding this process contributes to the attainment of evolutionary consciousness. Evolutionary consciousness raises the question: Can we give direction to the evolutionary process? As we continue to develop this work, I expect that there will be an increasing recognition of the notions that indeed we can and we must.

5.1.1.3.d Limits and Constraints. In the course of revisiting evolutionary core ideas in Part I, we explored Gould's metaphor of the left and right walls of the evolution of life. Life emerged from the left wall some 3.5 billion years ago. The right wall is in the future. It stands for the *possible limits* of how far evolution can go. Since its origin, the evolution of life moved toward ever increasing complexity. But how complex can it become? As the various life forms live their history, they run into right walls. As Goethe said: "It is ordained that trees cannot grow to heaven." Things created by nature or artifacts made by man have limits in their heights. The question we can raise is: Is there a right wall in our future evolution?

There are also constraints in the variability of life forms. We know that evolution has brought forth a rich variety of life forms. But evolutionary scholars have observed that variability is constrained. Only certain patterns of life forms appear to be viable. We also know from the evolutionary law of pruning-off that in certain life forms, some variations are more viable than others. An example of limits of variability are the 250,000 species of higher plants, around whose stems only three patterns of distribution are in evidence. Furthermore, all but 20% of these are of spiral forms. The same constraints are manifested in the formation of bones and, in general, in the morphology of all domains of life. In theory an infinite number of variations of life forms can be proposed but the evolutionary law of constraints permits putting up only a few for evolutionary experimentation. There is another side of constraints. Paradoxically, constraints at certain levels enable freedom from constraints at a higher level. For example, in language (Artigiani, 1999, personal communication), the limited number of sounds and morphological and syntactical rules enables great freedom and efficiency in the use of language.

Reflections

In the preceding section, we explored the notions of the design space in which evolutionary experimentation happens, reviewed the general model of design experimentation, revisited the concept of evolutionary potential, and understood the limits and constraints of evolution. It is in the systemic interaction of these four aspects that we can capture aspects of the process of evolution. We now understand that the design, and the design-based emergence, of new life forms are made possible by their enfolded evolutionary potential. This potential can unfold by a creative surge that puts up for experimentation new variations of species by which the viability of proposed life forms can be tested. New variations of a life form are limited and constrained by boundaries that are established in the evolutionary design space. The interactive systemic process described here can be defined as the *dynamics of creation in the evolutionary design space.*

Activity 13

This activity calls upon you to do some "knowledge discovery" of your own. Select one of the specific civilizations I described in the evolutionary story, for example, the ancient civilization, the Hellenic, or the scientific–industrial. In the preceding text we explored ideas on the generative order in societal evolution. We saw how the rise and fall, the growth and decay, the creation and destruction affect civilizations. A new civilization rises by transcending the one that preceded it. Then, following growth and development, it reaches a stage when it becomes rigid, inflexible, unwilling to change, even though newly emerged "realities" call for change and transformation. Thus, it enters the decay phase and eventually self-destructs. Your task is to speculate why and how these events happened in the life cycle of the particular civilization you selected. You might consider the current evolutionary state in which we live between the decaying Third Generation HSS and the yet to emerge Fourth Generation. Enter your findings in your workbook.

5.1.2 Formulating Propositions about Sociocultural Evolution

The term "sociocultural" captures social, societal, and cultural evolution as an overarching designation. The propositions are to be built upon generic core ideas about concepts and principles that are properties of the various evolutionary generations of HSS. To capture what we have learned about evolution in general and sociocultural evolution specifically, I introduce here a framework that designates the various evolutionary domains within which *you can summarize what you have learned about sociocultural evolution*. Your findings should take the form of a set of propositions formulated under a specific evolutionary domain. For example, in the domain of biological vs. cultural evolution, a sample proposition might say that: "With the emergence of Cro-Magnons, cultural evolution superseded biological evolution." Briefly explain this statement. You will present your propositions in Activity 14. The source of the development of your propositions is the core ideas that you have developed in working with the text and the activities.

5.1.2.1 Biological vs. Cultural Evolution

Issues around which propositions can be framed could include the following: definitions of biological and cultural evolution, the difference in their nature, the principles under which they operate, the nature of their process, their purposefulness or lack of it, the capacity of their (change) rate, their intensity to generate change in evolution, the role of learning in both biological and cultural evolution, the role of environment in affecting them, the role of selection and the kind of selection, their difference in the specialization of parts, the possibility of self-directed change in each, the relevance of the Lamarckian principle of the acquisition of inherited characteristic in both, comparison of the biological replication of genes and the cultural replication of "memes," and the differences in their flexibilities.

5.1.2.2 The Nature and Role of Cultural Transmission

Issues to address in formulating your propositions include the following: the definition of culture; the nature and the role of cultural transmission, (CT); the principles of CT; the general role of language in CT; the role of speech/writing/print/electronic communication, and the Internet in CT; the role of art in CT; the function of CT in building the "Cultural Pyramid" in sociocultural evolution; and the role of creativity in CT.

5.1.2.3 Social and Societal Evolution

A distinction can be made between social and societal evolution. Two examples follow. (1) The *social organization* of Cro-Magnons. They lived in kinship groups and occasionally in larger groups that included several kinship groups. The same organizational arrangements prevailed in the agricultural villages. (2) The *societal organization* of the agriculture-based ancient civilizations. They created and lived in the first societies in large settlements. From that point on in evolution societies became the typical organizational mode.

Developing your propositions could begin with a definition of social/societal evolution (SSE) and its principles, followed by a consideration of issues within the context of SSE such as: individuation vs. integration, the role of consciousness, the evolution and the role of "world view," the role of cooperation, the kind(s) of learning, the role of ethics and morality, complexity, the capacity of culture, the notion of progress, the role of technology, the evolution of a "common culture," the idea of the common good, the notion of "hominization," the nature and evolution of social behavior, the role of language, and spirituality.

5.1.2.4 Cultural Evolutionary Models

Cultural evolutionary models (CEMs) (of the various generations of HSS) are systemic organizations of sets of evolutionary markers that represent the identity of an emerged generation of HSS. The model as a system is organized from a system of "clusters." Clusters are organized from a system of evolutionary markers. A particular cluster integrates its component evolutionary markers which interact with and interpenetrate each other, and thus mutually affect and define each other. The same processes (interaction, interpenetration, mutual effect, and integration) operate at the whole level. Issues to consider in setting forth your propositions include the following: the function of CEMs in understanding sociocultural evolution, the role of evolutionary markers, the position and function of subsystems in the system of CEM, the substantive meaning of the terms interaction, interpenetration, mutual effect, and integration in the dynamics of developing a CEM, substantive differences in the CEMs that represent the three generations of HSS, the forces that bring forth a new generation of HSS and its CEM, and the relationship of CEMs to the evolutionary design space.

5.1.2.5 The Driving Forces of Sociocultural Evolution

Evolution would not be possible without a force to drive it. A "cosmic force" brought forth the geological evolution on Earth, and it then transformed into chemical evolution. Chemical evolution was then transformed into biological evolution that created life. We have seen how biological evolution was transformed into cultural evolution, which modern man brought into existence. We can see this evolutionary force as a system of forces that moves evolution. We can include at least the following component forces in sociocultural evolution: creativity, evolutionary potential, coevolution, selection, learning, humanization, and the influence of natural and social environments. You might identify some others. Your task is to describe in Activity 14 how these forces have shaped sociocultural evolution.

Section 5.1.2 provided us with a framework that will help in summarizing what we have learned about evolution. You are now asked to use the framework in presenting your responses to the issues raised in Section 5.1.2, by completing Activity 14.

Activity 14
This activity summarizes what you have learned about evolution in general and specifically about sociocultural evolution. You are asked to select a few issues introduced in each of the five domains of Section 5.1.2. The sources of your response are the texts of Chapters 2, 3, and 4, the core ideas you constructed by working with the texts and the activities, and the perspectives introduced in the first section of this chapter. Your presentation should be made in the form of a narrative. Enter your findings in your workbook.

5.2 *Images of Evolution*

The first image is presented in the form of a matrix, which presents the descriptors of the major stages of the evolution of our species in a space/time frame. Next, a generalized image of the cultural model shows the coevolutionary relationships of the various evolutionary markers. The third is a spiral image of the never-ending evolution. The fourth is a systems view of evolution that emerges as we look at evolution with three lenses: a systems environment, a functions/structure, and a process lens.

5.2.1 Image 1: An Evolutionary Matrix

The matrix presents a comprehensive summary image of the evolution of our species. (See Table 5.1.) The horizontal categories of the matrix define the various evolutionary stages. The vertical categories introduce the descriptors of the stages. What follows here is a narrative of the summary image of the stages of human and societal evolution.

5.2.1.1 Stage 1: The Era of our Ancestors

Stage 1 is specified by: (1) designating our ancestral species and (2) describing the time frame of their existence, then defining (3) their state

TABLE 5.1
Characterization of the Evolutionary Stages

Stage	Era	Time	Consciousness	Communication	Social	Technology
Stage 1	Ancestors	7 M–0.5 M	Contracted	Signs	Family, small group	Tools
Stage 2	Transition	0.5 M–35 K	Magical	Vocal signs	Small group	Tools
Stage 3	Cro-Magnons	35 K–10 K	Magical, reflective, sensory	Speech	Tarbe	Tool chest
Stage 4	Agricultural	10 K–0.5 K	Mythical, Reflective, emotional	Writing	Ancient civilizations	Agriculture, metal
Stage 5	Scientific/ industrial	0.5 K–today	Rational, reflective mental	Print, electronic	Nation-states	Industrial
Stage 6	Information/ knowledge ?	Future	Reflective, spiritual, ethical	Internet ?	Regional, global	?

of consciousness, (4) their mode of communication, (5) their social organizations, and (6) their technology.

5.2.1.1.a The Era of our Ancestors, the Evolutionary Era of the Humanoids and Homo species. In this era several variations of our species were "put up" for experimentation in the evolutionary design space, but all but one disappeared.

5.2.1.1.b Time Frame. The evolutionary saga of our species began some seven million years ago, when the humanoid race branched off from the common ancestral trunk it shared with the large primates. What we know about this era is revealed only by fossil finds and campsite evidence. The first fragmentary fossil dates back 4.5 million years. The first partial skeletons, Lucy's and her family's, were dated three million years with the paleontological definition of *A. afarensis*. There was an explosion of evolutionary experimentation with five humanoid and two *Homo* species between 2.5 and 1.5 million years ago. Of the seven, only the two *Homo* species, namely *Homo habilis* and *Homo erectus*, left as our ancestors.

5.2.1.1.c Their State of Consciousness. The consciousness of our ancestors can be described as a dreamlike state, a state that was "undifferentiated from nature." It was oneness with nature, with no separation from others. It was "oneness with wholeness."

5.2.1.1.d Their Communication. Their mode of communication included signals, gestures, and speechlike sounds. Their vocal anatomies as well as their consciousness have not yet reached the level of capacity required for speech. However, it developed gradually in *H. erectus* but did not yet attain speech status.

5.2.1.1.e Social Organization. The unit of social organization of the humanoid was the family. In the *Homo* species, however, it reached the level of complexity of kinship groups.

5.2.1.1.f Their Technology. It started out as basic survival technology of gathering, later scavenging, gradually developing tool use, and eventually toolmaking by *Homo habilis* and *Homo erectus*. By the time these emerged, the technology extended to camp-making and fire maintaining—but not yet fire-making. The last one million years of *H. erectus* evolutionary saga was static; his tools had not changed over time.

5.2.1.2 Stage 2: The Transition Stage to Modern Man

First, I define the two species of the transition stage, then describe their time cycle, their consciousness, their mode of communication, their social organization, and their technology.

5.2.1.2.a The Archaic sapiens and the Neanderthals. From the *H. erectus* line came two lineages of *Homo sapiens*: the Archaic *sapiens* and the Neanderthals. Recent fossil finds change this picture. They place *Homo heidelbergenesis* as the direct ancestor of the Neanderthals. In our evolutionary story revisions such as this can be expected as new evidence calls for it. For example, as I write the final manuscript, four new discoveries of the hominid species are reported in *Time* by De La Cal et al. (1999).

5.2.1.2.b The Emergence and Life Span of the Archaic sapiens and the Neanderthals. The Asian/European Neanderthal *sapiens* emerged more than 500,000 years ago and the African Archaic *sapiens* (AAS) about the same time. The AAS lived in the sub-Saharan region. They evolved into anatomically (but not culturally) modern humans around 200,000 years ago. Then, about 100,000 years ago, they migrated out of Africa, eventually reaching all of Europe, the Middle East, Asia, and Australia. It was from this lineage that culturally modern humans evolved.

The Neanderthals left Africa possibly 300,000 years ago. We have fossil and campsite evidence of their presence in Europe and the Middle East dating back 120,000 years. They flourished for tens of thousands of years through the Ice Age but disappeared after the arrival of the first *Homo sapiens sapiens*, the Cro-Magnons.

5.2.1.2.c The Nature of Their Consciousness. Both of these species still lived in the state of archaic consciousness in which the individual and his group—and man and his world—were bound up with one another. They did not have self-reflective consciousness. Magic was their main vehicle to assert themselves. While this state continued to characterize the consciousness of Neanderthals until their departure, the assumption is that some degree of reflective consciousness was awakened in our predecessors, the Archaic *sapiens*. As they migrated out of Africa into the Ice Age environment of Europe, they had to learn to make warm clothing, build shelters, and use fire, all of which indicate a limited degree of self-consciousness.

5.2.1.2.d Their Communication. There was another significant difference between the two species: their ability to use vocal communication. The vocal-track structure (for speech) of the Neanderthals was 5 to 6 on a scale of 10 for modern man. Their speech—if they had any—must have been extremely slow and strongly nasalized. On the other hand, the Archaic *sapiens* had the anatomy of modern man by the time they appeared in Europe. Their vocal track structure was developed to the point at which spoken communication became possible. Their speech most likely possessed some of the consonant features of modern speech, but the absence of various forms of human activities implies that their speech did not have the features of human language. We should note that vocal-track anatomy is a necessary—but not sufficient—condition of language use. Speech is in coevolutionary relationship with brain development, the development of consciousness, and the level of social complexity. In the Archaic *sapiens*, these capabilities were not yet present at the necessary level to enable modern speech. Full language performance had to await the emergence of the Cro-Magnons.

5.2.1.2.e Their Social Organization. In this domain, we can point to another difference between the two species. The Neanderthals were not organized in a nuclear family. There is no evidence of a biparental provisioning for the young or organized sharing of labor or food. Males lived separately from females and children. Their style of life was profoundly different from that of the Archaic *sapiens*. The Archaic *sapiens* lived in kinship groups and maintained contact with other groups. On the other hand, the Neanderthals avoided any contact with other groups. There is no evidence of interbreeding between the two species.

5.2.1.2.f Their Technology. In their technology, the Archaic *sapiens* ushered in innovations. They prepared the raw materials from which artifacts were made. For example, they removed flakes from a lump of rock to make a flat top from which many flakes were made. This reveals designlike thinking, a major "cognitive transition" from earlier technology. On the other hand, the tool technology of the Neanderthals was unchanged during their life cycle. Still, there is some evidence that by the end of their life cycle, they adopted some of the tool technologies of the Cro-Magnons. Innovation was anathema to them. They did not like change and lacked the ability to plan. Unlike the Archaic *sapiens*, they left their tools behind when they abandoned their campsites. They did not modify their environment for comfort in the caves.

To sum it up, the differences between the two species in their consciousness, communication, social organization and way of life, and technology indicate clearly that the Archaic *sapiens* and the Neanderthals were two markedly different species. It was Archaic *sapiens* who represented the transition between the *Homo* species and modern man. They were the parents of *Homo sapiens sapiens*. They were our parents.

5.2.1.3 Stage 3: The First Generation of *Homo sapiens sapiens*

Following a designation of the new species of HSS, the time and place of their appearance and their consciousness and world view are described, their mode of communication and their social organization are identified, and their technology is characterized.

5.2.1.3.a The Cro-Magnons: the First Modern Men. burst on the evolutionary scene as the descendant of Archaic *sapiens*. (The designation Cro-Magnons relates to the place in France where their first fossil remains were discovered.) They emerged with an unprecedented accelerated pace. Thus, their emergence is often called the "the Human Revolution." Following the numerous evolutionary experimentations with the variety of life forms of our species over six million plus years, the Cro-Magnons became the only human species left. They reached and fulfilled the evolutionary potential of our race.

5.2.1.3.b Time and Place. The Cro-Magnons appeared on the evolutionary scene some 35,000 to 50,000 years ago. It was the era of the last Ice Age, which started around 70,000 years ago. By 18,000 years ago much of Northern Europe was polar desert. But the southwestern regions of France and Spain and the river valleys of Central Russia were reasonably hospitable. Their life cycle—as modern man—has continued over the history of humankind, but as the First Generation of HSS their cultural life cycle ended when they became the first Farmers of the agricultural age.

5.2.1.3.c Their Consciousness. Awakened from the dreamlike state of contracted consciousness, they stepped over the threshold of self-reflecting consciousness. Metaphorically, as they left the "Garden of Eden," they acquired self-awareness. They began to create their own world as they became separated from nature and from their previous state of oneness with wholeness. Their self-reflective consciousness created the capacity to "know that they know" and they entered the world of feeling, mystery, awe, laughter and tears, and love and fear. Reflecting consciousness became the "priming force" of creative consciousness, which brought forth a revolution in the arts, in spirituality, communication, cognition and learning, social and institutional development, and in their technology.

5.2.1.3.d Their Mode of Communication. While all the previous species of the human race used signals that were tied up with their referents and pointed to action, the spoken communication of the Cro-Magnons was symbolic. It was separated from its referents. Their language ability enabled them to separate the past from the present. Speech extended their time and space horizons. They developed an oral tradition. Their life events and social interactions became increasingly more complex, putting more and more pressure on their language development. The refinement of their symbolic speech, intermeshed with their remarkable artistic symbolism and symbolic mythical spirituality, resulted in creating "mental models" that reflected their "reality" and shaped their view of the world.

5.2.1.3.e Social Organization. The Cro-Magnons represented the most fundamental change in human and social behavior and social organization since the emergence of our species. They established social order and intra- and intergroup relationships that were totally different from earlier times. They created a healthy balance between competition and cooperation. They established relational connections with neighboring Cro-Magnon groups and created extended trade relationships. All these social activities amplified the need to develop sophisticated political communication. They demonstrated that human groups can help each other without being threatened by each other; in fact there was a mutual dependence in the most real sense.

5.2.1.3.f Their Technology. After many millennia of hardly noticeable progress in technology, the Cro-Magnons brought forth a creating explosion. A host of sophisticated tools proliferated with many regional variations. Where before there was "global monotony" in technology, in Cro-Magnon technology change replaced stasis. From Spain to the Urals hundreds of useful and artistic creations and complex dwelling structures emerged as manifestations of a new age in the life of the human race. In the history of our species, the Cro-Magnon era was a true "modern age."

The Cro-Magnons are inspiring in what they showed us about the human spirit, how humans can love and create a compassionate society, and how humans can live in harmony with their fellow humans and with nature. Standing at the threshold of the twenty-first century, I wish we would create a contemporary Cro-Magnon society.

5.2.1.4 Stage 4: The Second Generation HSS: The Age of the Agriculture-Based Civilizations

5.2.1.4.a The Second Generation's Agricultural Revolution. The Second Generation's agricultural revolution emerged by the establishment of the first agricultural villages. The Farmers who tilled the soil and produced food surplus made possible the emergence of agriculture-based civilizations, such as the ancient civilizations and the Hellenic civilization which folded into the Byzantine and Roman Empires.

5.2.1.4.b Time Frame. The time frame of this stage is from 10,000 B.C. to 1500 A.D. The small village farming period lasted for about 6000 to 7000 years. The agriculture-based ancient civilizations emerged around 3500 B.C. and the Hellenic period 600 B.C. It was followed by the Byzantine and Roman empires during the early Christian period. The establishment of these civilizations was enabled by the surplus production of the farming communities, by writing, and by many innovative technologies.

5.2.1.4.c State of Consciousness. The state of consciousness of the people in the farming villages was mythical consciousness, coupled with worshiping the Earth Mother, and developing a two-dimensional, cyclic sense of time of the present and future, which was necessary for planning for planting, cultivating, harvesting, storing, and transporting. People in the ancient civilization created a three-dimensional, sky-oriented deep consciousness. In the Hellenic culture, ego-consciousness was coupled with historical consciousness.

5.2.1.4.d Communication. The most crucial development of the ancient civilizations was the invention of writing by the Sumerians around 4000 B.C. Writing severed the "psychological umbilical cord" from man's habitat. It enabled the manipulation of ideas and extended the boundaries of consciousness within the transmitting distance of the written text. It made possible the recording of the individual and collective experiences through time. Without writing the administration of ancient civilizations and the development of science, education, commerce, and engineering would not have been possible.

5.2.1.4.e Social Organization and Way of Life. The social organization and way of life of the Second Generation had four markedly different forms. The social organization in the agricultural villages was defined by the working requirements and way of life of farming. In the large settlements of the ancient civilizations, however, it was totally different from that of the low density farming communities. Each of the ancient civilizations had interconnected cities as centers of religion and trade and cultures, and was governed by a powerful priestly class and a strong autocratic central government. Life was organized in occupational classes of administrators, merchants, engineers, scientists, educators, artisans, builders, traders, and warriors.

In the third type, in the Hellenic civilization, life was organized in city-states as a revolutionary new form of way of life: democracy. Living a good and moral life and serving the common good was the collective norm. Arts, the humanities, philosophy, education, and science were the leading vocations. In both the ancient civilizations and the Hellenic city-states the farming communities that surrounded the cities supplied the food and raw materials. The Chinese, the Byzantine, and the Roman Empires were the fourth type of social organization in which autocratic rulers governed with large armies as the instruments of power and domination.

5.2.1.4.f Technologies of the Second Generation. The technologies of the second generation mirrored the various types of social organizations of the agriculture-based civilizations. The greatly increased complexity of technology invites a new definition. Tn the context of the Second Generation, we should view and understand technology as the integration of: (1) *what* to produce, (2) *with what* instruments, (3) *how to design* and make the instruments, (4) *how* to carry out the production, and (5) how to plan for carrying out the *production process*. The technology developed in the farming villages focused on the instruments and methods of farming and the intellectual technologies of planning the farming activities and designing the instruments of planting, cultivating, harvesting, storing, and transporting.

Technology in all its forms exploded in the large settlements of the ancient and the Hellenic civilizations and the Empires. Farming technology was married with engineering, creating irrigation systems. Larg-scale engineering, construction, and transportation technologies were engaged in planning, designing, and building large cities. The technologies of the Second Generation HSS were based on metallurgy, using iron, copper, silver, and gold for a large variety of uses, and unfortunately increasingly more for creating implements for war.

The era of the Second Generation HSS lasted more than 10,000 years. It was a period in the life of mankind that created ways of life and forms of social and societal organizations that are very familiar to us because they are still with us. Most importantly, the Second Generation learned how to create societies and, unfortunately, also how to destroy them. During this era we saw repeatedly the creation and decline—the rise and destruction—of a succession of societies and civilizations.

The next stage of building our matrix is aimed at developing the summary characterization of the scientific–industrial era of the Third Generation. It will be, however, *your task* in the next activity to provide the description in the matrix of the scientific–industrial era.

Activity 15

Follow the example of the characterization of the first five stages of human and societal evolution and describe the Third Generation HSS in terms of the following descriptors: (1) designation, (2) time frame, (3) type of consciousness, (4) mode of communication, (5) social organization, and (6) technology. Enter your findings in your workbook.

Reflections

What was the purpose of revisiting the various evolutionary stages of human and societal evolution? What have we learned from an exploration of the evolutionary matrix? A review and a careful study of Table 5.1—along the vertical as well as the horizontal axes—will help us to understand the evolution of our species as a phenomenon of: (1) the constantly expanding boundaries of the time dimension, first from the present only to the past, and then from the present to the future; (2) the constantly expanding boundaries of space and distance which enlarged the living space of our species; (3) the invention/creation of new systems of communication, namely, speech, writing, print, electronic, the Internet, which made (1) and (2) possible; (4) the continuous unfolding of new relationships among human systems, leading to their reorganization at ever higher levels of complexity (from the family to the global community); (5) the emergence of new paradigms of knowledge organization and utilization, and new ways of thinking; (6) the emergence of new levels of collective consciousness which were the priming forces that integrated all the above into an internally consistent system of collective human experience; and (7) the recognition of discontinuity between the various evolutionary levels. There was no way to predict or extrapolate from a specific evolutionary stage the nature and characteristics of the next stage. The seven markers of understanding evolution comprise an internally consistent system of the evolutionary phenomenon.

5.2.2 The General Image of Cultural Models

Life forms of evolution are created in the evolutionary design space. In the design space, evolution as an ongoing creating process proposes or "puts up" new life forms for experimentation. To the extent we can understand this process and have knowledge about the characteristics of a particular emerged species, we can "model" it. Based on such knowledge we can create a systems model of an emerged evolutionary generation, which is composed of the interacting "evolutionary markers" of the particular generation. Evolutionary markers of a generation are, for example, its collective consciousness, world view, way of life, social organization, technology, and mode of communication.

For our present purposes, in modeling the design of a specific generation of our species, we have focused on creating systems models from the point in time where cultural selection began to supersede biological selection, with the emergence of *Homo sapiens sapiens*. In Chapters 3 and 4 we have used the designation "Cultural Model."

Here, I portray a generic image of Cultural Models that describe the characteristics of the various emerged generations of HSS. For each generation, we differentiated three clusters: *consciousness, intellect, and way of life*. Each cluster is organized from several markers. Evolutionary markers of consciousness include: *consciousness, spirituality, ethics/morality,* and *world view*. Markers of the intellect cluster are: *cognition, mode of communication, learning/human development, science, and aesthetics*. The way of life cluster includes: *social organization, polity/government, technology, economics, technology,* and *relationship with nature*. In understanding the nature and characteristics of the various generations of HHS, we learned that the primary force that brought about the emergence of our species (and their evolutionary markers) at a new evolutionary stage was *a new kind—a higher level—of collective consciousness.* Figure 5.1 introduces a generic image of the Cultural Model.

Activity 16

You are asked to depict and describe the Cultural Model of the Third Generation of HSS. Working with Figure 5.1, your task is to describe all interactions within the evolutionary markers, (1) within each of the three clusters, (2) among the clusters, and (3) between markers across the boundaries of clusters. Enter your findings in your workbook.

5.2.3 The Systems Image

The systems image reflects a systems view of looking at, understanding, and portraying a particular entity at an evolutionary stage of our species. To create the image, we use three lenses: the systems-environment, the functions/structure, and the process/behavioral lens (Banathy, 1992).

5.2.3.1 The Systems-Environment Image

The creation of this image informs us about the space and time environment of an emerging evolutionary entity, for example, one of the generations of HSS species as well as its environment. The *time environment* tells us the life span of a particular evolutionary entity. The description of the *space environment* refers to the geographical environment, the prevailing or changing climates, and the plant and animal species environment. The flora and the fauna are always affected by climate changes which in turn affect the life of our species. Faced with drastic climate changes, a species can either try to adjust or move away. For example, the migration of Archaic *sapiens* out of Africa 100,000 years ago brought them in contact with the Ice Age in Europe and created for them conditions in which they had to learn to reflect consciously on their environmental conditions and create their own environment purposefully, by building shelters, using and making fire, storing food, making protective clothing, and so forth. In fact, their natural environment was instrumental to the advancement of their consciousness and the development of their communication capability and technological competence. There is a saying that "ice made man." This coevolutionary relationship between environment and the human species became reversed when our runaway technology over the last hundred years or so destructively affected the natural environment

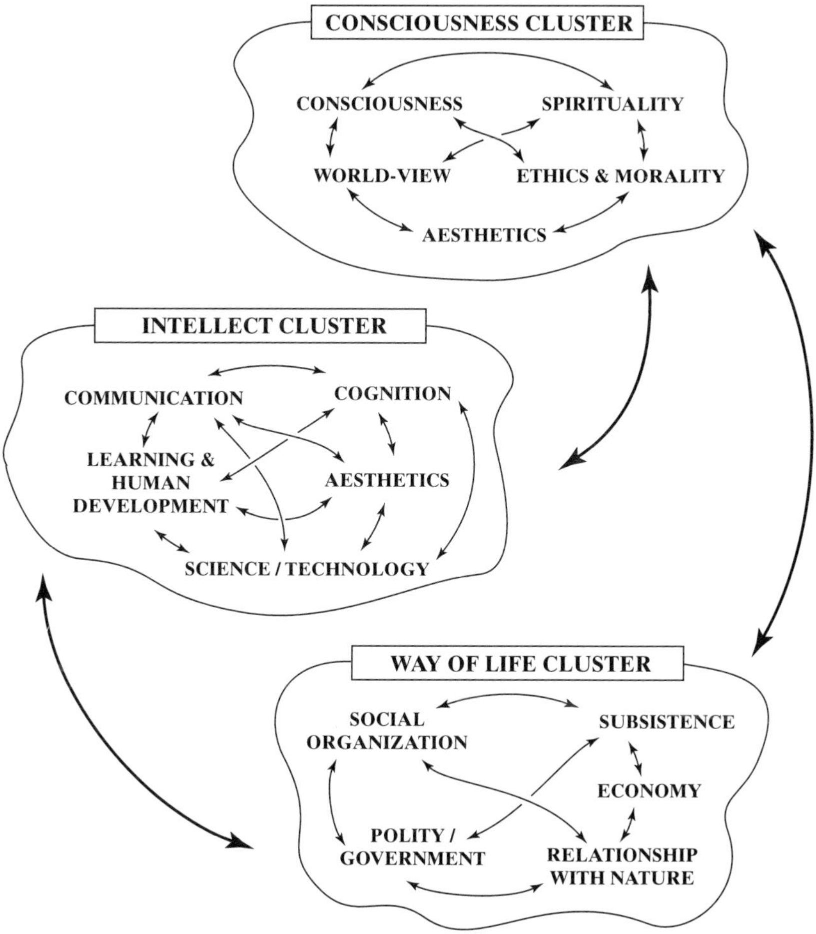

FIGURE 5.1 A General Image of Cultural Models

and the biosphere. The third type is the *social environment*. While there is no fossil evidence that would indicate social interactions among the various humanoid species who lived in the same time period, there are some speculations that there might have been some social interaction—but no interbreeding—between the Cro-Magnons and the Neanderthals. There is clear evidence, however, that trade relationships existed between various Cro-Magnon groups. From that point in time, social interactions among the various HSS groups became increasingly more significant from a social–environmental coevolutionary perspective. By the time the ancient civilizations emerged, some 3500 years ago, one form of social relationship attained a negative implication expressed in exploitation, hostilities, warfare, and destruction.

5.2.3.1.b The Function/Structure Image. The use of the functions/structure lens enables us to capture a "still picture image" of the evolutionary state of a species in a given moment in time. Discussing the Cultural Models that represent the various evolutionary stages of our species, we saw that these models are organized as the systemic arrangements of "evolutionary markers." These models stand for the various functions that an entity of our species carries at a given stage of evolution: functions, such as self-reflecting, believing, thinking, communicating, learning, providing, and creating. The systemic arrangement of the markers that carry out these functions provides the evolutionary image of our species at a given point in time at a specific evolutionary stage. It becomes a representation of the interactive relationship of the evolutionary markers. Certain markers organize themselves into clusters (subsystems) and clusters are organized in the evolutionary entity as a whole system. The markers are coevolutionary. They reinforce, mutually influence, and mutually define each other. The force that brings about these dynamics is *creativity*.

5.2.3.1.c The Process/Behavioral Image. The use of the process/behavioral lens presents to us a "motion-picture image," a series of pictures of the behavior of an evolutionary entity through time. This image arises from the interaction of the component markers of a cluster as well as the interaction among the clusters and among markers across the boundaries of the clusters. These interactions create the dynamics of the evolutionary process and the behavior of a given evolutionary generation. Each generation follows a generic process of its life cycle, namely: emergence, development, maturity, inflexibility and resistence to change, decline, loss of harmony, chaos, and self-destruction.

Applying a systems view approach and using the three lenses to describe an evolutionary generation is in no way sequential. It is unfortunate that in a text we are constrained by a linear presentation. We can see the whole picture of the entity only if we superimpose the three lenses so that they will manifest the three-dimensional image of the whole.

5.2.4 The Spiral Image of Human and Societal Evolution

The spiral image of human and societal evolution shows the ascendence of the emerging generations of HSS through time. This ascendence can be characterized by several considerations. (1) One is an ever increasing number of discernible evolutionary markers that describe the emerging human species. Any particular marker can evolve into subsystems of markers or become redefined based on new functions it carries out. (2) Another consideration is an increase in the complexity created by the interaction of the increase in the number of markers and by the dynamics of their interaction. (3) The horizontal widening of the spiral signifies the ever enlarging life space of our species, the constantly expanding boundaries of the space and time dimension of the human experience facilitated by new technologies of communication. (4) New and expanding relationships among human systems continuously unfold, leading to their reorganization at higher levels of complexity (as shown by the ascending spirals). (5) The horizontal spiral lines represent the successive emergence of new generations of HSS. (6) The vertical breaks between the horizontal spirals stand for "discontinuity" and "quantum jumps" between successive generations. There is no way to predict from a given generation the nature and characteristics of the next generation. (7) Finally, the infinity signs at the end of the vertical lines stand for the notion of a never ending—a forever ongoing—evolution. Figure 5.2 depicts the spiral of cultural evolution.

Reflections

In Section 5.2 our task was to summarize what we learned as we listened to the evolutionary stories of our species at the various stages of evolution. The first image was an evolutionary matrix that presented the markers of human and societal evolution in a time frame. This matrix captured a birds-eye view of the evolution of our species. The second image introduced a "generic" image of the cultural models of the various generations of HSS. The third image captured a systems view image and the fourth a spiral image of the ever-ongoing evolution—the ascendence of our species through time.

5.2.5 Another Systemic Discovery

My intent in this work has been and continues to be to take a "systems view" of human and societal evolution. As noted in Chapter 1, we have few if any works that have attempted to view our entire evolutionary story from a systemic perspective, even though many systems scientists have made significant contributions to evolutionary scholarship. Here, I provide an example showing how two scholars looking at the same evolutionary phenomenon viewed it very differently. One view presents a reductionist and the other a systemic picture.

Stephen Jay Gould (1998) reports on the life work of the two greatest scholars of Cro-Magnon cave art: the Abbe Henri Breuil (1906) and Andre Leroi-Gourhan (1967). They provided two profoundly different

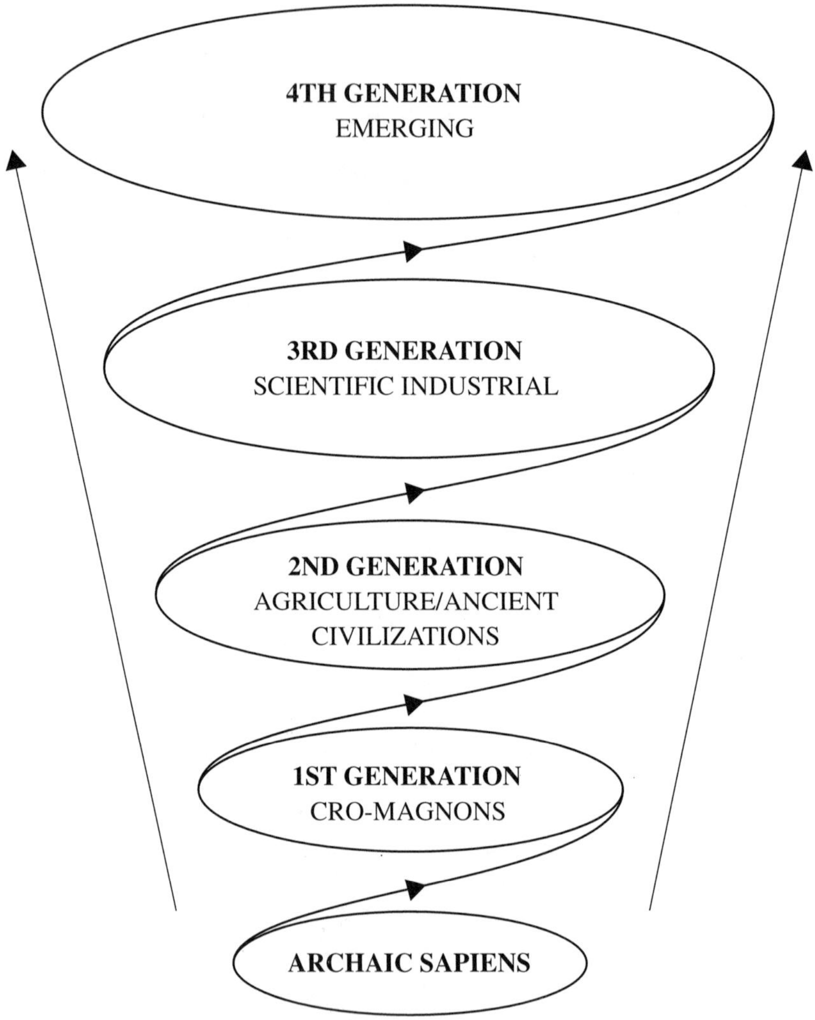

FIGURE 5.2 A Spiral Image of Cultural Evolution

interpretations of cave art. Breuil studied *parts—individual figures*. He read their meaning in piecemeal fashion rather than as components of an integrated composition of the whole cave art. Leroi-Gourhan approached the same subject from a totally different perspective. An avid student of French structuralism set forth by the anthropologist Claude Levi-Strauss he "viewed each cave as an integrated composition, a sanctuary in which the numbers and position of animals bore unified meaning" (1906, p. 167).

Breuil focused on the *part*, while Leroi-Gourhan captured the artistic images in the cave of the *whole picture. He saw the whole, the one that organizes the parts. He had a systems perspective; he was a systems thinker.*

5.3 *Organizing Principles of Evolutionary Consciousness*

In the first two sections of this chapter we reflected upon what we have learned from the evolutionary story. This learning was portrayed as sets of evolutionary propositions and images. Having been informed by these, a set of principles is presented here that can help us organize our thinking about the evolutionary design space, the workings of the generative order in evolution, the life cycles of the various generations of HSS, the creation of evolutionary models, the necessary conditions of evolutionary emergence, and the notion of increasing complexity in evolution.

5.3.1 The Principle of "Free Play" in the Evolutionary Design Space

Design creates novelty by exploring and testing design alternatives and selecting the most viable alternative for development. In the course of the biological evolution of our species, new life forms have been put up and tested (and selected in or out) in the evolutionary design experimentation space by the process of natural selection. Since modern man emerged, the process of selection became cultural. In the design space of cultural evolution, we can observe four main design operations: (1) transcending the current state of evolution; (2) creating an evolutionary image of the desired future; and—based on the image—(3) considering alternative future evolutionary states; and (4) selecting, describing, and modeling the most promising alternative. The notion of creating design alternatives is viewed by Bohm and Peat (1987) as "free play" in the design experimentation space. Such free play can and should become part of the process of conscious and guided evolution.

5.3.2 The Principle of Evolutionary Potential

A species evolves from "enfolded" evolutionary potential that unfolds as the species emerges in the evolutionary design space. Evolutionary potential is not fixed; it "can learn." New potential is built upon unfolded potential as the potential is realized in the emerged evolutionary markers that "can learn" as they evolve through time and as they create new potentials. Evolution always offers new possibilities. It is neither predictable nor predetermined. That is why a conscious understanding of evolution (evolutionary consciousness) offers the possibility of conscious and directed evolution.

5.3.3 The Principle of Generative Order

This principle is derived from Bohm's proposition. The generative order is the deep-seated inward order from which life forms can emerge or may unfold from the cosmic creating force of evolution. The principle of generative order takes issue with the Darwinian notion of life arising slowly and gradually as a fortuitous chance of mechanically determined small steps of random mutations. Arising from the inwardly enfolded generative order of the "whole," the whole organizes its evolutionary markers. *Thus, the notion of the whole organizing its parts into an integrated evolutionary model is another core organizing principle.*

5.3.4 The Principle of a Sudden Evolutionary Leap

It is from the enfolded generative order of the whole that the new life from "unfolds" through sudden evolutionary emergence. The evolution of a new generation of HSS *does not* come about from the gradual and step-by-step change (mutation) or improvement of evolutionary markers. It comes about as a sudden evolutionary leap, a sudden unfolding of the new generative order. The new generative order is *not* an improved version of the previous one. It is *discontinuous with it.* It transcends it.

5.3.5 The Principle of Evolutionary Modeling

Understanding this principle helps us to guide our thinking about the creation of sociocultural models of an evolutionary entity designed at a specific evolutionary stage. These models are organized from clusters of evolutionary markers, and the whole entity organizes the systemic arrangements and integration of the clusters. The Cultural Models of the three generations of HSS show a similarity in their systemic design and organization. Each is organized from three internally consistent clusters: consciousness, intellect, and way of life. And each cluster organizes its own set of markers.

5.3.6 The Wholeness Principle

The wholeness principle refers to the harmonious and balanced interaction among, and integration of, the evolutionary markers in a whole. It is the coevolution, the mutual effect and influence, the interaction, and the interpenetration of the markers that ensures the internal consistency, the integrity, and the wholeness of the evolutionary entity. When, in the life cycle of a generation of HSS one or more of the markers becomes out-of-sync with the rest, balance and harmony are destroyed, wholeness is violated, and decay sets in. Such is the case at the current evolutionary state in which technology, consumption focused economy, and the exploitation of nature have offset the balance of the whole, threatening the destruction of the social/societal order. Faced with this evolutionary predicament, we call for a new creative surge that will enable us to restore wholeness by envisioning a new evolutionary image and bringing it to life by purposeful design.

5.3.7 The Principle of Evolutionary Emergence

We can be conscious of evolution and engage in conscious evolution only if we are aware of the conditions that are necessary for the emergence of a new evolutionary entity. As we have explored the emergence of the various evolutionary generations of HSS, we have seen that a key necessary condition of emergence of a new generation is the emergence of a new *collective consciousness*; the surfacing of a new and higher level of consciousness. But this is not as simple as it sounds. Consciousness does not cause emergence; it is not emerging on its own while the rest is waiting. Its emergence is coevolutionary and conditioned by recursive and mutually influencing interactions with (1) an increase in the complexity of *cognition* and *language* and (2) the demand that is placed on both consciousness and cognition by an increase in the complexity of *social organization*. Another core condition is the availability of new evolutionary potential, manifested in a new generative order.

5.3.8 The Principle of Ever Increasing Complexity

The most popular definition of evolution is that it is a process of the reorganization of life at ever higher levels of complexity. This principle may not hold as a general principle of the biological evolution of all life forms (Gould, 1998). But it does hold for cultural/societal evolution. Csikszentmihalyi (1993) defines *harmony* as one of the core evolutionary principles. "Harmony is usually achieved by evolutionary changes involving an increase in an organization's complexity, that is an increase in both *differentiation and integration*" (p. 156). Thus, complexity can be defined as a state of integrated differences, which is the state of evolution as the "whole" organizes and integrates the evolutionary markers as the generative order of a new HSS generation emerges.

In the context of our discussion, differentiation in evolution refers to the degree to which the evolutionary markers differ from each other in function and structure. Integration refers to the state in which the different markers communicate, coevolve, interact, interpenetrate, and enhance one another. An evolved entity, which is more differentiated and integrated than another, is more complex than an entity that is less differentiated and integrated.

While complexity stands for integrated differences, complicatedness means unintegrated differences. Complicatedness is demonstrated at that state of an evolutionary generation when decay sets in, when harmony is destroyed as the integration of the evolutionary markers breaks down. At that point, the state of life of an evolutionary generation "feels" complicated, confused, anxiety ridden, and it becomes out-of-sync with new realities—a fitting description of the current state of evolution.

5.3.9 The Principle of the Life Cycles of Evolution

We have seen that the various evolutionary Generations of HSS have their own life cycles. They suddenly emerge and rise, fueled by a creative surge. They then attain maturity as they develop and establish themselves in a "standardized state" of habitual orders of customs, laws, and institutions. In time, these characteristics become rigid and inflexible. Change is rejected even in the face of newly emerging realities. Defending and sticking to "what IS," creativity becomes "out-of-order." The loss of creativity means not only stagnation but also the onset of decay and eventual destruction. In a twisted and paradoxical way, creativity is used to reject the creation of novelty and is mobilized to defend the status quo. In the course of our journey, we have seen stagnation, decay, and destruction in the life of the ancient civilizations, in the Hellenic and Byzantine civilizations, and in the scientific–industrial era.

Activity 17
In this section eight organizing principles of evolutionary consciousness were introduced. I assume it is an incomplete set. You are asked now to reflect upon what you have learned about evolution and seek to formulate at least one more organizing principle.

Reflections

Having arrived at the end of Chapter 5, we can say with some confidence that we have attained evolutionary consciousness from the insights we have gained, by exploring sets of evolutionary propositions, perspectives, images, and principles. At the same time, we have become conscious of our own troubling uncertainty about the present state of our evolution. While we are at the threshold of a major evolutionary transformation, remnants of the dying industrial machine age are still with us.

We therefore raise several questions. *Is the vicious circle of rising, developing, decaying, and dying inevitable in the life of an evolutionary generation? Can we avoid the chaos of decaying by attaining evolutionary consciousness, and, provided we develop it: Can we "step into" the evolutionary design space, develop evolutionary competence, and generate willingness to engage in conscious evolution?* In the chapters that follow, I begin to answer these questions.

6

From Evolutionary Consciousness to Conscious Evolution

> Intelligent life on a planet comes of age when it first works out the reason for its own existence. If superior creatures from space ever visit earth, the first question they will ask, in order to assess the level of our civilization, is: "Have they discovered evolution yet?"
>
> *R. Dawkins*

In Chapter 4, I gave an account of the "Human Revolution," the emergence of the First Generation of *Homo sapiens sapiens* (HSS). Stepping on the evolutionary stage the Cro-Magnons represented the first life cycle of three Generations of HSS. The second life cycle, the Second Generation of HSS, was the era of the Agricultural Revolution. During the 10,000 years in which it existed, it saw the emergence of three successive civilizations. The Third Generation HSS, the era of the Scientific–Industrial Revolution, emerged during the Renaissance. Each generation came forth as a major quantum leap, a major transformation in the existence of HSS. Today we again live at the very edge of another massive evolutionary transformation—to the Fourth Generation of HSS. We are at the exit door of the design space of the Third Generation and try to prepare ourselves to take a quantum jump into the design space of the Fourth Generation. But we can jump-start this transformation only if our answer to the question of the "fictional superior creature from space" is a resounding *yes*. Yes, we are conscious of our evolution. Yes, we have attained evolutionary consciousness.

In Chapter 5 we asked: *What have we learned from our evolutionary story*? In particular: What have we learned from an understanding and appraisal of the *process* of the evolution of our species? In conclusion we asked: What are evolutionary principles and characteristics that can define—and enlighten—our own consciousness of evolution? Answering the above questions is a necessary condition for active participation in the upcoming evolutionary transformation. Without an attainment of evolutionary consciousness we cannot engage in conscious evolution.

Now, in this chapter we ask: *How can we apply what we have learned from the evolutionary story?* How can we apply our conscious understanding of the principles, patterns, and processes of evolution to the evolutionary transformation from the Third to the Fourth Generation HSS? Providing an answer to these questions is the purpose of this chapter, which is organized in three sections.

In Section 6.1, I explore the unfolding of collective consciousness, by establishing its meaning and characteristics and by organizing it in a systemic framework. In addition, the idea of "the evolution of evolution" is explored.

In Section 6.2, I ask: What does it mean to be at the cutting edge—at the crossroad—of the evolutionary transformation from the Third to the Fourth Generation of HSS? I depict the current evolutionary landscape and describe the evolutionary predicament that has arisen from the collapse of the Third Generation. This collapse created an evolutionary imbalance and led to chaotic discrepancies between the evolutionary markers. Here we shall uncover critical evolutionary gaps between recently emerged "realities" of our era and the state of our systems and societal institutions that still behave—and attempt to continue to operate—in the context of the industrial machine age.

In Section 6.3, the question arises: *transition or transformation?* We shall use the insights gained from our evolutionary consciousness to recognize that we cannot change the current troubling evolutionary landscape—and cannot deal with the evolutionary predicament we face today—by merely continuing an analysis of what is wrong with the current state. We must *transcend* it and focus on the evolutionary principles processes that can guide us in taking a creative role in the evolutionary transformation to the Fourth Generation.

6.1 *Evolutionary Consciousness: The Springboard to Conscious Evolution*

In this section, we first explore the nature of human consciousness and review levels of collective consciousness at the various generations of HSS. Then, we explore the function of consciousness in sociocultural evolution. This review is followed by the framing of a systems model of evolutionary consciousness, which provides an organized presentation of the processes and products of societal evolution. Next, I propose a set of evolutionary qualities that have enriched the various evolutionary stages. The section ends with an exploration of the notion of "evolution of evolution," namely, evolutionary epistemology.

6.1.1 The Emergence and Nature of Human Consciousness

In Chapter 2, exploring the evolution of evolutionary ideas, we stated that the most crucial point in the evolution of our species was the time when our ancestors moved out from the blissful undifferentiated state of consciousness to the dawning light of consciousness. They asked: Who am I? Where am I? Who are you? Humans have asked these existential questions throughout the ages, and they remain today the most vexing of all questions (Wilber, 1981a). Wilber suggests that when we ask these questions we are describing ourselves as we come to know ourselves. We are trying to establish our identity, trying to define our "self." The definition of our identity depends upon where we draw the line between the "self" and the "non-self." We are describing what we are and what we are not. The boundary line between the self and non-self, however, might shift. This shift or redrawing of boundaries may enlarge the space of consciousness. Wilber (1981a) expressed this idea best:

> In a sense, the person can re-map his soul and find in it territories he never thought possible, attainable, or even desirable. As we have seen, the most radical re-mapping or shifting of the boundary line occurs in experiences of the supreme identity, for here the person expands his self-identity to include the entire universe. We might even say that he loses the boundary line altogether, for when he is identified with the "one harmonious whole" there is no longer any outside or inside, and so nowhere to draw the line. (p. 5)

Keep this inspiring statement in mind as we contemplate later the emerging collective consciousness of the Fourth Generation of HSS. Further, our evolutionary story informed us that every transformation from one evolutionary stage to the next involved a redrawing—an enlargement—of the space of collective consciousness.

Pankow (1976) suggests that "Consciousness is a meta-perception, the vindication of perception, the perception of perception" (p. 29). With the emergence of consciousness we widen the margin between what is now—the factual—and what might be—the possible. We understand ourselves by interaction with others. Plato said that "I understand myself only in conversation with another." Our personal consciousness is socially constructed. Pankow commented that we understand ourselves by understanding our environment, and we understand our environment by actively transforming it into our world. Among all the species we are the most open not only because we have the highest potential for development, but also because we recognize this potential.

Ian Tattersall (1998) suggests that HSS is "not simply an improved version of its ancestors—it is a new concept, qualitatively different from them in highly significant respect" (p. 189) The "human capacity" that first became manifested in the extraordinary art of the Cro-Magnons is not simply an extrapolation of the earlier trends of our ancestors that many of the studies of the paleoanthropologist elucidate. It is more an *emergent quality* whereby a new combination (and interaction) of evolutionary features produces "totally unexpected results." Human beings, says Tattersall, are so remarkable and so unprecedented from all other wonderful creations of evolution, that "only introspection will suffice to tell us who we are. And such self-examination has to start with our remarkable consciousness from which all else flows" (p. 190). In our evolutionary story—as we explored the emerging designs of various evolutionary stages—we found that it has been consciousness that has been infused in—and integrated with—all the other evolutionary markers. Tattersall's definition of consciousness best explains its essence. Consciousness is an "inner experience. It is not derived from the outside world, although it has everything to do with the way we perceive the world. Our consciousness is, if you will, the filter though which we view and interpret the environment around us"(p. 190).

6.1.2 *Consciousness as the Core Marker of the Various Stages of Our Evolution*

An understanding of human consciousness helps us to explore the phenomena of a changing and differentiated *collective consciousness* that animated the various stages—the various generations—of sociocultural evolution. An understanding of the nature of collective consciousness that operated at the various stages of sociocultural evolution constitutes a central component of attaining evolutionary consciousness.

In Chapters 3 and 4, we defined consciousness as the dominant marker of evolution. We differentiated several levels that characterized the various evolutionary stages. I review these stages only briefly. For a more complete picture you might go back to the sections in Chapters 3 and 4 that described the consciousness of the various evolutionary stages. A quick review of these levels is based on the earlier descriptions (Elgin, 1993) and provides another component of attaining evolutionary consciousness.

6.1.2.1 The Archaic, Contracted, or Embedded Consciousness

This prereflective type of consciousness guided the emergence of the Archaic *sapiens* and the Neanderthals, who lived in a primal, undifferentiated, dreamy state, in which they experienced oneness with wholeness and oneness with nature. Their consciousness was contracted and embedded within life. In this state the individual and the group were bound up with one another. The sense of self had not yet emerged as clearly differentiated from others. At this stage our ancestors lived in early dawn, waiting for the sunrise of reflective consciousness.

6.1.2.2 The Reflective (Magical) Consciousness: The Sun Has Risen

The emergence of reflective consciousness signaled the emergence of "modern man," namely "us." At this crucial and revolutionary state of evolution we stepped into the sunlight of self-reflective consciousness and begun to ask: Who am I? Where am I? How are you? The self-reflective capability of the Cro-Magnons led them to step back and observe themselves, others, and nature. It was the time of separation from wholeness and nature. What appeared to be strange was explained as magic. The Cro-Magnon consciousness was *sensory consciousness*: experiencing life through the senses. The power of self-reflective consciousness triggered creating consciousness, manifested in their beautiful cave art, which is yet to be surpassed.

6.1.2.3 Depth Consciousness

Depth consciousness characterized and animated the agricultural civilizations. Also associated with mythical consciousness, it created a growing capacity for imagination, empathy, and emotional life, coupled with a language to represent it. Depth consciousness enabled the description of emotional qualities and distinctions. There was a growing depth of conscious experience and a deepening sense of self and community.

The Second Generation evolved four ascending consciousness: (1) the Earth Mother consciousness of the agricultural villages, (2) the sky-focused consciousness of the ancient civilizations, (3) the Hellenic consciousness of ancient Greece, and (4) the consciousness created by the great spiritual leaders. At the last stage of this era, ego-consciousness emerged which marked the "mental-ego state" that became dominant during the era of the Third Generation HSS. Evolving with ego-consciousness, historical consciousness emerged, which is "today's reflection on yesterday."

6.1.2.4 Dynamic–Mental–Rational Consciousness

This consciousness animated the scientific–industrial era. This period brought forth the beautiful creations and aesthetics of the Renaissance, and extended the special boundaries through the print media. It stood for the logical and rational description of the world, for an unlimited belief and pursuance of material progress, for the domination and exploitation of nature, and for worship of science and technology as the means of solving all our problems.

6.1.3 The Consciousness of Evolution

The consciousness of evolution is portrayed here. First, I suggest that being aware of and knowing the stories of collective consciousness reviewed earlier contributes a great deal to our consciousness of evolution. In addition, to further explore evolutionary consciousness, I also examine the relationship between the process and product of evolution.

6.1.3.1 The Consciousness Stories Are Also Stories of Evolutionary Consciousness

At each stage of evolution a new level of consciousness emerged that animated and organized a new cultural model of evolutionary markers. Each level of consciousness is unique to that stage and discontinuous with the one that preceded it. One cannot extrapolate one consciousness story from the other. Even though the consciousness stories are unique, stand on their own, and are disconnected from each other, if we want to attain evolutionary consciousness, it is of *critical importance* that we understand the following: The sequence of the evolutionary stories tells us the story of *the ever increasing maturation of our species, the ever increasing complexity* of social organization of the various generations of HSS, *and most significantly, the ever increasing fulfillment of our ever increasing evolutionary potential.*

Here, we have reviewed the consciousness stories of the evolutionary stages and have gained an understanding of their role in evolution. In so doing we have built another component of the system of evolutionary consciousness, and have further advanced our *attainment of evolutionary consciousness.* What follows adds another component to evolutionary consciousness.

6.1.3.2 The Process of Sociocultural Evolution

In the course of our evolutionary journey we found that core ideas of a new species—as well as novel/advanced ideas of a new generation of

HSS—are designed and put forth for experimentation in the evolutionary design space. Here, I review the *process* of how a new evolutionary entity emerges, matures, attains stability, defends its stability, loses its ability to continue to coevolve with its environment, resists change, declines, and loses internal consistency. At the end of its life cycle an evolutionary generation disintegrates and faces entropy. Then, through a new creative force it might unfold new evolutionary potential, engage in coevolution with its environment, attain new consciousness, and create a new sociocultural system in the evolutionary design space. This stream of processes has been recurring in the course of the evolutionary journey of our species.

6.1.3.2.a The Emergence of the Core Idea of a New Generation of HHS. In the evolutionary design space, a novel form of the human species might unfold by the interaction of an enfolded core idea, the unfolded evolutionary potential, coevolutionary interaction with the environment, and the integrating force of an emerging new consciousness. Keep in mind that without new consciousness there is no possibility for the emergence of a new evolutionary (or for that matter any social) system. We have to attain new consciousness. A case in point is the declining and dysfunctional, collapsing Third Generation HHS of the scientific–industrial era. The Third Generation finds itself to be totally out-of-sync with the new realities of our current era. Given this condition, if our species wants to continue our evolution—rather than become a failed evolutionary experimentation—we have to take on the task of conscious evolution, attain a new level of consciousness, create a new core idea for the Fourth Generation of HSS, and bring the idea to life by self-organizing design. The evolutionary process described in the preceding triggers in the evolutionary design space the unfolding of the enfolded evolutionary potential that is inherited from the previous evolutionary generation. This new—yet enfolded—potential builds on the capacity and capability attained by the previous generation of HSS. Unfolded by the force of the emerged new level of consciousness the *core idea of the new generation of HSS is created.* For example, animated by the mental/rational dynamic consciousness and the evolutionary potential generated by the Second Generation, the core idea of the Third Generation of HSS emerged and was manifested in the scientific–industrial era.

6.1.3.2.b The Process of Self-Organization. Animated by the newly emerged consciousness and the unfolding inherited evolutionary potential, this process is one of selfcreation, the creative self-organization and integration of the various evolutionary markers into an internally consistent sociocultural system of a new generation of HSS. This process (Artigiani,

1996) brings forth by emergence a new level of reality in which the relationship between the components redefines their attributes. "An emerged self-organized system represents a reality that can only be described *using qualitatively new kinds of information. An emergent system, in other words, creates itself and it must be understood in terms of the constraints, laws, and phenomena it generates*" (p. 2). This quote supports a core process of evolution, manifested first in the emergence and the self-organization of the Cro-Magnons, the First Generation of HSS.

6.1.3.2.c The Process of Maturation, Stagnation, and Decline. This process is manifested in two opposite processes: (1) the process of evolution toward full maturation and (2) a process of devolution toward decline and eventual collapse. The first leg of this process is maturation in the course of which the guiding consciousness becomes increasingly more infused into the evolutionary markers, thus creating increasing internal consistency and wholeness. During the next leg of the life cycle of the system, we experience a stasis as institutions are firmly established. They expend much effort in maintaining their existing state, resisting change despite their changed environment and emerged new realities. If there is no creative/self-organizing response to these changes, if this protecting and defending the "way it is" continues and takes the upper hand, the process leads to devolution, decline, and eventually disintegration. It is particularly important for us to understand and become conscious of the processes of stasis and decline if we are serious about engaging in guided evolution.

In the previous sections we have made further progress toward attaining evolutionary consciousness by understanding the process and product of evolution. To continue our journey in acquiring evolutionary consciousness, we have to go deeper and try to unfold the *qualities* that are the properties of sociocultural evolution.

6.1.4 Evolutionary Qualities

Qualities that become manifested in sociocultural evolution are not explicit. They are enfolded in the generative order of evolution, and their unfoldment requires probing and reflection. As I worked with the theme of this book and stopped and reflected at times, these qualities slowly emerged in my consciousness. I can probably say that they *reemerged.* Because these are qualities I value in my life, these are qualities I seek to bring about in the systems in which I live and work, even though I have never "listed" them before as such. Plato said that "I understand myself in conversation with another." It seems that a conversation with our ancestors has helped me to better understand myself. But let us see what these qualities might be. First, I shall name a set of core qualities and then explore their relationships. The set presented is preliminary and will be revisited later.

6.1.4.1 Naming the Qualities

The qualities introduced here are generic to the various evolutionary Generations of HSS. They characterize a new evolutionary stage at the point when it has reached maturity.We can suggest three categories of qualities: *core qualities, enabling qualities, and meta-qualities. Core qualities* include *balance, harmony, internal consistency, synergy, symmetry,* and the "twin" qualities of *self-referencing and self-organizing. Coevolution* and *cooperation are enabling qualities, and collective consciousness, wholeness,* and *creativity are meta-qualities.*

6.1.4.2 Core Qualities and Their Relationships to Enabling and Meta-Qualities

The *core qualities* are in relational arrangement with each other as well as with *enabling qualities* and *meta-qualities.*

6.1.4.2.a Balance. Balance refers to a desirable qualitative state among the various evolutionary markers. It is a state in which the markers, animated by the new level of collective consciousness, have coevolved and mutually influence each other and attain a balanced state. *Coevolution* was the enabling quality that helped to attain such a balanced state. An out-of-balance state comes about when one or more of the markers dominate others. In the life cycle of an evolutionary generation an out-of-balance state indicates a declining stage. For example, during the declining stage of the Third Generation of HSS, the dominance of industrial age technology over nature and the health of people created many harmful effects.

6.1.4.2.b Harmony. Harmong is a quality that is enabled by intensive *cooperation* among the evolutionary markers. Harmony exists when the markers support each other. It is lost when there is discord among them. The more intensive the cooperation, the higher is the degree of harmony. Harmony also exists in a mutually defining *coevolutionary* relationship with the balanced state of the sociocultural system. The weakening of harmony can be mitigated by increased cooperation among the markers.

6.1.4.2.c Internal Consistency. Internal consistency is a qualitative state of a sociocultural system that is attained when the qualities of *balance* and *harmony* are manifested and more importantly when the meta-quality of *collective consciousness* is continuously infused in all the evolutionary markers. A lack of internal consistency moves the system toward a crisis of consciousness and a chaotic state.

6.1.4.2.d Synergy. Synergy is manifested in the interaction and the integration of the parts (components) of a system, orchestrated by the whole. It comes about in an evolutionary system as the evolutionary markers interact and integrate and mutually support each other. In a second sense, synergy has a special role. In evolution, synergy is achieved through the *coevolution* and integration of the collective consciousness of an evolutionary generation.

6.1.4.2.e Symmetry. Symmetry develops from *synergy*. The depth and dynamism of the material realm of evolution are mirrored by the corresponding depth and dynamism of consciousness, states Elgin (1993). "Material evolution provides the physical necessities of life that support the widespread realization of psychological and spiritual potential. Consciousness of evolution provides the insight, compassion, and creativity needed to orient material development into a sustainable pathway to the future. Co-evolution is a balanced, middle path that integrates a co-evolving spiral of mutual refinement" (p. 293). Consciousness and material evolution thus are mutually enabling. Because these two aspects are symmetrical, a conscious organism has the capacity to become fully self-referencing and self-organizing. One of the characteristics of the "declining" stage of evolution is the loss of balance and harmony, when the material side outruns the consciousness side, resulting in a symmetry break. A symmetry break also signals the decline of an evolutionary generation.

6.1.4.2.f Self-Referencing and Self-Organization. Selff-referencing and self-organization are complementary evolutionary qualities. A balanced, harmonious, internally consistent, synergistic, and symmetrical sociocultural system may be attained by ongoing self-reflection and self-referencing. When the system is aware of itself, it knows how it is doing, what it desires and aspires to attain. It is not only conscious of its evolution but is also engaged in conscious, continuous self-organizing and self-renewing evolution, continuously paying attention to and maintaining its evolutionary qualities.

6.1.4.2.g The meta-qualities of *consciousness, creativity, and wholeness* are directly involved in attaining the core qualities. The infusion of consciousness into the evolutionary markers is a necessary condition for the emergence of a new generation of HSS that will integrate the evolutionary markers so that the new entity will have the evolutionary core qualities defined previously. It is the ongoing *creating* process by which this infusion interpenetrates the evolutionary markers and by which the core qualities are attained in the evolutionary design space. The ultimate meta-quality, *wholeness*, becomes manifested in the life of an evolutionary generation when all the other qualities are attained.

6.1.5 The Epistemology of Evolution

In the most basic sense, "ontology" manifests what we think *IS* and "epistemology" *HOW* we think the *IS* came about. They are our own constructions. In Part I we told the *ontological story* of the evolution of our species. In this chapter we began to explore the HOW: the epistemology of evolution.

The term "epistemology of evolution" or evolutionary epistemology was introduced by David Campbell (1974). It is not of wide currency in the evolutionary literature. Evolutionary epistemology is not carved in stone. It is evolving and changing. First, we should consider evolutionary epistemology as an issue at a given time of our evolutionary (hi)story, for example, the metamorphosis from biological to cultural evolution. Second, we can consider the nature of change in epistemology itself, which we might call the *epistemology of the evolution of evolutionary epistemology*, which is therefore a meta-issue of evolution.

In addressing evolutionary epistemology, I limit the scope of exploration to the process of sociocultural evolution, beginning with the time when the Cro-Magnons appeared on the evolutionary scene. What follows can be considered a reasoned portrayal of evolutionary epistemology. It is based on the understanding we have gained from our learning about evolution and attaining evolutionary consciousness, and these are prerequisites to engaging in conscious evolution.

In the evolution of our species the foremost epistemological function was the bringing about of the crucial revolutionary metamorphosis from the Archaic *sapiens* to *Homo sapiens sapiens*. The subfunctions of this transforming epistemology: (1) enabled the transformation of biological to cultural evolution, (2) brought forth an acceleration of sociocultural evolution, (3) infused a new level of consciousness into the evolutionary markers of the new evolutionary entity, (4) provided for coevolutionary interaction among the evolutionary markers, (5) increased the complexity of emerged evolutionary entities, and (6) increased the complexity of their markers. These epistemological functions are described in the following subsections.

6.1.5.1 The Epistemology of Transforming Biological to Cultural Evolution

The first major metamorphosis brought forth by evolutionary epistemology came about when the Cro-Magnons burst on the evolutionary scene. Their emergence was the most dramatic and revolutionary "quantum jump" in the evolutionary saga of our species. This revolutionary "quantum jump" occurred when biological evolution ceased to be the key epistemological method of evolution of our species and cultural evolution took over as the driving force of sociocultural evolution.

6.1.5.2 The Acceleration of Evolution

For millions of years, biological evolution created the various species of our race, but the first six million plus years of the evolution of our race was a slow process. For example, for one million years there had been practically no change in the way of life of *Homo erectus*. But when *H. erectus'* grandchildren, the Cro-Magnons, appeared on the evolutionary stage, a lighting bolt acceleration in evolution hit the evolutionary scene that continues to this day. Another accelerating quantum jump occurred 10,000 years ago, and still another four hundred years ago. We are now at the threshold of another acceleration.

6.1.5.3 The Evolutionary Epistemology of Infusing Collective Consciousness in the Evolutionary Markers

A new level of collective consciousness is the core marker of an emerging sociocultural system. It is this new collective consciousness that is infused into all the evolutionary markers: it coevolves with them and integrates them into an internally consistent sociocultural system. Thus, the evolutionary epistemology that brings this infusion and integration about has to be in a synergistic relationship with the collective consciousness that animates the emergence of the new system. The epistemology should enable/enhance the infusing, coevolving, and integrating functions of collective consciousness.

6.1.5.4 Increase in the Complexity of the Operating Evolutionary Epistemology

During the evolutionary journey, we witnessed the reorganization of the various generations of HSS at an ever higher level of complexity. This phenomenon was made possible only by an evolutionary epistemology that was more complex than the one acting at the previous stage of evolution. Working with increasing complexity requires increasing complexity in the evolutionary epistemological process.

6.1.5.5 Epistemology Enabling a Change in the Nature of Evolutionary Markers

As the various generations of HSS emerged, there has been an increase in the evolutionary markers, in the complexity of their interaction, and in the refinement and elaboration of the markers. This is the result of two processes. One is the *evolutionary learning* attained by the newly emerged evolutionary sociocultural system. The other process is the infusion of (inherited) *evolutionary potentials* in the newly emerging evolutionary generation. To accomplish these enabling tasks in the evolutionary design space, a new and more sophisticated epistemology needs to operate that enables the increase, interaction, and refinement of the evolutionary markers.

6.1.5.6 Epistemology that Guides Evolutionary Transformations

This aspect of evolutionary epistemology is the most important and at the same time is the most mysterious. Operating in the evolutionary design space, this is the *creating process*, a process (of design) that creates a new sociocultural system, a system that is discontinuous with the system that preceded it. The emergence of the Cro-Magnons is the best example of an evolutionary epistemology that had to operate (in the evolutionary design space) to produce this miraculous revolutionary transformation from biological to cultural evolution. The evolutionary epistemology that brought forth this transformation was not known to the Cro-Magnons. They obviously did not know what evolution is about or how to guide it. But we know. By now we have attained evolutionary consciousness. Thus, having eaten the fruit of knowledge from the tree of evolutionary consciousness, the burden of this "knowing" is on us. It is like "leaving Eden a second time." We now have the responsibility to fend for our evolutionary future. Once we accept this responsibility, *we must create the evolutionary epistemology* that will guide the creating process of the Fourth Generation of HSS.

6.1.6 Our Epistemological Challenge

> "Two roads diverged in the woods—and I—I took the one less traveled by, and that has made all the difference."
>
> *Robert Frost*

Today, we are at the crossroad of societal evolution. We have a choice to make. We can continue our journey on the well-traveled road of unguided evolution and continue to be the spectators—and often the victims—of relentless undirected evolutionary changes. Or, we can choose the less-traveled road, become players on the evolutionary scene and guide the second crucial metamorphosis of the evolution of our species. If we elect the road less traveled, ***we have the enormous task of developing the evolutionary epistemology of guided evolution.*** *Thinking and speculating about this task is the challenge of our learning journey.*

We know for certain that the epistemology of guided evolution has to address at least the six epistemological functions described previously. But in our present case, an epistemology has to be created that will operate in the evolutionary design space as a purposeful, disciplined, and decision-oriented inquiry carried out by us all who are willing to create our own evolutionary future.

Reflections

By now our journey has led us from an *understanding of the story of the evolution of our species* to *learning from that story* and *attaining evolutionary consciousness*. In the first part of this chapter, I explained that listening to and understanding the evolutionary story of our species was a prerequisite to attaining evolutionary consciousness. Next, the nature of human consciousness was explored, followed by a discussion of collective consciousness as the core marker of sociocultural systems. This led us to consider the attainment of evolutionary consciousness as the "springboard" of engaging in conscious evolution. In conclusion, the epistemology of evolution was explored and we speculated about the role of evolutionary epistemology in conscious/guided evolution.

Activity 18
Capture the core ideas from the text of Section 6.1 and introduce them in your workbook. Then, write a text for a high school class on the journey from evolutionary consciousness to conscious evolution. Enter your presentation in your workbook.

6.2 *The Current Evolutionary Landscape*

The imaginary visitor from space asked us earlier whether we know how we evolved. The visitor, looking at the "landscape" of our current state of evolution, *most* likely would tell us that we have an enormous task ahead of us. He would tell us that he knows what he is talking about. He says that several millennia ago they faced the same predicament we are now facing. Still, he cannot give us any advice because they have a law of "non-interference." The task is enormous he would say, because—on the one hand—we are at a crossroads between two generations of our species and we are faced with great uncertainties. On the other hand—having "eaten from the tree of knowledge" by knowing how we evolved—-we have the gift and burden of this knowledge, and we have to make crucial and risky choices that will guide our evolution into the next millennia.

The above hypothetical story captures what this section is about. More specifically, I paint the current evolutionary landscape by sketching three of its features. First, I portray the current evolutionary predicament that has arisen from the collapse of the Third Generation of HSS. In many ways and in many aspects, however, we still live the life of the Third Generation, and this is the source of the manifold predicament in which we find ourselves today. Second, the new realities of our age show the evolutionary imbalance and the widening evolutionary gap between where we are now and where we should be. Third, I suggest that the only viable choice is to transcend the current evolutionary state and engage in an evolutionary transformation.

6.2.1 *Our Current Evolutionary Predicament*

The current predicament is a result of the fact that we find ourselves today standing with one foot in the decaying Third Generation of HSS, while trying to find a foothold in the Fourth Generation that we are to create. Even though we are at the tail end of the Third Generation, it seems that the "tail (still) wags the dog." In most cases we try to hold on (often desperately) to our established systems and institutions and the ways of doing of the Third Generation.

Volumes have been written about the current predicament, and I shall not dwell on it much. First, I revisit the learning we generated about the declining/decaying stage of the various evolutionary stages of our species, and then introduce statements made by a few evolutionary scholars.

6.2.1.1 The Life-Cycle Stages of the Evolutionary Generations

In the life cycle of the three evolutionary generations we observed and described several stages. The first is the emergence of the new sociocultural system in the design space of a new generation. Then comes a creative surge toward maturity, when the various evolutionary qualities, such as harmony, balance, synergy, symmetry, coevolution, cooperation, internal consistency, and wholeness become fully manifested in the life of the society. Then, stagnation sets in, when habitual customs, laws, rules, and institutions become fixed. Limits are firmly established, and creativity becomes dormant. The idea of any significant change is resisted. The evolutionary qualities that created the new sociocultural system gradually degrade into disharmony and imbalance, losing synergy and internal consistency. Wholeness breaks down, and a symmetry break moves the system toward a bifurcation point. At an evolutionary bifurcation point, the alternative to chaos and collapse is a quantum jump into a new evolutionary regime, by creating, through self organization and self-directed design, a novel, more complex state, a new generation of HSS. *This is precisely the evolutionary choice we have to make at the onset of the twenty-first century.*

There is an enormous difference between what happened at past evolutionary bifurcation points and what we face at the current one. Past transformations were brought forth by evolutionary forces operating in the evolutionary design space. They were not subject to the purposeful guidance of the various evolutionary generations. However, at the current evolutionary bifurcation juncture, having attained evolutionary consciousness *it becomes our task, our responsibility, and our burden to choose our evolutionary future.*

6.2.1.2 The Current Predicament: Decline and Decay

Toynbee (1947) suggests that civilizations, after they experience maturity and success, are vulnerable to disintegration. It is difficult for them to let go of their ways of life and their institutions that have worked so well in the past. They become inflexible and deny the need to change. They invest much effort to maintain the status quo.

Elgin (1993), characterizing the declining state of the industrial era, says that "the scientific-industrial era generated unprecedented social dynamism, moral relativism, intellectual absolutism, nation-state egotism, and technological gigantism" (p. 113). Living in the industrial society people became isolated in vast urban regions and alienated from nature and from others. The sense of community is lost. Life seemed to have little meaning beyond ever more consumption, ever more accumulation of material goods. But the gravest injury and destruction were inflicted on nature by the excesses of runaway technology.

According to Jantsch (1980), evolution creates wholeness. But in our current era wholeness is violated by social and societal fragmentation. Furthermore, he says that "We do not learn any longer, we are taught (with particular ready-made products), we do not design our environment, it is supplied by industry, we do not live in a healthy way, we are medically taken care of. We no longer determine the values of our life, they are prescribed by experts. Humans, who can no longer produce autonomous values, have to be supplied with them " (p. 261).

Salk (1983) observes that "humankind seems hopelessly sick." We need to bring wholeness to the collective mind in a situation when it is difficult to bring wholeness even to the single mind. "At this point in our development it is clear to me that the degree of pathology is such as to make the future inauspicious unless health-creating individuals and groups arise in sufficient number and coalesce into a powerful force" (p. 103). Salk suggests that only such a force will have the capacity to counter the effects of the internally arising sociocultural pathogenic influences that are responsible for the current state of humankind.

Hubbard (1998) remarks that during the past four decades, "Our basic social and economic systems have attempted to maintain the status quo despite the many warnings that the old ways where no longer sustainable. In many instances our existing systems are not humane. Homelessness, hunger, disease, and poverty consume the lives of hundreds of millions of people and the environment continues to degrade. We can view the reactive and conservative ways of the past few decades as a survival mechanism—as the caterpillar 's immune system rigidly holding on to old structures" (p. 11).

Our evolutionary predicament is shaped by the presence of two evolutionary gaps (Banathy, 1996). These gaps opened up as the result of the synergistic effect of the speed and the intensity of evolutionary developments during the last few hundred years, creating a perilous evolutionary imbalance. While our recently emerged communication capabilities created the potential and conditions for global human community, our consciousness is still locked within ethnocentric, racial, and national boundaries. This lack of internal consistency has created an evolutionary consciousness gap. Furthermore, the technological revolution, while giving us an earlier unimagined and power, has accelerated to the point where we have lost control over it. "We have simply failed to match the advancement of our technological intelligence with an advancement in socio-cultural intelligence, and advancement in human quality and wisdom. This situation has created the second evolutionary gap " (p. 315).

Activity 19

You are asked to continue to develop the societal evolution text for your high school students. Your new topic might be something like this: "How and why has evolution gone wrong?"—or any title you select to introduce to your learners the notion of "evolutionary predicament." First, review Section 6.2. Start out by noting/describing the core ideas. Then, draft the text by considering the current state of evolutionary markers and evaluating them from the perspective of the evolutionary qualities I introduced earlier. Enter your text in your workbook.

6.2.2 The New Realities

During the last decades of the twentieth century, a new evolutionary stage seems to emerge. It is called by many different names, such as: the "post-industrial society" (Bell, 1976), the "post-business society" (Drucker, 1989), and the "post-modern society" (Harman and Horman, 1990). The recurring label "post" indicates the onset of the transformation of our society into something very different from what it has been. But we are not sure yet what name to give to the emerging age. Whatever we call it, we know for certain that this new stage of societal evolution will unfold new thinking, new perspectives, new scientific orientation, and a new planetary world view. It will bring about massive changes and discontinuities in all aspects of our lives. In his speech to the U.S. Congress, Czech President Havel (1990) voiced his vision of a world in which history has accelerated, and once again "It will be the human mind that will notice this acceleration, give it a name, and transform those words into deeds."

What we can see today is a widening gap between the emerged new realities and our institutions, systems, and ways of living and doing, all still grounded in the collapsing Third Generation of HSS. In this section, my task is to introduce a limited set of "new realities." The set will be inclusive enough to indicate our current evolutionary crisis. Drucker (1989) calls the emerged changes the "new realities." It is of primary importance that we individually and collectively understand what these new realities are, and grasp their implications for conscious evolution, for the design of our lives and the design of the Fourth Generation of HSS.

In the course of the last few decades, observers of the societal landscape have described the beginning of a major societal transformation from the industrial machine age to the post-industrial information/knowledge age. Describing these transforming societal features, Bell (1976) distinguishes three discontinuous stages in societal evolution. Our current era, which is at the threshold of a new evolutionary stage, focuses on processing with the use of knowledge-based intellectual technology. Thus, the emerging era is vastly different from the declining industrial age, which focused on fabricating with the use of machine technology. And the industrial era was vastly different from the pre-industrial period of agriculture and mining, which was primarily extracting in nature. Bell (1976) suggests the unfolding of a major shift that is characterized by the centrality of knowledge, the creation of new intellectual technologies, the spread of the knowledge class, a massive change in the nature of work, a focus on cooperative strategy, and the central role of "new science." The emerging evolutionary transformation brings about a major shift from producing goods to generating information and knowledge.

Drucker (1989) says that the biggest change we experience today is a shift from the industrial worker to the knowledge worker. Knowledge becomes the true capital of our age. The dominant task of the day is by no means confined to high technology or technology in general. In fact, social innovation and its intellectual technology—social systems design—may be of greater importance and have much greater impact on life than any technological invention. Drucker also says that to respond to the new realities, within the next decade education must "change more than it has changed since the modern school was created by the printed book over three hundred years ago" (p. 232). Today only a focus on evolutionary learning leads to the attainment of evolutionary competence, which becomes the individual's and society's most important possession.

Reflections

As we reflect upon the new realities presented in the preceding, we can become increasingly more aware of the widening gap between the emerging realities and current state of our systems and our ways of living and doing things. In times of emerging new realities and accelerating and dynamic changes, when a new stage is unfolding in societal evolution, inquiry should not focus on the improvement or fixing of our existing systems. Such a focus limits perception to adjusting or modifying the old design of the industrial age in which our systems are still rooted. A design rooted in an outdated image of a bygone age is useless. In fact it creates far more problems than it solves. So, we should transcend the boundaries of the existing systems, explore change and renewal from the larger vision of an emerging collective consciousness, create a new image of the Fourth Generation of HHS, and bring that image to life by conscious and purposeful design.

6.2.3 Attitudes and Approaches toward Change

The new realities that are in evidence in our current evolutionary landscape call for a major evolutionary change, an evolutionary transformation. They call for the emergence of a new Evolutionary Generation of HSS. But, before we can consider conscious and purposeful change, we should reflect upon how people and systems relate to change. Ackoff (1981) describes four orientations toward change: reactive, inactive, preactive, and interactive. I shall interpret his ideas to orientations toward evolutionary change. Ackoff suggests that the four orientations seldom appear in pure form, but usually one of the four types dominates.

6.2.3.1 The Reactivist: Back to the Future

Reactivists are longing for the past. Their reaction to change is an attempt to try to unmake it. They romanticize about the "good old days" when life was simple. They are driving toward the future by looking in the rear view mirror, focusing on where they have been instead of where they want to go. Reactivists hold that experience is the best teacher. They do not like change. They are the modern version of the Neanderthals. If they find that they must change, they seek guidance from past experiences. They consider technology the main cause of change and reject it. They rationalize, by pointing to some of the negative consequences of technology applications. They rely on old, well proven, and familiar organizational forms and protect the current structure and operational procedures, often at any cost. In the reactionist style, planning is ritualistic, and is directed from the top. Given the official plan, other levels react to it, develop implementation plans, and pass it up for approval. Problems are "fixed" in a piecemeal fashion, believing that "getting rid of what is not wanted we can achieve what is desired." Ackoff suggests that the reactive orientation has three main attractions. It maintains a sense of history from which we can derive guidance; it maintains continuity and seeks to avoid changes; and it preserves tradition, protects familiar grounds, and maintains a feeling of stability and security.

Ackoff's portrayal of the reactivist orientation mirrors well the current state of trying to protect what has been and rejecting a major transformational evolutionary change. It seems obvious that the people and systems having a reactivist orientation to change will need to undergo a major change in their attitudes, consciousness, and beliefs before they can embrace the idea of conscious evolution. Again, we are reminded to Einstein's admonition: We cannot address a problem from the same consciousness that created it. We have to think anew. Yes, we have to develop a new collective evolutionary consciousness that will animate the emergence of the Fourth Generation of HSS.

6.2.3.2 The Inactivists: Don't Rock the Boat

The inactivists are satisfied with things as they are; therefore, they also resist change. The label "inactive," however, is misleading. A great deal of energy and effort is spent on keeping things as they are and preventing change from happening. The operating principle is preserving stability at any cost. It takes a great deal of work to keep things from changing. The inactivists say that things may not be the best today, but they are as good as can be expected. Having a dominant orientation toward the present, the inactivists want to preserve the status quo, rely on present practices rather considering new knowledge as a guide, and, like the reactivists, they are reluctant to use new technology.

Having a desire to keep things just as they are, the inactivists are bureaucratic. They rely on red tape to slow things down. They use committees and study groups in an endless process of gathering facts, pass on information from one group to the other, and revise positions and recommendations. This process goes on until they might resolve that there is no reason to change anymore. They value conformity more than creativity. The status quo is valued and "don't rock the boat" is the code phrase. Working with problems, their approach is similar to that of the reactivist. Problems are treated piecemeal. Inactivists change as little as possible; they muddle through. Ackoff suggests that the inactive style is dominated by the perception that even if there might be situations that call for change, doing nothing is better than doing something. They believe that problems fade away if left alone and those who act cautiously seldom make mistakes of catastrophic proportions. The inactivists will not be ready partners in engaging in conscious evolution. Like the reactivists, they are defendants of the status quo and believe that somehow we shall muddle through our current evolutionary crises if we are careful enough, if we just "don't rock the boat." But let me note that often people who resist change do it because it is imposed upon them and because they are denied the right to participate in making decisions that affect their lives.

6.2.3.3 The Preactivists Ride the Wave of the Future

The preactivists anticipate change and prepare for it when it arrives so that they can exploit its opportunities. Since preactivists believe that change is brought about by external forces, they do all they can to guess possible changes to come and say: When it comes, we will be ready for it. They hold that designing for the future is the job of the futurists, the experts. They put a great deal of energy and effort in finding out who are the best futurists or use science-based methods of predicting. They agree with the reactivist that technology is the principal agent of change. But unlike the reactivists, they want to promote technology as a panacea. Preactivists are more concerned about missing an opportunity than about making errors. For them, errors of commission are less costly than errors of omission. They build an organizational culture of anticipation of change and want to be the first to try out new things. They value growth, want to become bigger, want to capture the largest share, want to become "number one." Given several possible futures, plans are prepared for each. Ackoff (1981) suggests that the preactivists' close association with modern science and technology accounts for much of its great appeal as well as its prestige. Advocating change gives the preactivists a progressive stance at the frontiers of the future.

6.2.3.4 The Interactivists "Steer the Kayak Down the Rapids."

The interactivists believe that it is within our power to attain the future we envision by designing it. The past, the present, and the future are in an interactive relationship. The future depends more "on what we do between now and then than it does on what has happened until now" (Ackoff, 1981, p. 146). The interactivists create the desired future state by engaging in the intellectual technology of social systems design, which applies the concepts, principles, and models of the systems and design sciences. The interactivist believes that the value of technology is manifested in the way we make purposeful use of it as a tool. They seek idealized solutions in creating the future. The interactivists engage in two major operations: designing the desired future and planning ways and means for implementing it. Interactivists do not wait for the expert to create the future. This orientation "provides the best chance we have for coping effectively with accelerating change, with increasing organizational complexities and environmental turbulence" (Ackoff, 1981, p. 65).

Of the four orientation toward change, it is obvious that it is the interactivists' style that fits best conscious self-guided evolution.

Activity 20
Review the four styles of working with change and use them as a mirror. Ask yourself: Which style is my dominant style? And: Which style would I aspire to develop and why, as my preferred and dominant style for participating in conscious and guided evolution? Enter your findings in your workbook.

Reflections

If we compare the four styles in a summative way, it becomes obvious that the interactive style is the only one that learns from the past, values what is good in the present, and takes responsibility for the future. It is the only orientation that attempts to harness individual and collective aspirations and creativity for the purpose of seeking the ideal and, based on it, giving direction to change, and shaping our evolution by design.

6.3 *Transition or Transformation?*

Standing at the declining stage of the Third Generation of HSS, we face the choice between *transition* and *transformation*. As used here, *transition* means an attempt to stay in the evolutionary space of the industrial society and transit to the information society by trying to fix our problems while continuing to maintain and somehow improve our institutions, our way of life, hoping that by so doing we might transit into an "improved" Third Generation. *Transformation* means engaging in conscious evolution, by developing evolutionary competence, envisioning an image of a desired future state, and creating that future by collaborative and purposeful design.

6.3.1 Making a Crucial Choice for Transformation

"Conscious evolution will emerge from the evolution of consciousness and the consciousness of evolution. Conscious evolution will result from a sufficient evolution of consciousness and a sufficiently developed consciousness of evolution" (Salk, 1983, p. 111). Salk suggests that when these two develop effectively, they can enable conscious, creative sociocultural evolution. This development will require a change in perception, a change in consciousness. It requires a *transformation* from a system of beliefs that presently prevail to a system of beliefs that will guide *conscious, creative, and transforming sociocultural evolution. Unless we self-create that system of beliefs, unless we engage in evolutionary transformation, humankind will muddle along, seeking remedies, trying to correct the host of problems of the decaying Third Generation, searching for ways by which we somehow might avoid the continuing evolutionary decline and collapse.*

So it is up to us. We can continue our journey on the evolutionary path of the decaying Third Generation. We may try to fix the host of problems that have arisen with the evolutionary markers as they lost balance, harmony, internal consistency, synergy, and symmetry. Or *we can embark on the greatest venture humankind ever attempted, collectively developing the conditions necessary to engage in conscious evolution and guide our evolution by design.*

6.3.2 Conscious Transformation: A First in Evolutionary History

In the course of our evolutionary history, we have reached several critical events when a life form of our species transformed into another. During the time span of the last 40,000 years we have experienced three transformations and today we are stepping through the threshold of the fourth—if we so chose. Yes, if we so chose, because *it is the first time in evolutionary history that we have the opportunity and the burden to engage in conscious, deliberate, self-organized, and self-guided evolutionary transformation.*

The emergence of modern man, the coming on the evolutionary scene of the Cro-Magnons, was the result of such transformation. Since their emergence the biological characteristics of our species have not changed. But the sociocultural characteristics have gone through major transformations as manifested in the emergence of the Second and the Third Generations of HSS. But let us now explore more fully the notion of evolutionary transfomation.

6.3.3 The When and How of Transformation

In the course of the evolution of our species there comes a time when the state of evolution becomes pathological, when decay sets in and when the balance among the evolutionary markers is lost. Harmony is degraded to disarray, internal consistency becomes out-of-sync, synergy loses its power, symmetry is broken, wholeness becomes fragmented, and creativity is replaced by inaction. These pathologies, once recognized, call for evolutionary transformation.

The necessary conditions of engagement in transforming conscious evolution include the following: (1) the convincing presence of consciousness of evolution; (2) the emergence of a new collective consciousness that will animate the transformation process; (3) the understanding of the new realities of our transforming era; (4) a grasp of the implications of what those realities mean for a transforming evolution; (5) a readiness to leave behind our old ways of thinking and doing; (6) the development of a new collective consciousness; (7) the development of evolutionary competence by evolutionary learning; (8) the willingness to transcend the existing system; (9) a commitment to take responsibility for conscious evolution; (10) the full engagement of our creating potential; (11) the envisioning a new evolutionary image; and (12) the transformation of our systems by purposeful design.

6.3.4 Making It Happen

The main obstacle to making transformation happen is a lack of willingness to transcend and leave the present state behind. Metaphors often help us to capture the meaning of such concepts as leaving the past behind and embarking on a new venture. One such metaphor is the story of Cortes and his crew. When they came to shore at Veracruz, Cortes was aware of the great uncertainty and adversaries ahead of them. Many of his men wished that they never come. Cortes burned the ships. This dramatic mark of an ending became a new beginning, the beginning of a new journey. This is the type of resolve we need to make evolutionary transformation happen.

Another metaphor is a Native American story of leaping out from—and transcending—the "here and now." It is the story of the "jumping mouse." Hearing an enticing noise in the distance (of a better future), the mouse left his familiar home ground, and—encouraged by the frog at the river—dared to jump high (transcending) and captured the image of the sacred mountain. He then embarked on an arduous transforming journey toward the sacred mountain. During the journey he had to give up his sight (the old way of seeing) to the blind wolf (leaving the past behind), who was the only one who knew the way to the top of the sacred mountain.

I suggest that the ease or difficulty of transcending the existing state depends a great deal upon our attitudes and our orientation toward change. Earlier in the present chapter we reviewed four different orientations. Those who have a "reactive" (re-activating the past) or "inactive" orientation will have the highest degree of anxiety about change. Having a "preactive" orientation, on the other hand, would lead people to rush into change whenever certain "trends" indicate or "forecasters" predict a solution. I suggest that it is people who have a proactive and interactive attitude who will be ready to transcend their existing state and seek to engage in conscious and purposeful evolution.

We can offer two major strategies for initiating transformation by transcendence. One of the strategies offers an opportunity to those who need some reassurance, some support to arrive at the realization of the need to transcend the current evolutionary state. They need to develop a well-grounded understanding of evolutionary consciousness. They need to explore and understand the implications of the decline and decay of the Third Generation HSS. They should have a clear picture of the current evolutionary problematic, including an understanding that most of our current systems are still grounded in the precepts and practices of the industrial machine age. Most significantly, they need to answer the question: What would happen if we would not change, if we would leave things as they are?

The other strategy, on the other hand, is appropriate for those who have developed a well-grounded evolutionary consciousness, who have a clear picture of the evolutionary predicament, and who have an interactive and positive orientation toward change. They hold that it is a waste of time to analyze the problems of what is wrong now, and therefore they are prepared to transcend the current evolutionary state and leap out from their systems, leave the baggage of the decaying Third Generation HSS behind, and engage in the conscious and purposeful evolution of the Fourth Generation.

Reflections

"One does not discover new land without consenting to lose sight of the old shore" (French novelist Andrè Gide). The new land for us is the new land of the Fourth Generation of HSS. We are now vividly aware that we cannot discover this new land unless we are willing to transcend, we are willing to lose sight of, the evolutionary land of the Third Generation. For many of us it is hard to let go. But we really have no other choice. In this section we explored the conditions we have to meet to embark on the journey that we shall lay out so that it will lead us from the land of the Third Generation to the land of the Fourth. Knowing what these conditions are and having the willingness to meet them are the prerequisites to transcend the problem-laden current evolutionary stagnation and engage in conscious, purposeful evolutionary transformation by design.

Activity 21
Review Section 6.3 and capture the core ideas that constitute the mosaic of evolutionary transformation. Work with these core ideas and use them as you continue the creation of your evolutionary text for your high school students. Focus on the why and how of transformation. Enter your learning text in your workbook.

Summary

In Chapter 6, we embarked on a mental journey that led us from evolutionary consciousness to the threshold of conscious evolution. In the first section of the chapter, we looked at the many facets of consciousness. We defined individual and collective consciousness, revisited the various stages of the evolution of consciousness, and captured the meaning of evolutionary consciousness. Then, we traveled on the current evolutionary landscape, struggled with the notion of our evolutionary predicament, depicted emerged new realities of our era, and characterized various orientations toward dealing with change and working with these new realities. In the third section of the chapter, we used the insights gained from the previous sections to deal with the crucial choice between trying to fix the host of problems of the Third Generation of HSS or transcend these and engage in conscious evolutionary transformation. A quote from Salk (1983) appears to be a most appropriate concluding remark.

> The most meaningful activity in which a human being can be engaged is one that is directly related to human evolution. This is true, because human beings now play an active and critical role not only in the process of their own evolution but in the survival and evolution of all things. Awareness of this places upon human beings a responsibility for their participation in and contribution to the process of evolution. (p. 112)

If we accept this responsibility and engage creatively in the work of evolution we shall take part in a crucial and a first ever event in the seven million years of our evolutionary saga: We shall be the designers of our future, we shall become the guides of our own evolution and the evolution of life on earth and possibly beyond.

7

Calling for Conscious Evolution

In this chapter, we revisit several evolutionary scientists who called for conscious evolution in Chapter 2. In the second section, I introduce ideas of additional scholars who have explored a comprehensive approach to self-guided purposeful evolution.

7.1 Ideas about Conscious Evolution

In Chapter 2, we traced the evolutionary ideas of scholars and philosophers of the nineteenth and twentieth centuries. Now, we revisit some of the statements they made about conscious and guided societal evolution.

Bergson (1983) objected to the Darwinian conception of mechanistic, mindless, purposeless evolution. He saw evolution as a continuous creative process in which a vital impulse, working from within, aims to achieve perfection and freedom. He believed that this vital impulse is the source of evolutionary change. Obtainable only through intuition, this vital impulse *enables us to creatively guide evolution.*

Teilhard (1975) saw the evolution of humanity leading toward the development of a common and shared thinking; he called this thought system noosystem. The noosystem becomes the *mind of humanity,* which will realize new possibilities for evolving a planetary humanity. Evolution is an ever ongoing and unfolding "groping," says Teilhard. It is a thrust toward "the precise orientation of a specific target" (p. 110). Human and societal evolution is no longer a chance. It is "directed chance."

Csanyi (1989) holds that the highest organizational level of the human and societal evolutionary process on Earth is the global level. Global problems have been identified and efforts to solve them are being initiated. But "these problems cannot be understood and solved within the framework of former social paradigms" (p. 189). Systems thinking and the development of a new systems theory will be necessary as a new evolutionary paradigm for global applications. Furthermore, the design and introduction of new learning content and learning strategies are inevitable. They will enable and empower us to transform the idea systems of present societies into those of global humanity.

Gould (1996) highlights the difference between Darwinian natural evolution and cultural evolution. It lies in our explosive and collective capacity for directional and rapid change. The cultural mechanism of inheritance gives cultural evolution a "directional and cumulative" character that no natural selection can provide. It accumulates favorable innovation by the amalgamation of cultural features. It allows a culture and society in general to *choose* from the most useful innovations.

Chaisson (1987) holds that our challenge is to guide social, technological, scientific, political, and economic changes and become "the agents of change." Once we understand this, we can guide evolution so that it is beneficial to mankind and brings forth a new evolutionary stage, which Chaisson calls the *Life Era*. "Of all the implications of the Life Era concept, the most important to my view is that we, as the dominant species on Earth, develop—evolve if you will—and quickly too—a global culture. We need to identify and embrace a form of planetary ethics that will guide our attitude and behavior toward what is best for all humankind" (p. 176).

Dawkins (1989), exploring the power of the "selfish gene," suggests that if we wish to have a society in which people cooperate unselfishly toward a common goal, we should understand what our "selfish genes are up to" so that we might have a chance to upset their design. This means that we need to develop a culture in which we have altruistic idea systems of memes, and we can teach and learn to become unselfish in the service of the common good.

Dennett (1995) says that "We, unlike the cells that compose us, are not ballistic trajectories. We are guided missiles, capable of altering courses at any point, abandoning goals, and formulating new ones. For us, it is always decision time, and because we live in a world of memes, no consideration is alien to us, or forgone conclusion" (p. 460).

Huxley (1946) suggests that "man's true destiny emerges in a startling new form. It is to be the chief agent for the future of evolution on this planet. Only through man can any further advance be achieved" (p. 32). We must embark on a psychosocial stage of evolution to create a major advance, a radical change in the dominant idea systems. His new idea system, called *evolutionary humanism*, seeks to unify the spiritual with the material. It is a dynamic and directional process rather than a static mechanism. Evolutionary humanism affirms that knowledge and understanding can be increased, that social conduct and social organization can be improved, and that more desirable directions for individual and social development can be defined and attained.

Laszlo (1987) says that "The evolution of our societies and therewith the future of our species is in our hands. If we make use of our capacity to think rationally and act purposefully, we can defuse immediate threats and promote the creation of more mature, more autonomous, more dynamically stable societies. Only by becoming conscious of evolution can we make evolution conscious, and only by conscious evolution can we have the assurance that we can survive the awesome combination of highly evolved order and complexity in our brain and immature and underdeveloped order and complexity in our societies" (p. 102).

The orientation of the scholars who made the preceding statements ranges from the spiritually inspired, such as Bergson, Teilhard, and Huxley, to the Darwinian Dennett and Dawkins, and to systems scholars Csanyi, Chaisson, and Laszlo. It is rather remarkable that scholars representing such a wide range of orientations agree on the possibility—even the need—for conscious evolution.

Activity 22
Your first task is to capture the core ideas of the evolutionary scholars, reviewed in this first section of the chapter. Your second task is to write an essay that synthesizes those core ideas and reflects your own understanding of the possibility and significance of conscious and guided evolution. Enter your essay in your workbook.

7.2 Some Comprehensive Ideas about Conscious Evolution

In this section, I introduce a variety of ideas of scholars who explored in some depth the need for, the implications of, and the possible outcomes of conscious evolution.

7.2.1 Csikszentmihalyi

Here, I introduce two evolutionary core ideas of Csikszentmihalyi: directed evolution and the role of flow in evolution.

7.2.1.1 Directed Evolution

Csikszentmihalyi (1993) suggests that the way human life evolved up to the present cannot be considered to be a planned effort. Billions of more or less randomly interacting small and large events created the current evolutionary state that now binds us.

> And now we suddenly realize that, unless we take it in hand, this process of change will continue under the sway of relentless change, a change entirely blind to human dreams and desires. Like horrified passengers on an airplane, who are told that the pilot mysteriously vanished from the cockpit as the plane is cruising miles above the ground, we know that we must find a way to master the controls or the trip will end in disaster. But will we conquer ignorance and fear before the fuel runs out? *If there is a central task for humankind in the next millennium, it is to start on the right track in its effort to control the direction of evolution.* (p. 149) (Italics mine.)

Csikszentmihalyi holds that to engage in the task of directed evolution, we need to attain a genuine understanding of evolution itself. He further states that an understanding of evolution suggests, that there are two opposing tendencies in evolution. One leads toward harmony, which is sustained by cooperation. Its opposite tendency leads to entropy by exploiting other organisms and causing conflict and disorder. Harmony is achieved by evolutionary changes that increase both differentiation and integration. *Differentiation* is the degree to which a system is composed of parts that differ in function and structure. "*Integration* refers to the extent to which the different parts communicate and enhance one another's goals. A system that is more differentiated and integrated than another is said to be more *complex*" (p. 156). An example is the family system. In a differentiated family, both the parents and the children are allowed to play out their distinct individuality unconstrained by a common purpose. However, if they are also integrated, they are connected by mutual care and support. We can say that they are there for each other. As a family, they have collective consciousness and they live by it. The family example introduced here applies to all social and societal levels, up to the global. [Csikszentmihalyi's example and his focus on integration are excellent representations of the emergence of an evolutionary generation of *Homo sapiens sapiens* (HSS) by the interaction and integration of its evolutionary markers.]

Complexity, at any level of analysis, involves the optimal development of both differentiation and integration, according to Csikszentmihalyi (1993).

> If we are to direct evolution toward greater complexity, we have to find an appropriate moral code to guide our choices. It should be a code that takes into account the wisdom of tradition, yet it is inspired by the future rather than the past; it should specify right as being the unfolding of maximum individual potential joined with the achievement of the greatest social and environmental harmony. (p. 162)

Therefore, confronted with everyday issues as well as substantive evolutionary choices, we should always evaluate which alternative brings more harmony to life. This task is a core requirement of directed evolution. The attainment of this task is much enhanced by developing "quality of flow" in our individual and collective lives.

7.2.1.2 The Role of "Flow" in Evolution

Evolution is creation. It is a process of continuous creation and re-creation of species in the evolutionary design space. Csikszentmihalyi suggests that "we experience enjoyment when we take on a project that extends our skills in new directions and when we recognize and master new challenges. Every human being has this creative urge as his or her birthright. It can be squelched and corrupted, but it cannot be completely extinguished" (p. 175). *Creating is the core process of evolution.* And there are few things that are more rewarding, more enjoyable, and more gratifying than taking an active part in creating our future. A flow is "what the lover feels talking to the beloved, the sculptor chiseling marble, the scientist engrossed in her experiment" (p. 177). Imagine the creating flow that the Cro-Magnon cave artists lived and experienced as they painted those marvelous images! Or the flow that Michelangelo felt as he painted the ceiling of the Sistine Chapel!

Flow is a sense of discovery, finding out something new, creating what we aspire to bring about. These are activities we enjoy most because they enable us to express our potential, to stretch ourselves, to make valued contributions. It is for this reason that flow is such an important force in evolution. Without it, our genetic programs would instruct us to continue pursuing what has been "good for us" in the past, but flow makes us receptive to the entire world as a source of meeting new challenges. It becomes the arena for evolutionary creativity.

> The connection between flow and enjoyment may have been at first a fortunate genetic incident, but once it occurred, it made those who experienced it much more likely to be curious, to explore, to take on a new task and develop new skills. And this creative approach, motivated by the enjoyment of facing challenges, might have conferred so many advantages that with time it spread to the majority of the human population. (p. 190)

Over the last several decades, I have been working with others in ideal-seeking, self-organizing, collaborative, social systems design. As our work has called for deep collective involvement, commitment, and high concentration the sense of time passing became irrelevant and unnoticed. We truly experienced what Csikszentmihalyi calls "flow."

What is most relevant to our discussion is the *role of flow in conscious and guided evolution*. Those who possess the skills, the competence, the experience of *flow* will be instrumental in *transforming* the current evolutionary problematic, the paradoxes, the chaos and entropy into meaningful creation of the next evolutionary stage. "In so doing, they not only enjoy their own lives, but they contribute to the evolution of complexity for humanity as a whole" (p. 204).

This quotation has an important implication for engaging in conscious evolution. For conscious evolution to come about, it has to happen at all levels of the society, from the family to the systems in which we live and work, in our communities, and at all levels of the society. The "they," as used in this quote, is *US*. It means thinking about participation in the great venture of guided evolution by all of us in a collective and coordinated way. In closing, it is helpful to quote Csikszentmihalyi's vision of the inevitability of conscious evolution.

> Even if nothing were to change in our lifetime, even if signs of a new dark age proliferated, if chaos and apathy were on the ascendant, those who cast their lot with the future would not be disappointed. Evolution is not a millenarian creed event, expecting a Second Coming next year, the next Century, or the next millennium. Those who have faith in it have literary all the time in the world. The individual life span with all its woes and disillusions is only an instant in the awesome cosmic adventure. At the same time, our actions have a decisive impact on the kind of life that will evolve on this planet, and perhaps on other planets as well. (p. 293)

We shall revisit Csikszentmihalyi's work several times as we unfold ideas about conscious evolution and explore evolutionary learning and the role of creativity and flow in evolutionary epistemology.

7.2.2 Hubbard: Conscious Evolution

Barbara Hubbard is a leading voice in the social potential movement. In her recent work (1998) she calls for the activation of our individual and collective (social) potential for an evolutionary transformation by the creation of what we have called the Fourth Generation of HSS. In her book she reviews five lessons of what we have learned from a study of evolution. She develops a rather comprehensive definition of the meaning of conscious evolution. Then, she proposes an evolutionary agenda for the twenty-first century. In the following subsections I review these three aspects of her work as part of our continuing exposition and exploration of conscious, guided evolution.

7.2.2.1 Lessons of Evolution

Hubbard proposes that in looking back to past evolutionary transformations or quantum jumps we can learn and uncover the pattern and the process of evolutionary transformation. Once we have uncovered these, we can apply what we have learned to engage in the conscious evolution of the next generation of HHS. She presents us with five lessons, which are interpreted in the following paragraphs.

7.2.2.1.a Quantum Transformations. Quantum transformations are the rule of the game in the course of the evolutionary journey. A quantum jump, an evolutionary transformation, it is a discontinuous leap, a leap out from an existing evolutionary stage into a new stage. It cannot be attained by incremental changes or adjustments. Evolution works through radical change. "The capacity of evolution to produce radical newness is truly astonishing" (p. 48). As we have seen in Chapter 4, some 35,000 years ago we experienced such an astonishing radical newness, a discontinuous quantum jump with the emergence of modern man, the Cro-Magnons.

7.2.2.1.b Crises Precede Transformation. When an evolutionary stage reaches its limitation and begins to decline, it creates disruption, chaos, and social disorder. In such a state, it cannot adapt or change incrementally; it must leap into a new creating regime and transform into a new evolutionary entity. What we earlier called paradoxes and evolutionary predicaments become the driving forces toward evolutionary transformation. For example, "The environmental crisis is awakening us to the fact that we are all connected and must learn how to manage planetary ecology" (p. 48). Such a realization becomes an essential driving force for our current evolutionary transformation.

7.2.2.1.c Holism. Holism in evolution forms whole systems from parts. We have seen this repeatedly when at the emergence of a new evolutionary stage a higher level evolutionary consciousness organized the evolutionary markers into an new integrated system of an evolutionary generation. "Every tendency in us toward greater wholeness, unity, and connectedness is reinforced by nature's tendency toward holism. Integration is inherent in the process of evolution" (p. 48).

7.2.2.1.d. Evolution Creates Beauty. In evolution only the beautiful endure. "Every leaf, every animal, every body that endures is exquisite" (p. 45). Earlier we talked about the laws operating in the evolutionary design space. These laws favor the emergence of an elegant, aesthetic design of life forms. This gives us the courage, says Hubbard.

It has been the decline, disharmony, and decay of the Third Generation HSS that has created the current evolutionary predicament and produced the crude forms of so many modern cities, houses, and machines and *dysfunctional* social, political, social service, and economic systems and public and private institutions and organizations. It is the challenge of conscious evolution to bring forth aesthetic designs in all spheres of life.

7.2.2.1.e The Lesson of Consciousness and Complexity. Hubbard refers to Teilhard de Chardin's law of "complexity/consciousness." As in the course of the evolutionary journey a system becomes more complex, it jumps into higher levels of consciousness. Standing at the threshold of a new evolutionary stage, a new level of consciousness is awakening in us as an integrating, whole system, global consciousness.

7.2.2.2 The Meaning of Conscious Evolution

Hubbard suggests that we are seeking a new world view that calls for creative action and will engage our "immense powers" toward creating and attaining evolutionary purposes. "That guiding world view is, I believe, conscious evolution. It holds that through our unprecedented scientific, social, and spiritual capabilities we can evolve consciously and co-creatively with nature and with the deeper patterns of creation" (p. 57). She says that

> Conscious evolution as a world view began to emerge in the latter half of the 20th century because of scientific, social, and technological abilities that have given us power to affect the evolution of life on Earth. Conscious evolution is a meta discipline; the purpose of this meta discipline is to learn how to be responsible for the ethical guidance of our evolution. It is a quest to understand the processes of developmental change, to identify inherent values for the purpose of learning how to cooperate with these processes toward chosen and positive futures, both near term and long range. (pp. 57–58)

Hubbard also holds that through language—and cultural evolution I would add—each generation learns from the previous generation, thus vastly accelerating change and innovation. Conscious evolution is based on our ability to understand the evolutionary innovations of the past (we called this ability the attainment of evolutionary consciousness) so that we can consciously codesign the future. Conscious evolution is a recent, radically new concept. It has not yet been incorporated into our academic, political, and religious realms. It makes us aware of the need to design social systems that are in alignment with higher consciousness, greater freedom, and more synergistic order. With an increase of freedom should come an increase in responsibility for the use of our creative powers for the common good. The spiritual aspects of consciousness call for deepening our sensitivity to the patterns of creation. It calls for us to learn how to design social systems as vital parts of the society.

> If we can learn to combine our advanced technological capacities with an evolved spirituality and the ability to design synergistic social systems, we will be part of a quantum transformation, a jump as great as the jump from Neanderthal to *Homo sapiens*. Conscious evolution is the design innovation that empowers us to become a universal humanity. We are crossing the evolutionary divide between creature and co-creator. (p. 76)

Barbara Hubbard tells us that conscious evolution provides the context and the logic for working in alignment with the process of creation. This process has a direction that encourages us to attune to our creative potential and express it for our sake and that of others. "Conscious evolution offers the possibility of an open-ended, choiceful, glorious future. Within each of us stirs the mighty force of creation in the process of its next quantum transformation" (p. 94).

7.2.2.3 Plan of Action for the Twenty-First Century

Hubbard (1998) suggests that "Our goal for the next century should be the broad acceptance of an evolutionary agenda, which is supported by the world-view of conscious evolution and manifested through social innovations and social systems that lead toward and ever-evolving future" (p. 121). She further suggests the development of society as a whole system in transition. A view of current society depicts an incoherent state. (Earlier, we characterized this state as a lack of harmony, balance, symmetry, synergy, and above all, wholeness.) Today, social innovations are fragmented, disconnected from each other, dispersed, and often ignored. Still, breakthroughs are emerging where there are breakdowns, and solutions are emerging where there are problems. We need to seek ways to connect these breakthroughs and solutions and organize them into a comprehensive design of a positive future.

In contemplating the design of a new societal model, Hubbard suggests that we should ask—and I would add test—"What will work?" This question is crucial, and how we respond will guide the selection of social innovations. Our collectively defined values become the basis for answering the question. There are values that are inherent in evolution itself. The five "lessons we learned from evolution" reveal fundamental values.

From the perspective of conscious evolution we value any belief or intention that expands and elevates our consciousness toward a higher level, toward a more spiritual and more whole-centered evolutionary stage, toward having reverence for higher dimensions of human nature and unity with nature. We have to value greater freedom *from* hunger, poverty, oppression, and greater freedom *to* realize our yet untapped potential for self-actualization. But freedom should be coupled with compassionate responsibility for the benefit and well-being of others as well as commitment to act in the service of the common good.

Finally, says Hubbard, we select innovations for their value of bringing together greater wholeness and cooperation "in our intimate relationships, in our communities and among religions, nations, disciplines, races, and cultures. We value what joins us together to form a whole society that is different from, unpredictable from, and greater than the sum of its parts" (p. 127).

Hubbard designed an "innovation wheel," comprised of sectors in which innovations should emerge for social transformation and toward a peaceful, cocreative world. The sectors include: government and law, education, economics, business, philanthropy, health, relationships, and personal growth; science and technology; spirituality and religion; environment and habitat; and culture, media, and communication. Within each sector guidelines and goals are suggested by the author.

Hubbard suggests that social innovations should be mutually reinforcing and interconnecting (and, I would add, internally consistent to attain harmony). They should "foster intrinsic values by embodying greater cooperation, creativity, optimism, a tolerance for differences, a sense of reverence for life, and faith in the potential of all people, They share an emphasis of self-actualization rather than self-sacrifice. Ultimately, they form the basis of the co-creative society" (p. 131).

In closing, we can review criteria that Hubbard adopted for assessing social innovations. An innovation should move the society toward a just, humane, and regenerative future offering many choices. It should be guided by higher consciousness, greater freedom, and more systemic order. It should enhance integrity, nonviolence, inclusivity, and sustainability. It should foster personal responsibility and respect for self and others. It should assist in the positive evolutionary transformation of the society.

7.2.3 Salk: Conscious Evolution

Earlier, we presented several of Salk's ideas on evolutionary consciousness. Here we continue our conversation with him as he explores merging intuition and reason in conscious evolution, forming a new evolutionary cosmology, and setting forth the idea of individual mutualism.

7.2.3.1 Merging Intuition and Reason

Salk (1983), surveying the current evolutionary landscape, recognizes the many different visions, perceptions, and world views as sources of goals toward which people are drawn. Some of these represent evolutionary paths, while others do not. (The "others" continue to travel on the well-known roads of the past, worrying about and fixing the potholes of the old road.) The evolutionary paths are the less traveled. But how do we know which path to follow? How can we know which to choose, given the many forces that act upon us and the various impulses that work from within? There must be something to guide us, something that we sense through our *intuition* and *reason*. Salk asks: Do we possess the capacity to merge intuition and reason and—based on it—select the path that leads us toward conscious evolution? He suggests that we can do it, saying, *"It now remains for human beings to decide the ultimate course of human evolution"* (p. 14).

> By imagining ourselves inside the process of evolution and by imagining the process of evolution working inside our minds, we may discover how to deal with the problems as well as the opportunities arising from the uniqueness of the natural world, and we may learn how to empathize in a way that might influence the direction of evolutionary choices. (p. 14)

The challenges described in the foregoing require a commitment that is far greater than that existing today. There are signs, however, that this commitment indeed exists and is growing. If we are informed about what is happening around us and about the options that face us, "then, as individuals and members of the organism of humankind, we may be able to choose the most evolutionary advantageous path" (p. 3).

Salk clearly sees our current evolutionary predicament as a malfunction of the human condition. He suggests that the remedy lies in the reconciliation of our intuitive and reasoning powers. "Intuition is part of human nature; it is an expression of human nature. Its alliance with reason is an expression of what human beings have been given by nature to further the process of evolution. This innate capacity for intuition and reason emerged through the process of evolution itself. In that process qualities that serve life and knowledge have been selected and transmitted" (p. 107).

7.2.3.2 Calling for a New Evolutionary Cosmology

Both science and technology are essential to human life. It is impossible to imagine life without them because of the extent to which we have grown to depend on both, such that today we must to find a way "to allow the evolutionary force to help establish the balance necessary for the healthy evolution of science in relation to human life and the creative human spirit" (pp. 107–108). We need to evaluate the insights of great spiritual and intellectual leaders, which represent the 'mutational' events of cultural evolution as manifested in religion, science, and philosophy. "There is a need for a reconciliation of religion and science, just as it is necessary to reconcile intuition and reason, experience and knowledge" (p. 108).

Reason and science alone do not adequately equip us to deal with the present evolutionary predicament. We also need a new cosmology, a cosmology that is compatible with the epistemology of science and of human experience, and—very importantly—with the epistemology of evolution. Furthermore, an ethic and morality are required that are based on evolutionary principles for the establishment of a humane basis for human existence and for serving conscious evolution itself.

7.2.3.3 Individual Mutualism

"We need a new philosophy, a new ideology on the basis of which to organize ourselves in the future. This new philosophy or ideology might be called *individual mutualism.* It requires the collective to respect the individual and the individuals to participate mutually in the collective" (p. 109). Similar ideas have been expressed in different ways in various religions and cosmologies and many of these were appropriate in the past but are no longer relevant today, as our present needs are quite different.

> We are in need of an altogether new and different guiding philosophy if we are to use our energy and our resources efficiently and effectively to bring about the fundamental and essential changes called for in the present state of human evolution. The need is for a global change. Something as profound as this, however, cannot occur overnight, nor in a generation; it will require many generations and it will differ in different parts of the world. It will take a long time for well-established habits, practices, and belief systems to change, as it will take a long time for fears, hatreds, and mutual distrust to be alleviated. (p. 109)

Even if the overall process of transformation will take a long time, there are many things that we can do now to move toward the realization of a new global cosmology. We can begin by increasing our consciousness and understanding of where we are at the present stage of evolution, which means the attainment of evolutionary consciousness. Then, we must embrace a philosophy and ideology that will be expressed in our everyday life.

We now leave Salk by quoting the last sentence in his book, which most powerfully presents his core idea on conscious evolution. "It is in search of guidance from within ourselves that consciousness is invoked in the service of co-guiding and by co-authoring our own evolution" (p. 110).

We should keep the two concepts: *co-guiding* and *co-authoring* our own evolution in mind as we continue our journey toward the "how" of conscious evolution.

7.2.4 Eric Chaisson: Conscious Evolution; the Life Era

Chaison, an astrophysicist, sees in the universe constant change that has created three major evolutionary eras. The first, the Energy Era, was brought forth with the Big Bang. As the energy expended and cooled, it became transformed into matter, creating the Matter Era. The next transformation created the Life Era. In the Life Era emerged consciousness and intelligence. In his book, *Life Era* (1987), Chaisson reviews the story of the Life Era and concludes that we are at the point in time of this Era when conscious evolution could and should guide evolutionary change.

My review is focused on the last chapter of his book, in which he contemplates the implication of our cosmic heritage, the history of evolution of the Life Era, and, explicitly, the tasks of conscious evolution. Chaisson identified several aspects of evolution that integrate into the overall scheme of conscious evolution. These are the overall process of change; the evolutionary role of synthesis and consciousness; evolutionary humanism; the interface and integration of science, philosophy, and religion; global evolution; and ethical evolution. In each of the descriptions we can find the others embedded either explicitly or implicitly, revealing the wholeness of the evolutionary concept as well as the wholeness of Chaisson's evolutionary thinking.

7.2.4.1 Evolution = Change

Chaisson (1987) sees a now emerging scenario of the cosmos, based on the concept of change, change meaning "to make different the form, nature, and content of something" (p. 1). Change throughout time and space brought forth successively galaxies, stars, planets, and life. We call this universal process cosmic evolution. But "evolution is hardly more than a fancy word for change—especially developmental change. It seems that change is the hallmark for the origin, and development of all things in the universe" (pp. 3–4).

But how can change shape itself in impersonal evolution and how can we—or should we—shape it in conscious evolution? Chaisson accommodates these questions by saying that life forms eventually advance sufficiently to determine when and how changes occur. "Human actions become just as much part of evolution as the impersonal movements of atoms and galaxies" (p. 196). Hence we ask: How should we change? Should we change at all? These questions reflect our current predicament. Those of us who attained evolutionary consciousness know well that we should indeed bring about change and engage in conscious evolution. Those who defend the status quo try to keep things as they are.

Three more aspects of change are considered by Chaisson: the timing, the regulation of change, and the rate of change. Considering *timing*, premature—or forced—timing of change could harm our intended synthesis of evolutionary markers in the evolutionary design space. Still, can we afford to wait when, for example, even a single global issue, such as overpopulation—if it is unsolved—can overwhelm us? Another issue we should be aware of in contemplating conscious evolution is the *regulation* of change. We have the ability to prescribe cultural, social, political, economic, and social change and we are able to regulate these to speed them up, slow them, or even stop them. We are in the position now to usurp "the role previously played by inanimate nature, for we humans have become the agents of change—at least on the planet Earth" (p. 177). The third aspect of change that concerns Chaisson is the *rate of change*. Certain aspects the evolution—certain evolutionary markers—can change so rapidly the they can cause an onslaught of global troubles—as has been the case now when our sociocultural intelligence can no longer control or guide technological advancements. Chaisson suggests that the answer to the issue of change can be extended back to Aristotle, who said: "If human beings could be shaped by their environment, they can change themselves in equal measure by their own effort." And once we adjust our thinking to accept the permanence of change, we can change *change* itself so that it will lead to evolution that is beneficial to mankind, rather than causing devolution, entropy, chaos, and even extinction.

7.2.4.2 A Call for the Synthesis of Science, Philosophy, and Religion

Chisson begins his book by saying that "We are *entering* an age of synthesis." He notes that the evolutionary theme of synthesis was one of the core ideas of Teilhard, who, called for and worked for a religio–scientific synthesis. Today, in science, excessive specialization makes us myopic. It blinds us to the wider impact of our work. Chaisson states that nether science, nor philosophy, nor religion *alone* can answer the crucial issues we face at the current evolutionary turning pont.

> Granted, each of these institutions might *think* that they do, but my claim is that none of them individually can be counted on to provide an ethical standard needed for the human species to endure rapid, global, often self-induced changes in our political–economic and especially technological environments. Will Durant well articulated our predicament: We suffocate with uncoordinated facts, our minds are overwhelmed with sciences breeding and multiplying into specialistic chaos for want of synthetic and unifying Philosophy. (p. 199)

We are entering the age of synthesis, provided we can define the common denominator that can unify science, philosophy, and religion. That common basis, says Chaiasson, is *the concept of evolutionary developmental change*. It pertains to all objects, institutions, societies, and civilizations. Invented by philosophy and embraced by science the evolutionary concept is a core concept of all except the most fundamentalist religions. It can provide a guiding map for our future.

We must learn to think more broadly that we did in the past. Our evolutionary scientific philosophy must include and it should be synthesized with ethics, the humanistic subjects of religion, classical philosophy, economics, law, politics, and even sociology. Only the type of synthesis suggested here can help us to act "wisely," and always in the broad and long-range interest of humanity. Chaisson suggests that it is systems science that offers a meta-framework within which this synthesis can be attained.

7.2.4.3 The Global Perspective

Chaisson holds that unlike the problems faced by our ancestors, many of those that threaten our survival are global. They are large scale and debilitative. Unless confronted, they not only threaten the quality of human life but might even extinguish it over much of the planet.

The most important implications for the Life Era, says Chaisson, is that we, as the dominant species on Earth, must quickly develop a *global culture, inspired by a planetary ethics* that will guide us toward what is best for all *humanity*. We have to acknowledge that we are first citizens of the planet and only then members of national communities.

> The onset of globalism is destined to occur in some form whether we intend it or not. Increased speed in transportation, nearly instantaneous worldwide communication, international economic competition, arms control negotiations and treaties, environmental anxiety concerning acid rain and Chernobyl-like incidents—all suggest a natural congregation toward a planetary society. The world is being made whole, and it is good. (p. 205)

Opportunities to build a global system are everywhere. We are not powerless in this process. Everything that we do weaves a thread into the fabric of an emerging globalism: the work we do, the way we relate to others, what we learn, the causes we support. The world depends upon the choices we make. Our individual fate depends on the creation of a world community. "Not least, it is by means of such conscious choices that we can transform alienation into globalism, stagnation into learning, cynicism into caring, despair into hope" (p. 220).

7.2.4.4 Evolutionary Humanism

> We appeal, as human beings, to human beings, remember your humanity and forget the rest. If you can do so, the way lies open to a new Paradise; if you cannot, there lies before you the risk of universal death.
>
> *Bertrand Russell*

Earlier we noted Huxely's *evolutionary humanism*. Chaisson joins him in considering this concept as crucial in conscious evolution. Conscious evolution calls upon us to think big, to recognize the responsibilities of our planetary citizenship, to serve the need for more benevolent understanding. Knowledge and compassion are the "twin guides" to our future. To express it truly: conscious evolution is the scaffolding for a modern humanism, an evolutionary humanism. Therefore: "To forsake change, to reject the next great evolutionary leap forward, would be, according to the principles of the Universe that created us, most unconscionable" (p. 221). The survival of humanity involves the adoption of evolutionary humanism as both a theoretical concept and a guide for action.

7.2.4.5 Ethical Evolution

For years we have been in search of a name for the emerging evolutionary stage, calling it post-industrial, post-business, post-bureaucratic, post-modern, among others. Chaisson confronts this "naming" issue as he says that "if our species is to survive to enjoy the future, then we must make synonymous the words "future" and "ethical," thus terming our next evolutionary epoch "ethical evolution."

> Since we have recently become the agents of change on Earth, we must now begin playing an active role in the process of evolution. And I maintain that this active role must begin with a collectively recognized set of ethics or principles, suited to the preservation of all humankind. (p. 201)

If we wish to build a civilization of humane peace and order, we should establish—says Chaisson—a set of social norms and abide by it.

We have called these norms social or cultural evolutionary norms but they should be termed norms of *ethical evolution*. We must now thus behave ethically, more than just as social individuals. Exerting increasingly more influence over our own destiny, we must develop a learning and living agenda of ethical evolution.

Chaisson holds that the implications of the Life Era in conscious evolution is the attainment of global, even cosmic, ethics.

> We must now develop an integrated wordly culture, including a unified political–economic ideology—which is not just a hackneyed proposal for a world government—if we as a species are to have a future. What's more, only by thinking big and embracing change can we ensure the onset of a planet-wide program of ethics. Indeed, if we act wisely, quite beyond intelligently, then an epoch of something resembling "ethical evolution" should naturally emerge as the next great evolutionary leap forward in the overall scheme of cosmic evolution itself. Yet, it is an evolutionary advance dependent upon technologically talented life forms—namely, us the new agents of change—to plan and implement, lest we succumb to the seeds of our own destruction. (p. 8)

If we learn to appreciate the importance of cosmic evolution sufficiently to embrace global ethics, we shall prevail. Only if we welcome, even seek planetary citizenship as an aspect of "unitary globalism" shall we realize fully the Life Era.

7.2.5 Elgin: Reflective Consciousness in the Era of Communication and Reconciliation

Elgin (1993) describes the emerging evolutionary stage as the Era of Communication and Reconciliation. (Earlier, we named this emerging stage the Fourth Generation of HSS.) Elgin suggests that this now emerging stage is guided by reflective consciousness as our observing capability is "turned back upon itself." Reflective consciousness emerges as we are able to approach life with a measure of detachment. This new level of consciousness enables us to reflect upon the critical problems created by the industrial machine age society. A detached perspective "will support the process of local-to-global communication that can lead to reconciliation around a shared vision of a sustainable future" (p. 115).

It is the first time in societal evolution that humanity as a whole has to work together to create its future. It is for us an unfamiliar territory. But if we are willing and able to take on this task we shall become pioneers in guiding the evolutionary design of the Fourth Generation of HSS.

In venturing into the new territory of conscious evolution, we have two major challenges, namely, to develop a collective global consciousness that will heal the separation that now divides us, and overcome our separation from nature and reconnect with it by becoming its trustees rather than its exploiters. The collectively emerging reflecting consciousness "represents a powerfully evolutionary advance because it enables us to take charge of our social (and societal) development with a new level of intentionality. When we see our actions in the mirror of reflective knowing, we can become self-directed agents of our own evolutions. And this promotes accountability, responsibility, and follow-through in personal and social action" (p. 119).

It will be helpful at this point to remind ourselves that it was the emergence of self-reflective consciousness that brought forth our ancestors from a dreamlike state to the sunshine of "awareness of awareness." But the essential nature of self-reflective consciousness has changed through the various stages of societal evolution. Elgin (1993) explains this change in the larger context of societal evolution. He suggests that self-reflection was operating at the various evolutionary levels since the emergence of the Cro-Magnons. Their reflective consciousness was *collective sensing consciousness*. It was focused on their physical existence, their sensations, and on the here-and-now. Many thousands years later, to the people of the agricultural revolution, self-reflection meant reflecting upon their emotional life and the possibilities of the future. It was *collective feeling consciousness*. It took several more thousands of years before a new kind of self-reflective consciousness—the ego-focused mental consciousness emerged. It created and animated the scientific industrial area. And now we stand at the threshold of the era of communication and reconciliation, the era of *collective reflective consciousness*, when "the capacity of self-reflective consciousness takes another quantum-leap forward as humanity consciously recognizes the *existence* of consciousness" (p. 124). (When we attain evolutionary consciousness, it enables us to engage in conscious evolution.) Next, I review Elgin's coevolutionary view and his proposal for global reconciliation.

7.2.5.1 Elgin's Coevolutionary View

Elgin (1993) suggests that from a coevolutionary point of view, evolutionary development involves the interacting refinement of the material and the consciousness sides of life. Material refinement means "learning to do ever more with ever less" and refinement of consciousness calls for learning "to touch life ever more lightly and gently." He says that:

> The synergy achieved through the simultaneous development of the material and consciousness aspects of life is enormous. Material evolution provides the physical necessities of life that support the widespread realization of psychological and spiritual potentials. Conscious evolution provides the insights, compassion, and creativity needed to orient material development into

> sustainable pathways for the future. Co-evolution is a balanced, middle path that integrates matter and consciousness into a co-evolving spiral of mutual refinement. Rather than competing with each other, the material and consciousness aspects are mutually enabling. (p. 193)

The coevolutionary process described in the preceding quote is synergistic. (Earlier we described both coevolution and synergy as two of several essential conditions of a healthy evolutionary stage.)

As we bring the consciousness and material aspects of evolution into a coevolutionary and integrated state, we can enter into an evolutionary journey (of the Fourth Generation of HSS) that would not be possible by focusing on either aspect in isolation. Such an integrated approach promotes balance and harmony among the evolutionary markers. The coevolutionary approach to conscious evolution described here will unfold an "ever more diverse and simultaneously unified whole" (p. 294).

7.2.5.2 Global Reconciliation

"A world divided against itself is a recipe for global collapse" (p. 121). Recalling our portrayal of the declining end of the age of the scientific–industrial era, we recognized such a collapse as one that endangers human life on the global scale. Reflective consciousness that ponders consciousness itself enables us to reconcile what divides us and create integrated global systems that will unite us. [Earlier, Combs (1996) described the consciousness of the Fourth Generation of HSS as the era of "integral consciousness."] The notion of integral or integrated consciousness has relevance to Elgin's idea, namely, that in the era of global reconciliation, we can focus on cooperative action that reconciles and eventually integrates several areas of our existence. He defined six such areas.

7.2.5.2.a Spiritual Reconciliation. Spiritual reconciliation will overcome religious intolerance, which produced—and is still producing—some of the bloodiest societal conflicts. It is crucial to our future to reconcile the various religious traditions, and to learn to appreciate and respect the core values and ideas of each that together stand for the richness of the human spirit.

7.2.5.2.b Racial, Ethnic, and Gender Reconciliation. Racial, ethnic, and gender discrimination dangerously divide us. Thus, their reconciliation is an essential condition of building a common culture. We can build a global culture and a satisfying future only through genuine cooperation across racial, ethnic, and gender boundaries.

7.2.5.2.c Ecological Reconciliation. Ecological reconciliation calls for living with nature as our partner. The integrity of the Earth depends upon learning to live with nature in total harmony. "To move from indifference and exploitation to reverential stewardship will require reconciliation with the larger community of life on the Earth" (Elgin, 1993, p. 112).

7.2.5.2.d Generational Reconciliation. Generational reconciliation recognizes that we are indeed trustees of future generations. It means that we are to meet our needs without endangering or compromising the ability to limit the options of future generations. A remarkable example of trusteeship of future generations is the custom of the Iroquois, who made decisions by considering the impact of decisions on seven generations to come.

7.2.5.2.e Political Reconciliation. Political reconciliation means bringing true democracy to life as an expression of a human community of self-determining individuals and communities. It means the reconciliation of the different values and views among the various cultures so that we can develop a healthy conversation locally in the communities and systems in which live, as well as globally among societies and cultures.

7.2.5.2.f Economic Reconciliation. Today there are enormous gaps between the resources of the poor and those of the rich. A crucial task in the era of reconciliation is to narrow this disparity. Those who have much also have the responsibility to simplify their participation in the material side and "shift increasing amount of energy into psychological, cultural, and spiritual growth while at the same time assisting developing nations" (p. 122).

Elgin (1993) suggests that to advance reconciliation, we need to attain an unprecedented level of understanding of each other, and an ability to communicate with each other at all levels of the society so that we can work together cooperatively.

> Yet, it is often very difficult for a person, community, or nation to accept responsibility for their past excesses and to seek new and healthy relationships. The world is filled with instances of genocide, religious intolerance, racial and gender discrimination, oppression of ethnic minorities, abuses of political and economic power, environmental destruction, and the extinction of other species. Some of these tragedies have festered for many generations and it makes the bridging of differences very difficult. Nonetheless, without deep and authentic communication across the barriers of suffering mistrustful humanity, our collective future will be gravely imperiled. (Elgin, 1993, p. 123)

In setting forth an agenda for reconciliation, Elgin envisions our collective future by proposing an evolutionary agenda for conscious evolution. Guided by systems thinking, I believe it is worthwhile for us to explore Elgin's six aspects of reconciliation, described previously, as a conceptual system of interacting and internally consistent components of an evolutionary image of the future. Activity 23 addresses this exploration.

Activity 23
This activity has three tasks. The first challenges you to review the six components of Elgin's set of reconciliations and ask the question: Is the set complete? Are there other aspects that invite reconciliation? If yes: What are those and how would you define them? The next task calls upon you to place the components of the agenda in a circle and seek to define interactions among them by asking: Do these components mutually reinforce each other? Are they internally consistent? The third task is to make a projection. Ask yourself: What would happen if we would not address those six areas of reconciliation? Introduce your findings in your workbook.

7.2.6 Jared Diamond: The Global Issue

All five authors introduced earlier emphasized the need to consider the global system and the global systems of humanity as we contemplate conscious evolution. The work of Diamond (1998) sheds light on this consideration and helps us to appreciate the set of issues involved in global evolution. First, I briefly describe the genesis of Diamond's interest in addressing global evolution; then, I interpret his core ideas on the complex set of issues involved.

7.2.6.1 Yali's Question

> Since the end of the Ice Age, some parts of the world developed literate societies with metal tools, other parts developed only nonliterate societies with farming tools, and still others retained societies of hunter–gatherers with stone tools. Those historical inequalities have cast long shadows on the modern world, because the literate societies with mental tools have conquered or exterminated the other societies. While these differences constitute the most basic facts of world history, the reasons for them remain uncertain and controversial. (Diamond, 1998, p. 13)

Diamond confronted this puzzling predicament in a personal way when he was conducting evolutionary research in New Guinea. One day, walking on the beach of one of the tropical islands, he met Yali, a political leader of New Guinea. They had a long conversation about the inequality of white and black societies.

Background. Two centuries ago, the people of New Guinea lived in the Stone Age, inhabiting small villages in an egalitarian society, using stone tools. When the colonizing whites arrived they imposed centralized authority. They brought steel axes, matches, clothing, and other goods which the Guineans called collectively "cargo." Diamond and Yali knew well that New Guineans are at least as smart as the Europeans. So Yali asked: "Why is that you white people developed so much cargo and brought it to New Guinea, but we black people had little cargo of our own" (Diamond, 1998, p. 15)? Diamond suggests that it was difficult to answer Yali's question. Still, in his book, *Guns, Germs and Steel: The Fate of Human Societies*, written twenty-five years after his walk with Yali on the beach, he provided a comprehensive answer to Yali's question.

As we contemplate conscious evolution on a global scale, it is helpful to follow Diamond's reasoning. Yali's contrasting the life style of the whites and blacks can be extended in viewing the larger global context. People of Europe, North America, and Eastern Asia have evolved to dominate the world. Today we consider them collectively as the "developed" industrial society. Much of the rest of the world was subject to colonization, subjugation, and even slaughter. Given this picture, Diamond reformulates Yali's question: "Why did wealth and power become distributed as they now are, rather than some other way? For instance why weren't Native Americans, Africans, and Aboriginal Australians the ones who decimated, subjugated, or exterminated Europeans and Asians" (p. 15)? Or, looking at our present world's inequalities we can ask: Why did societal evolution produce such different rates of development on different continents? And, in the context of our present exploration: What are the implications of these inequalities and differences in the rate of development to conscious evolution on the global scale?

7.2.6.2 Diamond's Answer

In his book, through some four hundred pages, Diamond leads us on a journey through the continents and sets the stage for answering Yali's question. (I suggest that you get hold of the work and enjoy the journey. The book traces human and societal evolution as we did in Chapter 3, but from the special perspective of the question raised by Yali.)

Diamond suggests that the differences between the evolutionary development of people on the different continents have been due to many factors. Rather than providing us with a long list, he highlights the most important ones, organized in four sets.

7.2.6.2.a Environmental Differences. Differences in their environment rather than innate differences lead to the striking differences between the long-term evolutionary development of peoples of the various continents. These differences relate to the availability of wild plant and animal species for domestication. As noted earlier, it was the production of surplus food in the agricultural villages that made possible the establishment of ancient civilizations. Surplus production was critical to feed the non-food-producing specialist, to build large settlements, which enabled the attainment of military advantage and, then, to political and technological advantage. "All developments of economically complex, socially stratified, politically centralized societies—beyond the level of small nascent chiefdoms—were based on food production" (p. 406).

As it turned out, there were striking differences in the number of wild plant and animal species available for domestication among the various continents. This was coupled with the Late Pleistocene extinction of species. "As a result, Africa ended up biologically somewhat less well endowed than the much larger Eurasia, the Americas still less so, and Australia even less so"(p. 406).

Rejecting the issue of innate differences as a factor in developmental differences, Diamond tells us his own experience while working with the New Guineans for 33 years. He says that:

> They impressed me as being on the average more intelligent, more alert, more expressive and more interested in things and people than the average European or American is. At some tasks that one might reasonably expect to reflect aspects of brain function, such as the ability to form a mental map of unfamiliar surroundings, they appear considerably more adept than Westerners. Of course New Guineans tend to perform poorly at tasks that Westerners have trained to perform since childhood and that New Guineans have not. Modern European and American children spend much of their time being passively entertained by television, radio, and movies. In contrast, traditional New Guinea children have virtually no such opportunities for passive entertainment and instead spend almost all of their waking hours actively doing something, such as talking or playing with other children or adults. (pp. 20–21)

(It is well known that child development studies indicate the beneficial effects of intensive social interaction activities on mental development.) I now return to Diamond's answer to Yali's question.

7.2.6.2.b Rates of Diffusion and Migration within a Continent. Rates of diffusion and migration within a continent comprise the second set of factors that have differed greatly among continents. They were most rapid in Europe and Asia because the ecological barriers on the Eurasian continent were rather modest. This condition enabled the movement of crops and lifestock as well as the diffusion of technological and cultural innovations.

Diffusion was slower in Africa and even more so in the Americas because of the existence of north–south ecological and geographic barriers. The same has been true in traditional New Guinea on account of the "long backbone" of high mountain terrains and ecological barriers which prevented any significant progress toward cultural, political, and linguistic unification.

7.2.6.2.c Rates of Diffusion and Migration between Continents. Rates of diffusion and migration between continents is the third set of factors that influences evolutionary development. The rates of diffusion and migration depend upon the degree of isolation vs. ease of interaction between continents. For thousands of years, such interaction was easiest between Eurasia and Africa, enabling the supply of much of lifestock to Africa. Until a few hundred years ago, intercontinental diffusion between Eurasia and the Americas was blocked by wide oceans. In the same way, Aboriginal Australia was isolated from Eurasia by the Indonesian Archipelago.

7.2.6.2.d Continental Differences in Population Size. Continental differences in population size constitute the fourth set of factors of evolutionary development. "A larger era of population means more potential inventors, more competing societies, more innovations available for adoption—and more pressure to adopt and retain innovations, because societies failing to do so will tend to be eliminated by competing societies" (p. 407). The African pygmies and many other hunting–gathering societies were replaced by farming societies. Competing societies were most numerous and largest in Eurasia and smallest in Australia and New Guinea.

From Diamond we can learn that evolutionary development, human creativity, and its products invention and design are the hallmarks, the enablers of societal development. "Without human inventiveness, all of us would still be cutting our meat with stone tools and eating it raw, like our ancestors of a million years ago. All human societies contain inventive people. It is just that some environments provide more starting materials, and more favorable conditions for utilizing inventions, than do other environments" (p. 408).

Diamond has made a significant contribution to explaining evolutionary consciousness, particularly consciousness about factors that have led to significant differences in evolutionary developmental and inequalities among peoples of the various continents. As we contemplate the conscious evolution of our species, especially of a global society, we should be sensitive to Diamond's message and develop strategies that (further) remove the barriers that have produced those inequalities and enable the development of a global human community of the Fourth Generation of HSS.

7.2.7 Fukuyama: The Great Disruption

In Chapter 6 and in the present chapter we have become increasingly more aware of our current evolutionary predicament that finds us struggling to find our way between two generations of HSS. Today, we live in the twilight zone between the decline of the industrial era of the Third Generation of HSS and the emerging Fourth Generation.

Fukuyama (1999a) designates what we called the "evolutionary twilight zone" as the era of "The Great Disruption." He suggests that the shift away from the industrial society toward the post-industrial information knowledge age has created a great disruption, which is manifested in troubling social disorder. In his work, he explores the key markers of this social disorder. At the same time, he sees signs of the reconstitution of social order that emerges through our conscious action.

Fukuyama's work is highlighted as follows. First, I provide an overview of his general characterization of the Great Disruption. Then, I discuss the current shift from "Gemeinschaft" to Gesellschaft. Third, I focus on what Fukuyama called "unbridled individualism." Then, the contradictory desires of autonomy and community are discussed. In conclusion. Fukuyama's approach to the reconstruction of social order is considered.

7.2.7.1 An Overview of the Characterization of the Great Disruption

Fukuyama places the period of the great disruption between the mid 1960s and the 1990s. This period, says Fukuyama, was marked by seriously deteriorating social conditions, rising crime, and social disorder that made inner-city areas almost unhabitable. Kinship as a social institution has been declining over the last two centuries, a process that has accelerated sharply during the last forty years. Marriage rates have declined and divorce has soared. In the United States one out of every three children—and in Scandinavia more than half of the children—were born out of wedlock. Trust in governments and institutions declined over this time. People's associations with each other now tend to be less permanent, looser, and occurring in smaller groups.

These changes are dramatic and constitute a great disruption in social values. "The perceived breakdown of social order is not a matter of nostalgia, poor memory, or ignorance about earlier ages. The decline is readily measurable in statistics in crime, fatherless children, broken trust, reduced opportunities for—and outcomes from—education, and the like" (p. 56). Fukuyama's hypothesis is that all these negative social trends, which reflect a weakening of common values, have been intimately connected with the transition from the industrial to the information age.

> Although many blessings have flowed from a more complex information-based economy, certain bad things also happened to our social and moral life. The changing nature of work tended to substitute mental for physical labor, propelling millions of women into the market place and undermining the traditional understanding on which the family had been based. Innovations in medical technology, leading to the birth control pill and increased longevity, diminished the role of reproduction and family in people's lives. And the culture of individualism which in the laboratory and the market place leads to innovation and growth, spilled over into the realm of social norms, where it corroded virtually all forms of authority and weakened the bonds holding families, neighborhoods, and nations together. (p. 56)

The complete picture, says Fukumaya, is more complex than what was said in the preceding quote and varies from one place to the other. Schumpeter called the technological change "creative destruction" in the marketplace. But it also caused creative destruction in the realm of social relationships.

7.2.7.2 The Shift from Gemeinschaft to Gesellschaft

Fukuyama further elaborates the shift from accumulated social norms, habits, and customs. To highlight this shift, he refers to the distinction made by the sociologist Ferdinand Tonnis who distinguished between two contrasting styles of social systems: Gemeischaft and Gesellschaft. In the *Gemeinschaft* type, social relationships among members of a group or community are valued as ends in themselves. The actions of members are grounded in and express underlying communal identification and shared values and beliefs. Individuals are bound together in mutual interdependence that touches all their life experiences. The relationships in the *Gesellschaft* type of social organizations, on the other hand, are impersonal, atomistic, and mechanistic. Values are considered to be means to an end. They are regulated by laws and standard operational policies. The social order in the industrial societies was the Gesellschaft type which still dominates the current scene. Fukuyama wonders whether we can maintain social order—and what kind—in the face of the technological and economic transformation that has been brought about by the post-industrial information age. "Social norms that work for one historical period are disrupted by the advances of technology and the economy, and society has to play catch-up in order to establish new norms" (p. 59).

7.2.7.3 Unbridled Individualism

Fukuyama suggests that the tendency of current liberal democracies "to fall prey to excessive individualism" is perhaps their greatest vulnerability. It is based on the notion that in the interest of "political peace" government would tolerate pluralism of opinions about the most salient ethical and moral issues. "In place of moral consensus would be a transparent framework of laws and institutions that produce political order. Such a political system did not require that people be particularly virtuous. They need only to follow the law in their own self-interest" (p. 58). But although the existence of law and political institutions is important, they do not guarantee a viable and successful society. "To work properly, liberal democracies have always been dependent on certain shared cultural values" (p. 58). In the United States, left and right blame each other for the other's excessive viewpoints. Those who speak for reproductive choice want to restrict choice in guns and those who are for unlimited consumer choice cry out against laws of consumer protection or environmental restraint, and on and on.

But by now many of us recognize the pitfalls of "unbridled individualism" in which "the breaking of rules becomes, in a sense, the only remaining rule" (p. 59). Shared moral values and social order are preconditions for social cooperation. We refer to society's stock of shared values as "social capital."

> Individuals amplify their own power and abilities by following cooperative rules that constrain their freedom of choice, because these also allow them to communicate with others and coordinate their action. Social virtues, such as honesty, reciprocity, and the keeping of commitments are not worthwhile just as ethical values; they have a tangible dollar value and help the groups that practice them to achieve shared ends. (p. 59)

Another problem with excessive individualism is that it does not allow for the creation of genuine communities. Communities emerge when their members are bound together by shared values and norms, by shared goals, expectations, and experiences. The deeper these are the stronger is the community. A society that constantly updates its norms and rules in order to expand individual freedom of choice, "will find itself increasingly disorganized, atomized, isolated, and incapable to carry out common goals and tasks" (p. 60). In such situations, social ties are becoming less and less binding and increasingly less long-lasting. There is an escalating decline in trust of social institutions with a shift toward fewer shared norms and values. True communities are becoming fewer and smaller, having less hold on their members. All these lead to "moral individualism and the consequent miniaturization of community" (p. 72). The end result is great loss of "social capital." Social capital can be defined (Fukuyama, 1999b, p. 299) simply as a set of informal values or norms shared among members of a group that permits cooperation among them. If members of the group come to expect that others will behave reliably and honestly, then they will come to *trust one another*. "Trust is like a lubricant that makes the running of any group or organization more efficient."

7.2.7.4 Reconstructing Social Order

Fukuyama asks: What can we do in face of ever increasing disruptions, disorder, atomization, moral decline, and anarchy? He suggests that the solution is the reconstruction of social order and the rebuilding of social capital. His line of reasoning follows.

We humans are "designed" to formulate moral rules and create social order for ourselves. There is an increasing body of evidence, says Fukuyama, that "human beings are born with pre-existing cognitive structure and age-specific capabilities of learning that lead them naturally into society. There is, in other words, such a thing as human nature. We are political and social creatures with moral instinct. Having this insight is extremely important, because it means that social capital will tend to be generated by human beings as a matter of instinct" (p. 76).

As we look back to our exploration of our evolutionary journey, we can reflect upon Fukuyama's reasoning. We are social beings with a human and a social nature that was created in the evolutionary design space of our species and that is our evolutionary gift. We have repeatedly called this gift our "evolutionary potential." It is this potential, which through our evolutionary history—once accessed, activated, and realized—enables us to advance evolution. Periodically, in the course of our evolution great disruptions occur such as the one we suffer today. In such times we lose our evolutionary balance and harmony and our sense of bearing; we lose our "soul," our collective consciousness. We find ourselves in such an evolutionary predicament today. But we can regain our humanity and evolutionary balance and harmony by forging collective consciousness, by reconstructing social order, and by rebuilding social capital. These are core tasks of conscious evolution.

In addition to the recognition of human nature as a source of supporting the (re) building of social order and (re) creating social capital, Fukuyama names human reason as another key source of social reconstruction. It is human reason's ability to spontaneously generate solutions through social cooperation.

> Mankind's natural capabilities to create social capital do not explain how social capital arises in specific circumstances. The creation of particular rules of behavior is the province of culture rather than nature, and in the cultural realm we find that order is frequently the result of a process of horizontal negotiation and dialogue among individuals. Order does not need to proceed from the top down—from a lawgiver (or, in the contemporary terms, a state) handing down laws or a priest promulgating the word of God. (p. 77)

Self-organizing groups of individuals have been continuously creating social capital for themselves throughout our evolutionary history. They have managed to respond creatively to the challenges of past disruptions.

Today, says Fukuyama, two parallel processes seem at work. "In the political and economic sphere, history appears to be progressive and directional, and at the end of the twentieth century has culminated in liberal democracy as the only viable choice for technologically advanced societies" (p. 80). However, in the social and moral realm a cyclic dynamics seems to operate with social order ebbing and raising. *Today, we have reason to hope that the very powerful innate capacity for constructing social order, our self-organizing capability to generate social capital, and our desire and aspiration to forge a new collective consciousness will guide us into the future.*

I firmly believe that Fukuyama's reasoning is well justified in that we can claim that our attainment of evolutionary consciousness, our awareness of evolutionary laws, and our understanding the overall pattern of evolution, coupled with our collective willingness to engage in conscious evolution, will empower us to create the new social order and a new society for the Fourth Generation of HSS.

Reflections

In this chapter we visited with a good number of evolutionary scholars, We have listened to scholars of the systems sciences, evolutionary scholars of spiritual orientation, philosophers, scholars of the social potential movement, palaeontologists, and social scientists, all calling for conscious or guided evolution. As I reviewed their statements, I found ample calls for conscious evolution in the evolutionary literature. Most of the statements provide a definition of conscious evolution, explain the reason for it, or project expected evolutionary outcomes. They tell us *what a proposed evolutionary transformation is and what it should lead to*. Their statements address the *"what's"* and *"why's"* of conscious evolution. With few exceptions, I find a lack of attention paid to exploring strategies and methods of *"how" to engage in and how to pursue conscious evolution*. This lack of attention to the *"how"* has motivated me to focus my present work on the *"how" of conscious evolution, on the evolutionary epistemology of guided evolution.*

Activity 24

In the present chapter we have visited with scholars, who—grounded in an understanding of evolutionary consciousness—testified to the need for and an urgency of engaging in conscious and directed evolution. You are now asked to engage in two tasks.

Task 1. Capture the core ideas of scholars presented in this chapter, work with their ideas, formulate your own understanding of conscious evolution, and create a text that presents your proposition of conscious evolution.

Task 2. Select out from the content of the chapter those ideas that pertain to the issue of *the epistemology of conscious evolution,* the issue of *"how" to proceed with conscious evolution.* Enter your findings in your workbook.

III

Self-Guided Evolution

> *The right of people to guide their destiny, to take part directly in decisions affecting their lives; to create authentic, caring, nurturing, and healthy communities; to control their resources; to govern themselves; and design their future* ***is a most fundamental human right.***

Evolutionary consciousness can motivate action toward conscious evolution. To attain evolutionary consciousness has been the purpose of working through Parts I and II. When I developed the idea of this book, I asked myself: How can we engage in conscious evolution unless we know how evolution works? So, we should consider evolutionary consciousness as being the stepping stone toward conscious evolution.

Conscious, self-guided evolution enables us to use our creative power to guide our own lives and the evolution of the systems and the communities in which we live and work. It is a process by which individuals, groups, organizations, and societies envision and create images of "what should be" and bring the image to life by design. This "creating" is based on the belief that, while the future is influenced by the past and present, it is not determined by it. Its shaping remains open to individual and collective conscious and purposeful action. The only constraint on our evolutionary design is a conscious consideration of the effect our decision shall have on future generations. We must avoid any decision that would close out options for them. They should be free—unencumbered—to *design their own lives and shape their own futures.*

Conscious, self guided evolution is a process of giving direction to the evolution of human systems and developing in these systems the human capability and organizational capacity to:

- nurture the physical, mental, emotional, and spiritual development and the self-realization ethics of individuals and their systems;
- extend the boundaries of social and economic justice and genuine civic participation;
- increase cooperation and integration among societal systems;

- honor and integrate diversity by creating healthy and authentic communities; and
- engage in the design of social and societal systems that serve the common good.

In Part I, I gave an account of the evolutionary journey of our species and told the story of the first three generations of *Homo sapiens sapiens*. In Part II, I retraced the evolutionary story of our species and asked: What can we learn from it? Answering this question helped us to attain evolutionary consciousness. We became conscious of the ways and means of human and societal evolution and—more importantly—we have learned and understood how evolution works in the evolutionary design space. Such understanding, the knowledge of the laws and the principles of evolution we attained, and the grasp of the workings of evolution have given us powerful insights. These insights will guide us now as *we move from reflection to creation* as we design and develop ways and means by which we can engage in conscious evolution.

Earlier, I used the biblical metaphor of eating the fruit from the tree of knowledge and its consequence of being ordered out of Eden. The first *"out-of-Eden"* event of our species happened some 35,000 years ago when we stepped out from a dreamless, slumberlike state into the sunshine of *"knowing that we know"* and received the gift of self-reflecting consciousness. We became *Homo sapiens sapiens: the double wise man*. But the burden of this knowledge and gift was to *take responsibility for our lives*. It took our species 35,000 years more to experience the second *"out-of-Eden"* event, when we attained the *new knowledge of "knowing that we know how we evolved." And, we attained the gift of self-reflecting evolutionary consciousness.* This knowledge and gift compel us to *take responsibility for our future*. We are standing at the threshold to become *Homo sapiens sapiens sapiens: the triple wise man, provided we use our wisdom and accept the responsibility for guiding our own evolution.*

The overall theme of Part III is *not* the story of the *what* the future of the Fourth Generation HSS should be. No one has the right to tell that story. This *new story* of humankind will be woven in families, in neighborhoods, in the systems in which we live and work, in our communities, in our myriad of social systems, in our societies, and ultimately, in our global community. Weaving the tapestry of the story of the future will be accomplished in the course of our continuing evolutionary journey. But *where we go there are no roads yet*. We have to build those roads.

The topic of this work thus is not about the *what is* or *what should be* our evolutionary future. It is about the *how* we might build the road toward that future, how we might write the *new story of evolution*. It is about how we might go about designing and testing evolutionary ideas and images in the evolutionary design space of the Fourth Generation of HSS. It is about exploring and *creating an evolutionary epistemology* that we might consciously and purposefully apply in the design the Fourth Generation of HSS.

Part III is developed in five chapters. In Chapter 8, I present *organizing perspectives* that may guide our thinking about conscious evolution. I will suggest that we have already stepped over the threshold of creating the Fourth Generation HSS. I will propose that conditions of acting upon conscious evolution is the acquisition of individual and collective *evolutionary competence,* and mustering the will and determination to engage in conscious evolution. In Chapter 9, I develop the four domains of evolutionary inquiry, explore the knowledge base for building an evolutionary epistemology, and ask: Who should be the evolutionary designers? In Chapter 10, our exploration will lead us to introduce the three major strategies of evolutionary epistemology: *transcending* the existing state of evolution, *envisioning an image* of our evolutionary future, *and transforming* our systems by *bringing the image to life by the design of evolutionary guidance systems* (EGS). In Chapter 11, I elaborate the approach and methodology of *the design of EGSs* and engage in open-ended exploration of ways and means by which we can *add value to evolutionary epistemology, increase its power, and enhance the design of EGSs.* In Chapter 12, I ask: How would an evolutionary community work in the twenty-first century? What happens if the idea of evolutionary guidance systems becomes reality? What happens if our families, our social systems, and our communities, our societal institutions would purposefully engage in the design of their own evolutionary guidance systems?

8

Organizing Perspectives and Conditions of Engaging in Self-Guided Evolution

In Section 8.1, I introduce key evolutionary perspectives that have the potential to guide our thinking about creating an evolutionary epistemology, and selecting approaches and methods of guiding evolution. In Section 8.2, I propose that we have already stepped over the threshold of conscious evolution. We now see at many places and in many forms the emergence of groups and communities, whose ideas and work manifest conscious evolution. However, these groups are disconnected from each other. In Section 8.3, I propose that conditions of acting upon conscious evolution include the individual and collective acquisition of evolutionary competence, and the development of will and determination to engage in conscious evolution.

8.1 Perspectives that Help Us to Organize Our Thinking About the Process of Evolutionary Design

The role of organizing perspectives is to help to shape our thinking about the conscious design and development of the Fourth Generation *Homo sapiens sapiens* (HSS). These perspectives are derived from the evolutionary knowledge base, introduced in Parts I and II. Working with these parts, you have attained evolutionary consciousness by interpreting the evolutionary knowledge base, and by exploring statements of evolutionary scholars who called for conscious evolution. The perspectives are relevant to both the *what* of evolution—to the general characteristics of evolution—as well as to the *how* of evolution, namely, to evolutionary epistemology. Here I focus on the *how*.

8.1.1 Complexity

The most often used definition of evolution is that evolution is a process of reorganization at ever higher levels of complexity. Sociocultural evolution has unfolded in the larger context of cosmic evolution as biological and cultural constructions. This occurred through a series of changes that brought forth reorganization at ever higher levels of *complexity*. As we have seen, each successive evolutionary advancement has produced an increase in *differentiation* (having differentiated evolutionary markers) and an increase in *integration* of the differentiated, emerged elements (evolutionary markers) of the evolving entity. Complexity, thus, can be defined as a state of integrated differentiation.

The first large-scale increase in complexity in the evolution of our species happened when our biological evolution came to end and cultural evolution began to guide our evolutionary advancement. This quantum leap brought forth explosive differentiation that, then, became integrated in the life experiences of the Cro-Magnons, the First Generation of HSS. Now we are at the threshold of another quantum leap: The conscious evolution of the Fourth Generation, where we not only face, but will also create the second large-scale increase—possibly the largest increase—in the complexity of our species.

What are the implication of complexity for evolutionary epistemology? First, we need to not only understand the working of complexity (in evolution) but also have to learn to be comfortable with it and how to create it. If we take over the work in the evolutionary design space, we have to develop an evolutionary epistemology that is capable of creating complexity. We have to learn to work with and bring forth diversity and integrate it as an emerging quality of the Fourth Generation HSS.

8.1.2 Nonlinear, Discontinuous Transformation by Transcendence

Evolutionary change brings forth a new entity. This change is *nonlinear* and is *discontinuous*. As we have seen repeatedly, change between two evolutionary generations of HSS is *transformative*, not transitional. We cannot extrapolate the characteristics of an emerging evolutionary entity from the one that preceded it. There is "big change," a quantum jump, between levels that results in discontinuity. Fossil records show that evolution had been static for most of the time, but this state of equilibrium is punctuated by rapid change that occurred suddenly over a brief period of time. Such a swift change cannot be predicted from earlier evolutionary states. The implications of such punctuated equilibria, the transformative nonlinear nature of change, and the discontinuity between two generations have far reaching consequences for conscious evolution and for evolutionary epistemology. Only an ideal seeking approach can envision an ideal future image of an evolutionary quantum jump.

Today, we witness large-scale and desperate attempts of many—at all levels of the society—to hang on to the dysfunctional and outdated institutions and the destructive ways and means of the declining Third Generation. Most of us are cursed with *trained incapacities* (Margulis, 1999) to embrace change. It gives us security to stick to old ways of thinking and acting. It is easier to cling to the familiar and to be comforted by the conformities of the past. If it becomes a choice to change or not to change, people find all the reasons and arguments why we should not change. If we are pressed for change, at best we try to improve our systems that are grounded in the dying industrial machine age and are totally out-of-sync with new realities.

Standing at the threshold of the Fourth Generation, one of our most important and most difficult initial tasks is *to transcend and leave behind us* all manifestations of the machine age thinking of the Third Generation, so that we can engage in the collective creation of the Fourth Generation.

8.1.3 Life Cycles of Evolutionary Generations

An evolutionary quantum jumps bring forth a new evolutionary entity—a new generation of HSS. We have learned from the evolutionary story that these generations have their own *life cycles*. They suddenly emerge, fueled by a creative surge. Guided by a new level of consciousness, they *create their wholeness* by integrating their evolutionary markers. They attain maturity and establish a "standardized state" of habitual orders, customs, laws, and institutions. Then, *stasis sets in*. Change is resisted. Creativity becomes dormant. Much effort is exerted to maintain the status quo, despite changing forces in the environment. If these forces are disregarded, if there is no creative/self-organizing response to change, then devolution, decline, and eventually disintegration follows. The evolutionary qualities that created the new generation gradually degrade into disharmony and imbalance, and a loss of synergy and of internal consistency set in. Wholeness breaks down and a symmetry break moves the system toward a bifurcation point. At this point, the alternative to collapse is a quantum jump into the into a new evolutionary regime, by creating—through self-organization and self-guided design—a novel, a more complex, a new generation of HSS. And this is precisely the evolutionary state we find ourselves in at the end of the twentieth century, but with a big difference.

Past transformations were brought by evolutionary forces operating in the evolutionary design space. They were not created by the purposeful actions of people in the various generations of HSS. Those generations did not yet have evolutionary consciousness. But today we have attained it, and having attained evolutionary consciousness, it becomes our task, our responsibility, our burden and privilege to choose and create our evolutionary future. This again has far-reaching implications for conscious evolution. The most salient implication is that we have to enter into the evolutionary design space and with a purposeful creative surge give birth to the life of the Fourth Generation of HSS. But to do so, we have to devise an evolutionary epistemology that, coupled with a new level of collective consciousness, will enable us to design the new sociocultural systems of the Fourth Generation. A very important challenge of this epistemology is to bring forth processes that would avoid the set-in of stasis and rigidity and that would avoid decline and the loss of balance, harmony, internal consistency, synergy, symmetry, and wholeness. Another challenge (of the epistemology we are to create) is to ensure continuous creative action by which to create our evolutionary systems. *Evolutionary epistemology works in the evolutionary design space, and as we now enter into that space, we should understand what it is and how to create it, so that we can operate in the design space.*

8.1.4 The Evolutionary Design Space

Evolutionary design is a creating activity that brings forth a potential-driven, intended novel sociocultural system in the evolutionary design space. In this space, alternative design ideas are proposed and tested for their viability and for their "goodness of fit" with their enfolded potential and the environment which becomes their life space. In the evolution of our species, for more than six million years, the design experimentation was primarily biological, which resulted in the pruning-off of more than a dozen variations of our species. But since the emergence of modern man (the Cro-Magnons), the evolutionary process transformed from biological into cultural. Over the last 35,000 years, cultural evolution produced three generations of HSS. In cultural evolution idea systems are put up for testing in the design space and are confirmed or pruned off. For example, the idea system of representative democracy that has been so successful in the Third Generation will be pruned off in the Fourth, when new emerging idea systems of participative democracy, enabled by Internet technologies, will be designed.

8.1.5 Evolutionary Potential

We can regard the evolutionary bifurcation point, where a new variation emerges, as an evolutionary threshold, where a new generation of HSS bursts forth, which has the "enfolded" *evolutionary potential* of a new design idea. The idea is offered for evolutionary experimentation. The design idea can become "unfolded" into the design of the new evolutionary entity. If a new offering fails the experiment, it is pruned off. The pruning-off happens because the new offering was unable to realize its enfolded potential. Evolutionary potential is *not* fixed. It evolves through time. *New potential builds upon unfolded potential.* Potential "can learn" even in its enfolded state. The unfolded potential creates new (enfolded) potential, which at the next stage of evolution is manifested in a new, more advanced, and more complex entity. This new entity then can create more advanced enfolded potential. Evolution always offers new possibilities. It is neither predetermined nor is it predictable. From ensembles of possibilities of identical initial conditions, different variations/alternatives may emerge and are tested in the evolutionary design space. It is of utmost importance to understand the significance of evolutionary potential for conscious evolution. Today, we should be able to decipher potential that we have had embedded in the Third Generation, so that we can take advantage of those that are useful for the design of the Fourth Generation. For example, we can ask: How can we capitalize on our highly advanced technological intelligence so that it can become beneficial to evolution rather than becoming a destructive force as it often has been by bringing forth products that endangered life. Then: What *new potentials can we develop* by learning so that their employment will be beneficial to conscious evolution?

8.1.6 Limits and Constraints

The next perspective speaks to the *limits and constraints* operating in the evolutionary design space. Gould's metaphor of the left and right walls helps us to understand limits. Life emerged from the left wall some 3.5 billion years ago and the right wall is in the future. The right wall stands for the *possible limits* of how far evolution can go. Since its origin, the evolution of life moved toward ever increasing complexity. But how complex can life become? The question is: Is there a right wall in future human and societal evolution? Earlier, we observed constraints in the variability of life forms. As we contemplate conscious evolution, what limits and constraints should we consider? Or what now existing limits and constraints should we try to overcome that would limit conscious evolution? Should we consider any constraints at all?

8.1.7 Evolutionary Epistemology

In the life cycles of previous HSS generations, epistemology was a manifestation of evolutionary forces. But what these forces were was unknown to man. Now, as we became conscious of evolution and learn about these forces, we have to create an epistemology by which we can take over the functions of these forces so that we can engage in the conscious design of our own evolution. I am reminded of Erich Jantsch's statement: "We humans are the integral agents of evolution, we spearhead it on our planet and perhaps in our entire solar system. We are evolution and we are, to the extent of our power, responsible for it" (1981, p. 4). The last sentence of the quote is the core theme of this work.

If we engage in conscious evolution, we assume the role of guiding evolution. We can play this role and we can assume responsibility to the extent we acquire the power of evolutionary competence. Evolutionary competence has several interacting components, including evolutionary consciousness, the knowledge and understanding of the "*what*" of evolution, and the knowledge and understanding of "*how*" evolution works. Such knowledge becomes the basis for developing evolutionary epistemology which we can apply as we engage in conscious evolution. (In the last section of the present chapter, I introduce a proposition for the evolutionary learning agenda.)

The emergence of the image of a new generation of HHS unfolds by the surfacing evolutionary potential, by coevolutionary interaction with the environment, and by the integrating force of a newly emerging collective consciousness. Keep in mind that without new consciousness there is no possibility for the emergence of a new evolutionary generation of HSS. An example is the evolutionary unfoldment of the Third Generation. Animated by the new mental/rational ego-centered dynamic consciousness and the evolutionary potential generated by the Second Generation, the evolutionary image of the Third Generation was brought to life and became manifested in the scientific–industrial era.

The process of emergence is followed by the *process of creative self-organization* and integration of the various evolutionary markers into internally consistent sociocultural systems of a new generation of HSS. In the design of the evolutionary systems of the Fourth Generation the task of the organization of the evolutionary markers becomes the responsibility of the evolutionary designing communities. Carrying out this task will require the creation of an appropriate epistemology.

Evolution from one stage to the next happens by reorganizing life at a higher level of complexity. An increase of complexity is manifested in three domains. First, the number of evolutionary markers increases. (Revisit Chapter 4 and see such an increase between the Third and Fourth Generations.) Second, the complexity of the markers themselves also increases. Third, as the markers increase in number and complexity, as they interact with each other and creat synergy, the complexity of their relationship increases significantly. In contemplating the conscious evolution of the Fourth Generation, we can expect nothing less than an increase in the number and complexity of the markers as well as in the complexity of their relationships.

8.2 Organizing Perspectives of Conscious Evolution

In Chapter 6, ideas about conscious evolution were explored. In Chapter 7, we had a conversation with several evolutionary scholars who have called for conscious evolution. I reviewed those chapters to capture and assemble organizing perspectives that can be applied to explore their relevance to the formulation of evolutionary epistemology. As you read the following text, contemplate the implication of the statements to the creation of an evolutionary epistemology.

We can realize new evolutionary possibilities for evolving a planetary humanity by developing a common, shared thinking, said Teilhard. Gould suggests that we have an explosive, cumulative, and collective capacity for directional and rapid change. This capacity produces favorable innovations by the amalgamation of cultural features. It allows a society to choose from the most favorable and useful innovations. In the scheme of evolution, says Dennet, we are not ballistic trajectories, but guided missiles, capable to alter course, abandoning goals and formulating new ones. Huxley's evolutionary humanism seeks to unify the spiritual with the material. It is a dynamic and directional process in which we are engaged collectively. Laszlo implores us that only by conscious evolution can we have the assurance that we can survive the awesome combination of highly evolved order and complexity in our brain and immature and underdeveloped order and complexity in our societies. Dimond brings to our attention factors that led to significant differences in evolutionary development and created inequities among peoples of the various continents. He calls for the development of strategies that remove the barriers that have caused those inequalities and enable the just and equitable development of the global human community.

Csikszentmihalyi headlines the theme of this work by saying that our societal way of life has been the result of random change of events and now we realize that unless we take things in hand, the process of change will continue to be entirely blind of human desires. If there is a central task in the next millennium, he says, it is to start on the right track in our effort *to control the direction of evolution*. He developed the concept of "flow," which is a process of discovery and producing novelty. It is the creation of what we aspire to bring about. It challenges our potential; it invites us to stretch ourselves. Flow can become an important force in the arena of evolutionary creativity. Those who possess the skills and the competence of flow will be competent to engage in conscious evolution.

Hubbard speaks about *quantum transformation* in evolution, a leap out from the existing evolutionary stage into a new one. When an evolutionary stage reaches its limitations and begins to decline, it creates disruption, chaos, and social disorder. At the decline state an evolutionary system cannot change incrementally; it has to transform into a new evolutionary entity by conscious evolution. I suggest that we can be part of such a quantum transformation by combining our technological capacity with an evolved spirituality and the ability to design synergistic social systems. Today, our social innovations are fragmented, dispersed, and often ignored. So we are to seek ways to connect them and organize them into a comprehensive design of a positive future.

Jonas Salk calls upon us to merge intuition and reason and, based on it, select the path that leads to conscious evolution. It remains for us to decide the ultimate course of human evolution. He also suggests that it is individual mutualism that is the basis on which to organize ourselves for the future, which requires the collective to respect the individual and the individual to participate in the collective.

Chaisson holds that our challenge is to guide social, technological, scientific, political, and economic changes and become the agents of change. Chaisson proposes several aspects that integrate into the overall scheme of conscious evolution, such as the overall process of change; the evolutionary role of synthesis and consciousness; evolutionary humanism; the interface of science, philosophy, and religion; global evolution; and ethical evolution. As the dominant species, he says, we must develop a global culture, inspired by a planetary ethics that will guide us toward what is best for humanity.

Elgin suggests that venturing into the new territory of conscious evolution, we have to develop global consciousness that will heal the separation that now divides us, overcome our separation from nature, and reconnect with it by becoming its trustees rather than its exploiters. A main agenda of conscious evolution is multiple reconciliation; global, spiritual, racial/ethnic/gender, generational, and political and economic reconciliation. To advance reconciliation, we need to understand each other and communicate with each other at all levels of the society, so that we can work together cooperatively.

Fukuyama suggests that today we suffer a great disruption in our societies, coupled with a crumbling social order. He explores the key markers of this disruption, while he also sees signs of a reconstruction of social order. A key solution to reconstructing social order is the rebuilding of social capital. Fukuyama (1999b) defines *social capital as a set of values and norms shared among members of a group,* such as truth telling, the meeting of obligations, cooperation, and reciprocity. If members of the group come to expect that others (in the group) will be honest and reliable, they will come to trust each other. Trust is the lubricant that makes the work of the group more effective and efficient. Human reason is another key source of social reconstruction. Reason has the ability to generate solutions through social cooperation. Order is frequently the result of a process of horizontal negotiation and dialogue among individuals. It does not need to proceed from the top. Self-organizing groups have been constantly creating social capital. *To sum this up: Today we reason and hope that our powerful innate capacity for constructing social order, our self-organizing ability to generate social capital, and our desire and aspiration to forge a new collective consciousness will enable us to guide our evolutionary future.*

Activity 25
Develop your own essay on the characteristics of conscious evolution, without any reference to the authors. You are not limited to the specific statement made by the authors. Reflect upon them and construct your own text. As you do, speculate on the implication of these characteristics on the creation of an evolutionary epistemology. Enter your essay in your workbook.

8.3 The Dawning of the Fourth Generation

The seeds of change of the Fourth Generation have been already planted. At many places and in a great variety of forms and ways—but yet not connected—we can witness the beginning of a sea change, which—once connected and systemically integrated in the evolutionary design space—will bring forth the new evolutionary generation. In this section, I introduce "signs of the sea change" and tell the story of an emerging population of "culture creators."

8.3.1 Signs of a Sea Change

Much has been written in the course of the last decade about the dawning of a new era. We already heard about it from several evolutionary scholars in Chapter 7, who gave voice to conscious evolution. Here, I continue to explore the ideas of cutting-edge thinkers who see the dawning of the coming of a new evolutionary generation.

8.3.1.1 A Great Turning

Today, says Goerner (1999), a *great turning is going on*, affecting all spheres of life. We can see the manifestation of a "big change" all around us. But what we cannot see is how the various manifestations of this big change connect. "In a specialist world, no one has time for integration. As a result, we are in the midst of a major change which is both obvious, and yet still largely invisible and hard to understand" (p. 9). What we see is a large-scale but fragmented movement, a movement that is larger than any previous historical shift. But most people are unaware that they are part of this movement.

> Individuals and groups rally around specific issues and work on rethinking in specific fields. As a result, they talk in the language and concerns of a hundred different fields. This make them unintelligible to one another and invisible to the outside world. It also makes their power diffused. We see only a few heretics pounding their heads against the walls of various complacent establishments. We hear only a Tower of Babel. Yet, world civilization is now filled with millions and millions of such voices, all aimed at the same direction. (Goerner, 1999, p. 10)

Individuals, addressing problems in disparate fields, for example, government, education, business, science, health, and spirituality (to mention a few) are coming up with remarkably common insights, finding themselves in harmony with a vast common effort. The "*great hive of mind we call humanity*" is at work and it reaches a critical mass of a new shared view. "This view was not invented by one great person, but many great people, some famous and most obscure. It is an act of the great *we* or, as the Chinese would say, of 'the web that has no weaver'" (pp. 16–17).

Reflecting upon the by now visible great turning I suggest that what we have learned heretofore about how evolution works helps us to understand the current state of the massive global movement toward creating a new society. It is truly remarkable to discover how the various initiatives and efforts that strive to build a new society can be associated with the various *evolutionary markers* of an emerging evolutionary generation. Everything is in place for the creation of a new generation of HSS but yet in an unconnected state. As I describe these initiatives next, I use *italics* to indicate those that stand for an evolutionary marker. We shall see that the whole range of markers are represented but are not yet connected into the evolutionary system of a new generation of HSS.

We can see various manifestations of the emergence of *a new level of consciousness*—beyond the still prevailing mental/rational consciousness of the Third Generation. But these manifestations have yet to merge into a collective consciousness. We can see many forms of a *spiritual* awakening. We can see an already operating new *world view of connectedness, synthesis, interdependence, and wholeness.* We see people calling for and advancing public *morality*, working for an *ethical* civil society, and exploring truly *participatory democracy*. We can see the forging of a respectful new *relationship—partnership—with nature.* We can see the signs of a major shift toward a *new science*, a *humane science*, and a thrust for *a technology* that serves the common good. New *educational* theories and practices are emerging. We experience the application of a new *communication* regime that can tie us into community-based evolutionary design groups as well as a web of global interdependence. We can hear the calls for social and economic justice and witness in many forms the forging of a new *social* and *economic* order.

What is remarkable is that we can already point to many small-scale self-organizing efforts that already manifest the characteristics described in the preceding. We see many attempts to create healthy, authentic, and nurturing communities. We might consider these as *experiments in the evolutionary design space of the Fourth Generation.*

But what we should be cognizant of is that the newly emerging characteristics described in the precedings are not yet organized into a new evolutionary entity at the societal level. They are not yet animated by a collective consciousness. They are not yet coevolving. They are not yet integrated. Furthermore, what is missing from the current picture of our evolutionary state is the explicit institutionalization of the set of evolutionary qualities, namely balance, harmony, internal consistency, synergy, symmetry, and wholeness. *It is the job of the creating force of evolution to organize the whole and bring all these qualities to life in a new evolutionary generation.*

But the emerging characteristics described in the precedings will not be brought together in internally consistent sociocultural systems by some unknown evolutionary forces. The sea change will not just happen. All of us need to make it happen. Having acquired evolutionary consciousness we became the *agents of conscious evolution*, the designers of the sea change. We are destined to be the creators of the Fourth Generation. But we cannot just declare this as something that should happen. We have to learn how to become the agents of evolution by *developing evolutionary competence.* We have to devise ways and means of the *"how" of evolution*: we have to create an evolutionary epistemology—and we have to learn to apply it, so that we can make it happen. And such creation and such learning is what the rest of our conversation is about.

8.3.2 The Social Potential Movement

An example of the dawning of the Fourth Evolutionary Generation is "the social potential movement." Barbara Hubbard (1998) describes this movement as a manifestation of conscious evolution in action.

> Conscious evolution gives us the context and logic for the emergence of the social potential movement. We are not working in a dying or meaningless universe. We are working in alignment with the whole process of creation. This process has a direction that is animating our heart's desire for greater freedom, union, and transcendence. The new meme encourages each one of us to attune to our creative urge, to express our potential for the sake of ourselves and the world. It urges us to find others and co-create together. Conscious evolution offers us the possibility of an open-ended, choiceful, glorious future. Within each of us stirs the mighty force of creation in the process of its next quantum transformation. (p. 94)

The social potential movement is an expression of conscious, guided evolution. It is longing for a more just, free, and compassionate society. It seeks out innovations working in many fields—in health, education, communication, the environment, science and technology, and in the public and private sectors. The movement seeks to design social systems that manifest the qualities the movement represents.

Reflecting upon the "dawning" of the Fourth Generation, we recognize that we are becoming "coevolutionary" with the evolutionary process. Participating in creation, we bring forth our spiritual essence as well as our scientific and social capacities. But we have to find ways by which we can accelerate the connections—and integration—of emerging evolutionary innovations. An increased attainment of collective consciousness combined with an intensified interaction among innovations and the development of evolutionary epistemology are critical to engaging in conscious evolution.

8.3.3 The Cultural Creatives

Paul Ray, a sociologist, conducted a national survey (1996) and found that a decade ago the U.S. population was split between two main cultural groups: *traditionist* and *modernist*. But within a decade something remarkable happened. A new, large cultural group has emerged. By now, one quarter of the population forged a third cultural group, which Ray calls *cultural creatives*. Cultural creatives seek to bring about a new society in which all societal groups and institutions work together for the *common good*. We can consider the existence of the cultural creatives as both a *qualitative* and a *quantitative* measure and manifestation of the sea change, described earlier. Ray estimates that in the United States today we have 88 million modernist, 56 million traditionist, and 44 million cultural creatives. Now, let us see what these groups are about.

8.3.3.1 The Traditionist: The Past

Ray calls the traditionist *heartlanders*. Their "values and worldview is validated by a nostalgic mythical image of the past, substituting for an image of the future" (p. 20). They are longing to return to the past. They are the back-to-the-future people, the type we described in Chapter 5. They resent and resist change. They find meaning in their family, in their church, and in their community. They seek to live by traditional values. They try to restore traditional values, often by autocratic means. They reject modernists who consider their yearning and values as irrational. An increasing number of the heartlanders are (right-wing) extremists, who hold that only their way is *the one right way*. At the fringe of this group we have the racists, the bigots, and some of the fundamentalists, all of whom want to exercise authoritarian control.

8.3.3.2 The Modernist: The Present

The modernist's "values and worldview is validated only by day-to-day concerns." They have "no image of the future" says Ray (1996, p. 20). The modernists' world view is marked by unconstrained technology, industrial capitalism, rampant consumerism, and materialism.

While in its early period, the modernist view was grounded in the principles of liberty, equality, and fraternity and promoted universal education, religious freedom, and humane tolerance, by the eighteenth century, "modernity, including science, capitalism and imperialist nation states gained a reputation for uncaring efficiency, and abuse of power. Modernism is now enamored of monied interest and well bolstered by 'scientism,' the dogmatic form of science. It is noticeably low on heart and soul" (Goerner, 1999, p. 420).

8.3.3.3 The Cultural Creatives: The Future

Ray says that the cultural creatives possess "integral values and worldview, validated by ideals and a new view of humanity." They are "developing a new image of the future" (Ray, 1996, p. 20). (By now you see why I suggest that the consciousness of the culture creatives are shapers of the Fourth Generation HSS.) Cultural creatives hold that we are all members of one planet and should be concerned about the environment. They hold that we are all members of the human community and should be concerned about social and economic justice. Their notion of systemic relationships is *horizontal* rather than hierarchical. They are *consciousness creating* in all aspects of their lives, personal, social, and planetary. While Ray (1996), the originator of the notion of cultural creatives, holds that cultural creatives are the fastest growing cultural group, he also suggests that they are not well connected in a movement and are not yet self-organized for a collective change. But now, their time has come to enter into the evolutionary design space of the Fourth Generation.

Interpreting Ray's work, Goerner (1999) suggests the spiritual roots of the creatives reaches back to "the revolutionary spirituality that started during the Renaissance and burst forth again in the nineteenth-century Romantic and transcendental movements. Yet, their wide-ranging spiritual interests from Native American to Eastern to early Christian and Islamic roots makes it clear that their concern is for deep meaning that inspires all religion" (p. 422). Creatives are also deeply involved with science and technology, reason, and critical thinking. They hold that technology and commerce should be tools not only for prosperity, but also for economic justice and common well being. Creatives try to go back to the very deep roots that "reflect the original thrust of both Modernist and Heartlanders. This is why they try so hard to bring head, heart, soul, and common wisdom back together. Bound together in a living world, Cultural Creatives believe that we must all learn the essential ecological truth, not separate, but together "(Goerner, 1999, p. 422).

Activity 26

In Section 8.3, we saw signs of the evolutionary dawning of the Fourth Generation. Your task is now to reflect upon the content of Section 8.3 and describe the core ideas that indicate this "dawning." Then, explore your own life experiences and look for and describe signs of the big change, signs of groups who are like the culture creators. Also tell us which of the three culture groups you identify with and why. Enter your findings in your workbook.

Reflections

In the first two sections of the present chapter we explored evolutionary perspectives and characteristics that help us to think about conscious evolution and explored ways that we make conscious evolution work. In the third section, we have been looking for signs that tell us the coming of the big change, the coming of the Fourth Generation of HSS. The overall impression one can gain from studying the promising signs of the dawning of the Fourth Generation is that we have many people and groups who are becoming conscious of evolution and are ready to engage in conscious evolution. Next we consider an item from the news.

Recently the networks and the news media reported the decision of the Kansas State Board of Education. The decision practically eliminates the teaching of the science of evolution and prescribes the teaching of creationism according to a literary interpretation of the Bible. A CNN/Times survey revealed that 85% of the U.S. population supports the teaching in the public schools of *both* creationism and the science of evolution. This is sobering, and even more, alarming, news. It reveals a high degree of science illiteracy. It seems we have a long way to go before dawn arrives. We are still living in the twilight zone of the industrial machine age, in the twilight zone of the world view of the clockwork universe. Most of our energies are still wasted in trying to defend and shore up the ways of life of the dying Third Generation. It seems that we *have a lot to understand, a lot to learn, before we can join the forces of* conscious evolution.

8.4 Conditions of Engaging in Conscious Evolution

We are the *first generation of our species* that has the privilege, the opportunity, and the burden of responsibility to engage in the process of our own evolution. We are indeed *chosen people*. We now have the knowledge available to us and we have the power of human and social potential that is required to initiate a new and historical social function: conscious evolution. But we can fulfil this function only if we develop evolutionary competence by evolutionary learning and acquire the will and determination to engage in conscious evolution. These two are core requirements, because *what evolution did for us up to now we have to learn to do for ourselves by guiding our own evolution.*

In this section, I first consider issues related to the development of evolutionary competence. Then, I highlight the condition of having the will and commitment to engage in evolutionary inquiry. Third, the self-reflecting and creating consciousness are brought into focus.

8.4.1 Developing Evolutionary Competence by Evolutionary Learning

Evolutionary learning is learning that has to be mastered in order to attain evolutionary consciousness and the competence to engage in conscious evolution. I forged the terms evolutionary learning and evolutionary competence in the early eighties, as I prepared for a presentation at a conference on future learning agendas. Later, I wrote an article, "The Characteristic and Acquisition of Evolutionary Competence," which was published in 1987 in *World Futures*. In two additional articles (1989, 1993b) and in a recent book (1996) I further elaborated the concept.

Conscious evolution is activated as we acquire evolutionary competence through evolutionary learning. Evolutionary competence enables us to give direction to our individual and collective evolution through purposeful design, provided we individually and collectively learn specific knowledge, ways of thinking, skills, and dispositions that jointly and interactively constitute the domain of evolutionary learning. The key point made here is that the hope for a future of promise lies in individual and societal learning of understandings, ways of thinking, skills, and dispositions that are necessary to *attain evolutionary consciousness and engage in conscious evolution*. In the next section, I elaborate on my earlier publications on the aspect of evolutionary learning and the conditions necessary to enhance evolutionary learning.

8.4.1.1 Evolutionary Learning

Evolutionary competence can be attained by evolutionary learning. The evolutionary learning agenda includes specific knowledge, ways of thinking and viewing, skills, and dispositions that jointly and interactively constitute the content of evolutionary learning. The greatest source of change in social systems is learning (Boulding, 1985), both the development of new knowledge and the know-how that the human race never had before. It is this source that we have to activate for learning and human development. Evolutionary learning is first explored here in the context of current practices in education. It will be shown that we face a major evolutionary task in education itself. The current educational agenda not only fails to provide for evolutionary learning, but in some cases it does not even provide for learning about evolution.

8.4.1.1.a A Major Hindrance to Evolutionary Learning. A major hindrance to the development of evolutionary competence is inherent in our current practice of education, which focuses on "*maintenance learning*" (Botnik and Maltiza, 1979). We focus on the acquisition of fixed outlooks, methods, and rules of dealing with known events and recurring situations. We are promoting already established ways of working in systems that are still grounded in the industrial machine age thinking of the Third Generation of HSS.

In an age of slow motion changes, when small and incremental changes suffice, maintenance learning is indispensable for the functioning of a society. But *today*, in times of turbulence, rapid change, discontinuity, and the massive transformations that characterize our current era maintenance learning has to move to the background and change learning, evolutionary learning must come into focus.

At the current evolutionary stage, evolutionary learning becomes survival learning. Evolutionary learning first and foremost leads us to attain evolutionary consciousness. Then, having attained evolutionary consciousness, we learn how to engage in conscious evolution. This learning enables us to *cope with change and complexity*. It enables us to capture new perspectives and attain higher level consciousness. Evolutionary learning empowers us to anticipate and face unexpected situations. It will help us to learn the know-how, the epistemology of conscious evolution, so that we can progress from unconscious adaptation to our environment to *conscious innovation and coevolution with the environment*. And most significantly, it enables us to become cultural creatives and engage in the conscious evolution of the Fourth Generation of HSS.

8.4.1.1.b Contrasting Maintenance Learning with Evolutionary Learning Maintenance learning leads us to operate in a "negative feedback" mode of error detection and correction. This type is adaptive. It reduces deviations from existing norms and is useful in maintaining the existing state. But we live in an era when we must learn another type, innovative/evolutionary learning, which operates in a positive feedback mode. It amplifies deviation from existing practices as it moves us in a double-loop learning mode (Argyris and Schon, 1978). In this mode, we become open to examining and changing our purposes and perspectives; we are transcending our existing state and redefining and recreating our systems. We now speak of the most significant learning of our age which enables us to engage in conscious/self-guided evolution.

While maintenance learning reinforces already learned ways of responding, it leads to reluctance and resistence to change and makes us unable to guide change, because we have not yet learned how to. Evolutionary learning enables us face unexpected situations and, even more, engage in purposeful change. We develop the will and capability to shape change rather than just cope with it or becoming its victims. Evolutionary learning calls upon and *nurtures our creative potentials* as it enables us to envision images of the future and bring those images to life by design.

Competition is rewarded in our current educational practices. Students compete against each other for grades. Conscious evolution places a *high premium on cooperation*, upon the shared envisioning of desired futures and collective action for bringing the vision to life. Evolutionary learning involves both cooperation as a mode of learning and purposeful learning and development of cooperative group interaction skills (Banathy and Johnson, 1977).

In our current educational practices the student is placed in subject matter boxes and is taught in a lock-stepped and reductionist mode. In evolutionary learning, we seek to learn to think and act systemically, to seek and understand integrated relationships, grasp the patterns that connect, and recognize embeddedness and interdependence of systems. In evolutionary learning we transcend the subject matter boxes and integrate them in functional contexts that are real and important to the learner. *Synthesis becomes the primary mode of inquiry.*

The contrast developed here can be best summed up by a metaphor that I heard from Simon Nichols (1979) of the Open University. He was guiding a seven-year multinational project of "children designing the future." He said that in our conventional educational mode we are driving children into the future while making them look into the rear-view mirror. The windshield is blacked out for them and teachers are doing the driving. Is it not time, he asks, to clear the windshield and enable students to do the driving? Evolutionary learning opens up for us an unlimited horizon and develops competence in driving toward the future with a purposeful destination in mind.

8.4.1.1.c Acquiring Evolutionary Competence. The acquisition of evolutionary competence enables individuals, families, groups, organizations, communities, and the society to guide their evolution by purposeful design. An evolutionary learning agenda includes:

- The *core evolutionary competence*, namely the acquisition of *evolutionary knowledge*: the knowledge of the evolutionary story of our species. This learning domain includes an understanding of how evolution has worked in the evolutionary design space, an understanding of the core principles and perspectives of conscious evolution, the appreciation of the current evolutionary predicament, and acquiring capability to engage in conscious evolution by mastering evolutionary epistemology.
- The nurturing of such *evolutionary values* as cooperation, trust, benevolence, altruism, love, and harmony and the development of a universal set of values that generate evolutionary consciousness and an ever-maturing vision of the future.
- The fostering of *evolutionary ethics* that include self-realization, social, and ecological ethics.
- The attainment of *cooperative group interaction skills* by which we can increase our capacities for entering into ever-widening human relationships and managing conflicts in a nonviolent manner.
- The learning of communication methods and techniques that enable us to attain consensus in group communication.
- The acquisition of competence in *systems thinking and practice*, by which to understand complexity, grasp connectedness and interdependence, and perceive wholeness.
- The development of a *systems view of the world* and the attainment of the capability to relate functionally to the ever enlarging social systems, and ultimately the planetary system in which we are embedded.

- The development of competence that enables the *creation of desirable images of the future* and the learning of the skills to *generate and evaluate alternatives* by which to bring those images to life.
- The mastering of *complexity*. Csikszentmihalyi (1993) holds that evaluating complexity and understanding what it adds to our lives is an important learning task. In selecting among experiential choices he suggests that we should choose the one that is more complex. By choosing the more complex experience we will be more likely to learn something new (differentiation) that will add meaning to our experience (integration). *Learning to understand and evaluate complexity and learning to seek out and prefer complex experiences becomes an essential component of the evolutionary learning agenda*. Such learning will enable us to not only cope with ever increasing complexities, but also create complexities in the course of our evolutionary design.

8.4.1.2 Creating Conditions for Evolutionary Learning

It is proposed that meeting the following set of conditions will enhance evolutionary learning: (1) a nurturing climate, (2) multiple learning types, (3) relevance to the learner's functional contexts, (4) broad-based learning resources, (5) self-created meaning, (6) creation of evolutionary images, and (7) application in real world contexts what has been learned.

8.4.1.2.a Creating a Nurturing Climate. Evolutionary learning can flourish only in a climate in which nurturing and caring relationships are created and support and trust flows both ways between those who learn and those who foster learning. As Elise Boulding (1981) noted, our current educational practices expect compliance, which often engenders insecurity, resistance, and even fear. Nurturing builds confidence, encourages exploration, and secures conditions for creativity and evolutionary learning.

8.4.1.2.b Offering Multiple Learning Types. Learning types that are conducive to evolutionary learning include: (1) socially supported individual learning in which the learner is guided by others; (2) self-directed learning in which the learner has access to learning resources; (3) team-learning arrangements in which learners cooperate and share experiences in joint mastery of learning tasks; (4) learning facilitated by the electronic media, primarily by the use of the Internet.

8.4.1.2.c Providing Learner-Relevant Functional Contexts. To become meaningful to the learner, evolutionary learning should be provided in the context of social systems, such as the learning group, family, organized youth groups, organizations, and the community. These are functional contexts for learning in which the learner is a participant and that offer actionable task environments for applications of what has been learned. Only in such contexts can we expect the development of knowledge, understanding, dispositions, and skills that enable the emergence of evolutionary values and competencies by which to guide one's own evolution and contribute to the purposeful evolution of systems in which one lives and works.

8.4.1.2.d Providing Broad-Based Learning Resources. Education is much more than schooling, and learning is much more than education (Banathy, 1991). Development of children and youth and continuing development through life meshes intricately with learning opportunities available in all facets of life. Beyond the boundaries of schooling and formal educational settings, learning opportunities and resources are offered in the home, in religious organizations, in youth and civic groups, in cultural and community agencies, through various media, in high-tech networks, in the world of work, and in many everyday situations. A powerful potential resides in the notion of creating an alliance of all the societal sectors that have the capability to support learning. The design and development of the sociocultural systems of the Fourth Generation of HSS can become the focus for creating such a learning alliance that can tap into a vast reservoir of resources for nurturing evolutionary learning and developing and applying evolutionary competence.

8.4.1.2.e Exploring Self-Created Meaning. Whatever learning task is offered to the learner, it can be "owned" by the learner only if the learner can self-reflect on it, if the learner can make sense of it, and if the task becomes truly meaningful to the learner. The learner should have an opportunity to internalize and integrate what is being learned into his or her cognitive map and should be enabled and encouraged to "construct" from what is learned his or her own meaning and understanding. The process described here is an essential condition to providing meaning to the learner in his or her learning experience.

8.4.1.2.f Creating Evolutionary Images. Learning is not completed without its application in contexts and situations that are meaningful and important to the learner. Having met all the conditions (of learning) described previously, the application of evolutionary learning is accomplished by challenging the learner to create an evolutionary vision of the future, a vision that is then elaborated in an evolutionary image. This creation is the product of the evolutionary consciousness and evolutionary learning.

8.4.1.2.g Bringing the Image to Life. The last task of the evolutionary learner is to create the system that brings the evolutionary image to life.

Reflections

What is emerging from an understanding of societal evolution and evolutionary consciousness and the acquisition of evolutionary competence is that all these appear to be essential prerequisites for engaging in conscious, self-guided evolution. Beyond this primary role, evolutionary learning will also have a powerful impact on education itself. The fostering of evolutionary competence through evolutionary learning has the potential to change the purposes, the content, and the method of education. Education, enriched by evolutionary learning, will become a means to develop a *creating culture* in a learning society, which can evolve a new societal purpose, a new societal way of life, and empower us to guide societal evolution by purposeful design.

Activity 27

Task 1. Review the preceding text and speculate about and describe learning content and learning experiences that you would add to those I introduced.

Task 2. Devise for yourself an agenda for evolutionary learning.

Enter your findings of the two tasks in your workbook.

8.4.2 Acquiring the Will to Engage in Conscious Evolution

In evolution the most advanced state of existence is humans endowed with consciousness. It is best manifested in those who are most developed in terms of their relationship to others and in their ability to interact harmoniously with all else in their sphere of life. The *highest form of consciousness is evolutionary consciousness,* which enables us to collaborate actively with the evolutionary process and motivate action toward *conscious evolution.* Conscious evolution enables us to guide our future, provided we have attained the competence and the will to engage in self-guided evolution. Self-guided, conscious evolution enables us to use the creative power of our minds to steer our systems and our society toward the fulfillment of their potential. Laszlo (1987) highlights our evolutionary responsibility as he says: "*The evolution of our societies, and therewith the future of our species, is now in our hands.* Only by becoming conscious of evolution can we make evolution conscious" (p. 122). Conscious evolution is enabled by "self-reflective consciousness" and it is activated by "creating consciousness" (Banathy, 1993a).

8.4.2.1 Self-Reflective Consciousness

Self-reflective consciousness is a process by which individuals, groups, organizations, and societies contemplate and make representations of their perceptions of the world—and their understanding of their place in the world—in their individual and collective minds. These representations are developed on the basis of values we hold and the ideas we have about how the world works, leading to the creation of a cognitive map of "what is." Cognitive maps are developed, confirmed, elaborated, tested, disconfirmed, changed, and redrawn. They are "alive." They affect our behavior and they are affected and influenced by it. This mutual affecting/influencing is recursive and it is constantly evolving.

Self-reflective consciousness enables us to *create the cognitive map of our present evolutionary state*. It is of crucial importance to understand that we cannot just "borrow" a cognitive map from others, regardless of how "highly acclaimed evolutionary scholars" they might be. They and others can inform us and share with us their cognitive maps, but if we wish to participate in the great venture of conscious evolution, if we want to make our unique contribution to it, if we wish to take charge of our future, we must engage in the self-reflective process of drawing our own cognitive map of the present state of evolution. Then, we can engage in self-reflective conversation with those with whom we share our life, starting with our family, with people in our neighborhood and our community, and people in the systems with whom we live and work. This conversation is expected to lead to a collectively painted picture of the "what is"; it will lead to a collectively drawn cognitive map of the present state of evolution.

8.4.2.2 Creating Consciousness

Creating consciousness is consciousness that *creates the future*. The genesis of creating consciousness is self-reflection. Now that we have understood the current dysfunctional and untenable state of evolution, we shall transcend the current state and engage in evolutionary creation. This creating is a process by which individuals, groups, organizations, and societies envision *"what should be." This* creating thrust is based on the belief that while the future is influenced by the past and present, it is not determined by what was or what is. It remains open to conscious and purposeful intervention that can be guided by an evolutionary image of the future. A representation of that image is the *first normative cognitive sketch of a desired future*, which we can create individually and collectively.

Summary

Chapter 8 provided us with a bridge that connects Parts I and II with the rest of the text. Based on what was presented as our evolutionary history, what we have learned as an understanding of how evolution has worked, and how we might work with evolution, in Chapter 8 we have explored perspectives that help us to organize our thinking about creating and selecting approaches and methods that we can apply in engaging in conscious and self-guided evolution.

I also proposed that we have already stepped over the threshold of conscious evolution. Thousands of flowers bloom on the emerging evolutionary landscape. We can see at many places, in many forms the emergence of ideas and actions that manifest conscious evolution. However, these ideas and actions are fragmented and often contradictory. They are disconnected from each other.

I also suggested that there are two essential conditions of acting upon conscious evolution. One is the *individual and collective acquisition of evolutionary competence*. Evolutionary competence can be attained by evolutionary learning. An evolutionary learning agenda was introduced and ways and means by which evolutionary learning can be implemented were proposed. The second core condition is *the mustering of will and determination to engage in conscious evolution*. I also explored the role of self-reflective and creating consciousness in conscious evolution.

In this chapter, working with the text and the activities, we have further developed knowledge, understanding, and competence that will enable us to create a system of evolutionary epistemology.

9

Self-Guided Societal Evolution

Our study of human and societal evolution has helped us to capture and understand the evolutionary story of our species and appreciate the intricate way evolution has worked in the evolutionary design space. In the course of working through the chapters, you have acquired evolutionary consciousness that lead you to grasp the role we have to play in guiding the continuing evolution of our species. Up to this time the evolutionary forces that have been working in the evolutionary design space were beyond our control. But now that we have acquired evolutionary consciousness we have the opportunity to enter the evolutionary design space and guide our own evolution. So, the crucial questions are: *What is this guiding role? What is the evolutionary process we have to master in order to engage in self-guided societal evolution? Who should be involved?*

To be of any use, an evolutionary process needs to be grounded in a clearly articulated theory and the theory has to be grounded in explicit underlining philosophical assumptions. If conscious evolution aims to go beyond speculation and haphazard trial and error, it needs to become a disciplined inquiry and has to work with these three domains: *philosophy, theory,* and *epistemology.* The fourth domain is the domain of *application in the evolutionary design space.*

In the present chapter, I first define the four domains of evolutionary inquiry and discuss their relational arrangement. Then, I introduce some ideas—as *examples*—of philosophy and theory that might be explored and considered in developing the philosophical and theoretical bases of evolutionary inquiry. Next, I explore the meaning of self-guided conscious evolution. In conclusion, I respond to the question: Who should be involved?

9.1 Evolutionary Inquiry

Evolutionary inquiry is a *disciplined inquiry by which evolutionary knowledge and evolutionary competence are developed and applied in engaging in conscious evolution*. In this section, I introduce the four domains of evolutionary inquiry, explore their relationships, and define the modes of evolutionary inquiry as disciplined inquiry.

9.1.1 The Four Domains of Evolutionary Inquiry

Evolutionary inquiry incorporates four interrelated domains: philosophy, theory, epistemology, and functional application. *Evolutionary philosophy* explores philosophical precepts that express fundamental assumptions about what evolution is and how it works. *Evolutionary theory* articulates interrelated concepts and principles that apply to evolution as a human and societal experience and activity. It seeks to offer plausible and reasoned general principles that explain evolutionary process as a disciplined inquiry. *Evolutionary epistemology* has two domains of inquiry. It studies the process of evolution in the evolutionary design space to generate knowledge and understanding about how evolution works. Based on this understanding and other insights gained from evolutionary consciousness, an evolutionary epistemology needs to be forged that is appropriate to implement conscious, self-guided evolution. Here, the task is to select/create/describe approaches, models, methodologies, methods, and tools that are applicable for use in the evolutionary design space. Thus, a more correct designation of what we up to now have called evolutionary epistemology is: an epistemology of design or evolutionary design. These models, methods, etc. of evolutionary design are then applied in the *functional contexts of evolutionary inquiry communities*.

9.1.2 The Interaction of the Four Domains

Evolutionary philosophy, theory, and epistemology come to life as they are used and applied in the functional contexts of conscious evolution. It is in the context of application of evolutionary inquiry that evolutionary philosophy, theory, and epistemology are confirmed, changed, modified, and reconfirmed. *Evolutionary philosophy* presents us with underlying assumptions, beliefs, and perspectives that guide us in defining and organizing in relational arrangements the concepts and principles that constitute *evolutionary theory*. Evolutionary philosophy and theory then guide us in developing, selecting, and organizing approaches, strategies, methods, and tools into the scheme of epistemology of evolutionary design, which then is applied in the functional context of creating the Fourth Generation HSS. But this process is not linear or circular in one-direction. It is *recursive and multidirectional*. One continuously confirms and/or modifies the other, as described next.

As *theory* is developed, it receives confirmation from three directions: the underlying assumptions of philosophy, epistemology, and its application in the functional context of the evolutionary design space. *Epistemology* is confirmed or changed by testing its "faithfulness" and relevance to its philosophical and theoretical foundations and its success of application in the evolutionary design space. The *functional context*—the various systems of evolutionary communities—from the family on up to the global system—is the primary domain which places demands on evolutionary inquiry. Figure 9.1 shows the interactions described here.

The four domains constitute the conceptual system of evolutionary inquiry. It becomes a system as the domains interact in a mutually influencing recursive relationship. It is important that the relational influencing links become fully explored and explicit. If they remain implicit, we

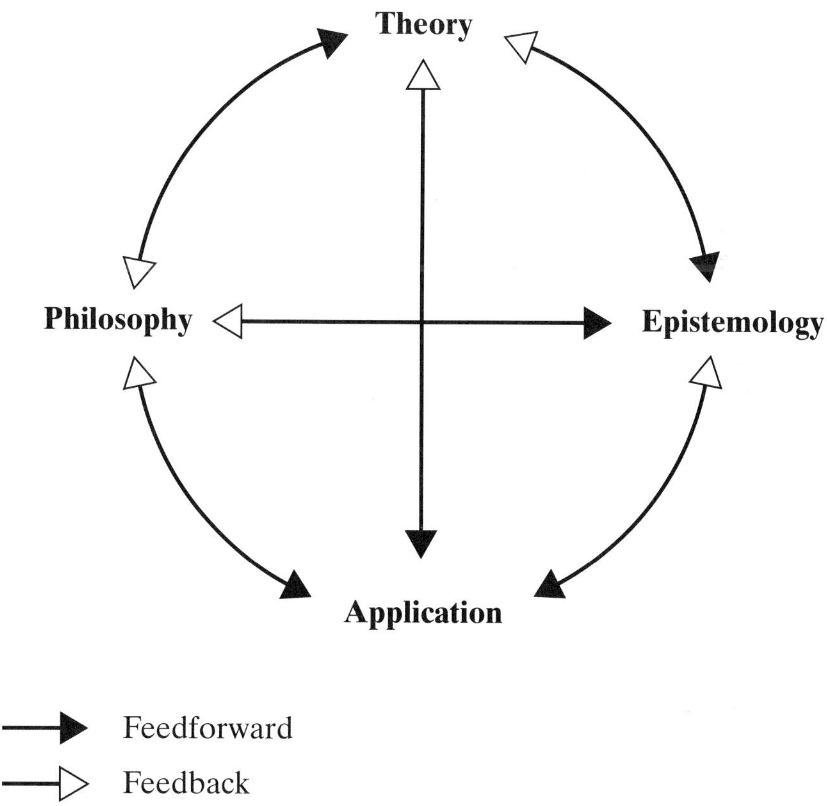

FIGURE 9.1 The Interaction Among the Four Domains of Evolutionary Inquiry

shall not be able to reason and support the decisions made in the evolutionary design space. As we work in the evolutionary design space and if we are unaware of the linkage with the philosophical and theoretical bases, we can never answer the question of why we have made a specific decision.

9.1.3 The Two Modes of Evolutionary Inquiry

Evolutionary inquiry is disciplined inquiry. It comes to life as the four domains interact recursively. In the course of this interaction, two modes of disciplined inquiry are operating: *decision-oriented disciplined inquiry* and *conclusion-oriented disciplined inquiry* (Cronbach and Suppes, 1969). Conclusion-oriented inquiry produces knowledge about a domain of inquiry. In the present case, in the first two parts of this work, our inquiry focused on *what* evolution is and *how* it works. This conclusion-oriented inquiry has produced knowledge about evolution. We apply this knowledge—in the decision-oriented mode—as we engage in conscious evolution in the evolutionary design space. Figure 9.2 depicts the relational arrangement of the two kinds of inquiry.

In evolutionary design, *the two modes of inquiry blend into each other.* As we operate in the evolutionary design space, and evaluate the appropriateness of the linkages among the four domains, we learn and gain new insights and produce new knowledge about the four domains of the evolutionary inquiry. Furthermore, we learn from the ongoing design activity and use what we have learned, as new knowledge in conducting the inquiry. As we keep on working in the evolutionary design space, evolutionary philosophy, theory, and epistemology become ever more refined and gain increasingly more power for application and our application becomes increasingly more powerful.

In this section, I defined the four domains of evolutionary inquiry, explored the dynamics of relationships among these domains, and introduced the two modes of disciplined evolutionary inquiry: one that produces knowledge about evolution and one that applies the knowledge as we conduct evolutionary inquiry.

9.2 The Philosophical and Theoretical Bases of Evolutionary Inquiry

At the present time, we do not have clearly articulated, comprehensive philosophical and theoretical bases of conscious evolutionary inquiry. There are at least four reasons for this.

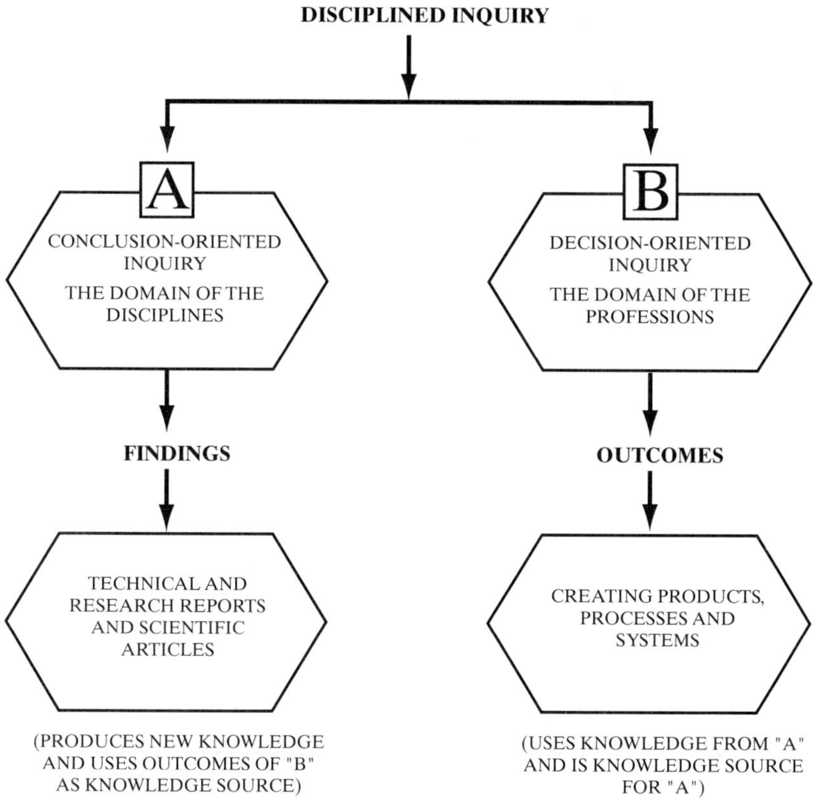

FIGURE 9.2 The Relational Arrangement of the Two Modes of Inquiry

- First, as one looks at the current evolutionary literature, one finds it to be *highly distributed, fragmented, and disconnected, often contradictory.* This condition is typical in newly emerging fields of inquiry and it invites respectful conversation rather than debate among those who take differing positions.
- Second, evolutionary thinking is *conceptually split* between two camps. One camp questions if we can—or should try at all to—engage in conscious evolution. The other camp—to which the present work aims to give a voice—holds that we not only can, but we should. It is our responsibility to enter into the evolutionary design space and engage in conscious evolution. There is probably a third camp, sitting on the fence between the two listening to the arguments.

- The third reason is that evolutionary inquiry is *grounded in a great variety of philosophical and theoretical positions in want of synthesis.* It will be the task of the evolutionary scholarship community to address this task and "sweep in" from the disciplines and the various fields of inquiry knowledge that have relevance to evolutionary inquiry and than synthesize those knowledge sources into an internally consistent "meta-knowledge base" of evolutionary inquiry.
- Lastly, the movement of *conscious evolution is a recent development.* It is in the process of developing, and maturing. There is a great deal to be done to create a clearly articulated and comprehensive theory and philosophy as bases of conscious evolutionary inquiry. This is a mega-task, the task of the *evolutionary inquiry community.* I see this community as one that is more interested in learning about and actively participating in conscious evolution than just pontificating about it. What follows is a modest contribution to this task. It is accomplished by a limited exploration of the potential contributions that the systems and design science community might make to evolutionary inquiry.

9.2.1 The Potential Contributions of Systems and Design Theory

What follows may contribute to the philosophical and theoretical bases of evolutionary inquiry as I explore the relevance of systems and design inquiry to evolutionary inquiry. It might give us a clue and indication of the value of exploring and learning from various philosophical and theoretical bases in order to articulate the philosophical and theoretical bases of evolutionary inquiry. I shall also explore briefly the relevance of ideas from what is called "new science."

By studying and working with human activity systems in the course of the last several decades, we developed an increasing appreciation of *the inquiry power we can gain from systems theory and systems philosophy and their application through systems methodologies.* We have liberated ourselves from the constraints and limitations of the analytically oriented and reductionist inquiry mode of traditional science. Systems inquiry enables us to orchestrate the findings of various scientific disciplines within the framework of systems thinking and to develop and apply systems approaches, models, and methods in working with social and societal systems. In this section. I explore the relevance of systems inquiry to evolutionary inquiry. This exploration addresses the relevance and potential contribution of systems philosophy, systems theory, and systems and design inquiry.

9.2.1.1 The Relevance of Systems Thinking

Systems thinking is the field of inquiry of the systems philosopher, who organizes internalized systems ideas, systems concepts, and principles into an internally consistent arrangement. This arrangement establishes for us a frame of thinking, a *systems way of viewing and understanding.* As we observe what is "out there," this way of viewing and thinking enable us to reflect upon what we experience; thus we construct our own meaning. We create our own cognitive map which is our own interpretation of the "out there." As we view and work with social and societal systems in the evolutionary arena, systems thinking enables us to *create our own cognitive map of guided evolution.* If we acquire a systems view and develop systems thinking we shall be empowered to participate in the evolutionary design of the various systems in which we live and work.

Contemplation of systems thinking is the primary interest of systems philosophers. Here I review the contribution of several systems philosophers, whose ideas will enlighten our understanding and appreciation of the potential contribution of systems thinking to evolutionary thinking.

9.2.1.1.a Churchman's Contribution. The most salient contribution to systems thinking is that of West Churchman. His contributions present an internally consistent set of core ideas that have direct relevance to evolutionary inquiry. His major works (1968, 1971, 1979, 1982) are grounded in the ideas of the philosophers of the Enlightenment, such as Kant. Churchman sets forth several core themes of systems thinking that are briefly described here. He suggests that in systems inquiry problems are *unbounded and tightly connected.* Every problem is an aspect of all others. This notion, he says, was already advanced in the fifth and sixth centuries by Greek scholars. One of them, Anaxagoras, said that "in everything is everything." But the reductionist Western scientific community rejected this notion and promoted bounded inquiry. Churchman (1982), a student of the Greek scholars, says that: "We need an *'unbounded'* systems approach which *sweeps-in all that is relevant to our inquiry*" (p. 8). The ideas of unboundedness and sweeping in all that is relevant to the inquiry of our interest should guide conscious evolution. The *ethics of the whole system,* says Churchman, includes a study of the *ethics of humanity,* not within a problem area, but *universally.* All issues in the social domain are *inherently ethical* and *not factual.* They are first of all *prescriptive* and not descriptive. He calls for determining the *ethics of the whole system.* He gives us an example: someone concerned about education should first *consider the nature of an ideal society* and then ask how education can serve it. This notion of seeking the ideal must be considered a core generative idea of conscious evolution. He also suggests that in our arrogance we often assume in our inquiries that we—the stakeholders of the present—are what counts. And he passionately asks: What are the *implications of our inquiry for future generations*? This forward-looking concern is one that we should honor in evolutionary inquiry. Sharing Churchman's concern, I suggest that whatever evolutionary design solution we create, we should never close out options for future generations. (A metaphor for this concern is the *"open chair."* At the conversations of the International Systems Institute we always have an open chair in the middle of the circle, as a reminder of generations not yet born and our responsibility toward them.)

Finally, Churchman calls for a focus on the *invariant of humanity*: the aspirations, the values, and the hopes of our collective humanity. All the core ideas mentioned here are germane to future building evolutionary inquiry.

9.2.1.1.b Mitroff and Linstone. Mitroff and Linstone (1993) follow Churchman's thinking as they elaborate the notion of unbounded systems thinking. Unbounded systems thinking holds that *all branches of inquiry depend fundamentally on one another*. The widest possible array of disciplines, professions, and branches of knowledge—capturing distinctively different paradigms of thought—must be consciously brought to bear on our inquiry. This is obviously true in evolutionary inquiry, which is grounded in all of the disciplines and fields of inquiry that are relevant to human experience.

9.2.1.1.c Geoffrey Vickers. The systems philosopher Geoffrey Vickers' (1981) main interest lies with the ecology of systems ideas as they relate to social systems. This focus tends to correct some of the "illicit" extensions of ideas (to the social realm), which have been derived from the natural sciences. In Newton's world, says Vickers, inert objects stayed put unless moved by some 'force.' By contrast, our world is one of active and dynamic reactions "in which *stability, not change, requires explanation*" (p. 19).

The interface between a social system and its environment *distinguishes inner relations from outer*. Inner relations hold social systems together and enable the system to act as a whole in the context of its environment. The scope of the system's external relations depends on the coherence that its internal reactions secure with the environment. In our current evolutionary state it is this coherence that is breaking down, as the Third Generation disintegrates.

Exploring identity, continuity, and change, Vickers asks: When does a system retain its identity and continuity through change and when does it itself vanish or become something new? These questions are of great practical concerns in conscious evolution. Human systems, says Vickers, are the only kind which can succeed or fail. "A human system fails if it does not succeed in doing what it was designed to do; or if it succeeds but leaves everyone wishing it never tried" (p. 21). This observation is a lesson for evolutionary inquiry. So is the understanding that social systems are governed by the demands of stability on the one hand and by criteria of success on the other. Vickers says "I find it *ridiculous to try to reduce the second to the first as so many people try to do*" (p. 21). Now contemplate the implication of Vickers' ideas for sociocultural evolution.

9.2.1.1.d Jantsch's Contribution. Erich Jantsch (1980) made a unique and powerful contribution to the evolution of systems thinking and most significantly to the evolution of evolutionary thinking. He is among those who shaped the intellectual process that has had a profound *concern for self-determination, self-organization,* and *the openness and plasticity of systems and their freedom to evolve and coevolve.* He says *process orientation* is primary, in contrast with an emphasis on solid components and structures. While "a solid structure determines the processes which it can accommodate, the interplay of processes may lead to the open evolution of structures. *Emphasis is then on becoming*—and even the being appears in dynamic systems as an aspect of becoming" (p. 6). Social systems are coherent, evolving, interactive processes that have nothing to do with the equilibrium and solidity of technological structures, which are geared to the output of specific products. Social systems are concerned primarily with *renewing and evolving themselves,* which are essentially learning and creating processes. "When a system, in its self-organization, reaches beyond the boundaries of its identity, it becomes creative. In self-organization, *evolution is the result of self-transcendence.* At each threshold of self-transcendence a new dimension of freedom is called into play for the shaping of the future" (pp. 183–184). Evolution is open through the self-organization of social systems, by the dynamic interconnectedness and the coevolution of the systems and their environment. *Evolution determines its own meaning.* This meaning is the meaning of life. "We are not the helpless subjects of evolution—we *are* evolution" (p. 8). In the course of our present unfolding discussion we often brought in the notions that Jantsch articulates so powerfully, such as self-determination, self-organization, openness, self-transcendence, and the freedom to evolve and coevolve by design. Evolutionary inquiry is a process of becoming, learning, and creating.

9.2.1.1.e Laszlo, a systems and evolutionary philosopher, suggests (1996) that a systems view of the world and sociocultural evolution is nonanthropocentric, but it is humanistic. All systems in nature have value and intrinsic worth.

> They are goal-oriented, self-maintaining, and self-creating expressions of nature's penchant for order and balance. The status of the human being is not lessened by admitting the amoeba as his kin, nor by recognizing that sociocultural systems are his suprasystems. Seeing ourselves as a connecting link in a complex natural holarchy cancels our anthropocentrism, but seeing the holarchy itself as an expression of self-ordering and self-cresting nature bolsters our self-esteem and encourages our humanism. (p. 91)

As human beings, we have a unique opportunity to get to know ourselves and the world in which we live. It is shortsighted not to seize this opportunity and confine ourselves solely to the business of living. "Our species may not be capable to exist for long without the use of rational insights in guiding its own destiny" (p. 92). Our knowledge (of evolution) has made us increasingly autonomous, it has freed us from the bonds of biological existence, it has enabled our species *to create the world of culture,* and it is enabling us now *to determine our own evolution.*

Here is where the holistic vision of systems thinking becomes so important to us. When properly articulated, says Laszlo, it can give us both factual and normative knowledge. "Exploring such knowledge and applying it in determining our future is an opportunity we cannot afford to miss. For if we do, another chapter of terrestrial evolution will come to an end, and its unique experiment with reflective consciousness will be written off as a failure" (p. 93).

Contemplating guided evolution, we should take to heart Laszlo's admonishment. If we fail the task of establishing a strong philosophical and theoretical foundation for our work in the evolutionary design space, our attempt to create the Fourth Generation will be built on shifting sand and will not stand the winds of the negative forces that are destroying the Third Generation.

Even a quick review of a few examples of statements made by systems philosophers shows us the unquestionable relevance of systems thinking to evolutionary inquiry. Systems thinking becomes not only a source but also a power of shaping evolutionary thinking and guiding conscious evolution.

9.2.1.1.f Kant's Comment. Kant was a systems thinker. His notion of *enlightenment* reflected many of the ideas we associate with systems thinking . He was concerned with man's *release from "self-incurred tutelage,"* with *people freely thinking for themselves, and deciding for themselves.* Kant's statement inspired the generative idea of this work as stated in the front part of the book. It is reintroduced here as the basic underlying assumptions of conscious evolution.

The right of people to guide their destiny; to take part directly in decisions affecting their lives; to create healthy, authentic, and nurturing evolutionary communities; and to control their resources and govern themselves is a most fundamental human right.

If people learn how to exercise this right, then they have the power to create a civil society, a true democracy, in which they can design their own lives, participate in the evolutionary design of the systems in which they live and work, and organize their individual and collective lives in the service of the common good.

9.2.1.2 The Relevance of Systems Theories to Evolutionary Inquiry

Since the mid-1950s, we have seen the emergence of several systems theories. Here, I only name members of the first set of theories, and discuss briefly recent theories which, in my judgment, have relevance to evolutionary inquiry. The exploration that follows is only an incomplete indication of the potential contribution of systems theory to evolutionary theory.

During the first two decades of the systems movement, we saw the emergence of *general systems theory* which aimed to establish relevance to all kinds of systems. Its orientation was organismic. *Systems analysis, Operations research,* and *systems dynamics* had a rather technical and mathematical orientation. We learned to call this orientation "hard systems, "which also includes *systems engineering* with a mechanical focus. *Cybernetics* has gone through several generations, for example, first order, second order, and third order. While the first two focused on information science, third-order cybernetics has a human system orientation. *Living systems theory* is an advanced elaboration of organismic systems theory. I suggest that *soft-systems theory* and *critical systems theory,* briefly described next, have relevance to evolutionary inquiry.

9.2.1.2.a Soft Systems Theory. It has been during the last three decades that systems inquiry became focused on the human experience, on human betterment, and the improvement of the human condition. West Churchman initiated and spearheaded this movement (1968), which has become know as the "soft-systems" approach, in contrast with "hard systems" approaches. Churchman has been the most articulate and most effective and committed advocate of a *humanistic oriented systems science,* an advocate of *ethical systems theory* and *morality* in *human systems inquiry.* Systems inquiry has to be *value oriented,* and it has to be *guided by the social imperative,* which dictates that *technological efficiency has to be subordinated to social efficiency.* He speaks of a science of values and the development of methods by which to verify ethical judgments.

Many of us have learned from West over the years. Among others, Checkland has worked on the continuing development and application of Churchman's ideas over two decades. He proposed a set of "constitutive rules" (Checkland and Scholes, 1990) of soft systems inquiry. Soft systems theory is interpreted in a design epistemology governed by these constitutive rules. These rules are relevant to evolutionary inquiry and are introduced and interpreted here. (It should be noted, however, that most of the work of Checkland and his followers has focused on the improvement of existing systems and the context of these applications has been primarily the business/corporate environment.) Thus, I limit my description to aspects that I judge to be relevant to our present concern.

The soft systems approach is based on ideas that come from a *conscious reflection on the world by using systems ideas.* These ideas yield the epistemology of the approach. The front-end part of the epistemology includes an exploration of the *sociocultural characteristics* of the situation, which includes the *exploration of values* and the creation of a rich picture of the design situation. The main body of epistemology includes constituitive rules, such as making a distinction between *unreflective involvement* in the everyday world and *conscious systems thinking* about it. In our inquiry we move back and forth between the two. The approach is to be *used by different people differently in different situations.* Conscious thought has to be given to adapting the approach to a particular design situation. The approach should be taken as a methodology and not a technique. *Every use should be considered as a learning experience,* as a methodological lesson. The lesson awaits discovery by conscious reflection on the use. As discussed earlier, such reflection creates new knowledge as we explore the relationship among philosophy, theory, methodology, and application.

9.2.1.2.b Critical Systems Theory. A new trend in systems thinking has emerged in recent years. Pioneered by Jackson, Ulrich, Flood, and others, it is called critical systems theory (CST). This orientation challenges some of the earlier aspects of systems inquiry. Critical systems theory is reflected in Ulrich's critical systems heuristic and Flood's and Jackson's total systems intervention. CST embraces a set of core commitments, such as critical awareness, social awareness, human emancipation, and complementarity (Jackson, 1992). *Critical awareness* closely examines the values and assumptions that enter into systems—in our case the evolutionary—inquiry. It provides tools that are useful to apply critical awareness, such as Ulrich's (1983) critical systems heuristics. *Social awareness* recognizes social/societal pressures that should be considered in systems (evolutionary) inquiry. It aims to guide us as we contemplate the social consequences of our inquiry. This commitment also calls for an open and free conversation on the justification of the use of our inquiry. The purpose of *Human emancipation* is to ensure the well-being of all individuals and the full development of their human potential. It aims to prevent coercion and the exercise of power that would prevent open and free conversation. *Complementarity and informed development* of all varieties of systems approaches is another commitment of CST. Various systems trends express various rationalities and theoretical positions. CST suggests that the positions and the methodologies that arise from these theories should be respected and their development should be encouraged. CST has a commitment to the complementary and informed use of all the various systems approaches whenever their use is appropriate to the context of various social conditions and situations. The relevance of these commitments of CST to evolutionary inquiry is more than obvious.

Activity 28

Even though the preceding section provides us with limited exploration of systems thinking and systems theory as potential bases to formulating evolutionary inquiry, enough systems ideas, concepts, and principles have been presented to enable you to judge their relevance to evolutionary inquiry. Your first task is to review Section 9.2.1.2 and make a list of systems ideas, concepts, and principles that you find relevant to evolutionary inquiry. Your second task is to reflect upon your finding, transform it into the context of evolutionary inquiry, and formulate an internally consistent set of principles that might guide evolutionary inquiry and work in the evolutionary design space. Introduce your formulation in your workbook.

9.2.1.3 Relevance of Design Thinking to Evolutionary Inquiry

Here we turn to a group of systems and design scholars and practitioners who have accomplished extensive work in systems design and revealed ideas that enlighten our understanding of the relevance of design thinking to evolutionary inquiry.

9.2.1.3.a Churchman's Design Thinking. Churchman's ideas on design (1971, p. 79) are reflected in his complex strategy for dealing with ill-structured design issues. The markers of his thinking include: (1) the necessity of *reflecting upon the viewpoints of the individuals and the collective viewpoints of the designing community*; (2) the consideration of the *nonhierarchical nature of the system's complexity*; (3) a *respect of wholeness* of the system and its *irreducibility*; (4) the critical consideration of the *purposefulness of the system*; (5) the *nonseparability* of a system's components; and (6) the *uniqueness* of the system. In Churchman's view, the designer's *main tool is subjectivity*, which includes social practice, community interest and commitment, ideas and ideals, the ethics of the system, the moral idea, faith, and self-reflection. For Churchman the issue is not whether we can design systems that are wholes and unique, but *whether we can design systems that make us more whole, unique, and self-motivated*. Churchman's strategy provides a powerful guide for working in the evolutionary design space of conscious evolution.

9.2.1.3.b Ackoff's Design Thinking. In Chapter 6, we explored Ackoff's (1981) four styles of how people think about and work with change. Of the four styles, his choice is the interactivist or proactivist who believes that the future is largely subject to our creation. Thus, design, says Ackoff, is the creation "of a desirable future and the invention of ways to bring it about" (p. 62). Ackoff's design thinking is reflected in three operating principles. The *participative principle* holds that the learning and using the design process (individually and collectively) is an outcome that is more important than the actual product of design. The other implication of the participative principle is that *one cannot design effectively for someone else*. In fact, no one should. (I call this an ethical principle of systems design.) The second, the *principle of continuity*, challenges the well-known routine of making a plan or design and updating it periodically in an on-again, off-again cycle. There are two reasons for design continuity. Continuous reflection upon and valuation of the operating design and its systemic environment enables us to be continuously in the design mode. Second, as we design something of value and pursue it, the value we place on it might change as time goes by, because our design values might change. This, then, indicates changes in the design or might call for redesign. The third, the *holistic principle*, has two aspects. First is the principle of coordination. It tells us that no part of the system can be designed independently from others that operate at the same system level. The second aspect means that all system levels should be designed itegratively.

Ackoff (1994, 1995) points to the antisystemic nature of such current popular panaceas as total quality management, continuous improvement, process reengineering, and right-sizing. Most of these *fail "to whole the parts."* They manipulate the parts without focusing on how they effect the performance of the whole. Second, they *fail "to right the wrongs."* Focusing on efficiency, they do the wrong things righter rather than the right things. Designers often say, as does Ackoff, that *getting rid of what is not wanted does not give you what is desired.*

9.2.1.3.c Nadler's Design Thinking. Nadler's design thinking is reflected in a set of propositions (1981). *The genesis of design:* Specific human and organizational attitudes, desires and behaviors are the starting points of systems design. *Uniqueness:* Individuals and systems are unique. There are no identical entities; therefore the design approach adapts to the specific situation rather than the situation to an approach. *Purposefulness:* Design is a purposeful human activity to be carried out in the context of human values and objectives. Design is *prescriptive.* Design is *grounded in theory:* It is based on some truths or axioms, supportable assertions, and empirical evidence about effectiveness of design approaches. For example, research shows that people fail to use solutions that were invented somewhere else. *Focus on what is desired, not on what is wrong:* In defining a design situation, the focus should be on what is desired rather than on what is wrong. *Systems design is holistic and interdisciplinary;* it is greater than the sum of its parts.

9.3.1.2.d Weisbord's Design Thinking. Weisbord (1992) presents a future search, whole systems design approach that brings people together to *achieve breakthrough innovations, empowerment, shared vision, and collaboration.* His main theme is that the world is moving from experts designing for people toward everybody, experts included, designing whole systems. He suggests that getting everybody involved is "the best strategy if you want long-term dignity, meaning and community" (p. ix).

Designed by Weisbord (1989, 1992), the Future Search Conference integrates the past, present, and future, as these are experienced and appreciated by the participants of the conference. This process uncovers shared values and leads to congruent action. The approach is based on seven assumptions: (1) The *world is knowable to ordinary people and their knowledge can be collectively organized by them.* "In fact, ordinary people are extraordinary sources of information about the real world." (2) It is believed that *people can create their own individual future* and, *collectively, the future of their systems.* (3) People want to have the *opportunity to engage their heads and hearts in determining their future.* They want to—and are able to—join the creative process, rather than leaving such creation as the sole domain of the "organization's elite." (4) The *nature of participation is egalitarian. Everyone is equal* and has an equal voice and right to make a contribution. (5) Given the chance, *people are much more likely to cooperate than fight*. The task of those who coordinate (facilitate) the process is to *structure opportunities to cooperate*. (6) The process should *empower people* to feel more knowledgeable about and in *control of their future*. (7) *Diversity is appreciated and valued*.

Weisbord sets forth a set of "minimum critical specifications" for success. These are as follows: (1) "Get the whole system in the room." This means to involve in the process the broadest possible community that represents maximum variety and diversity of interdependent people. (2) Have this community look at itself in a "global context." Explore all the system-relevant events, trends, relations within the wider world, and the institution/issues in focus. This means the exploration of the broadest possible knowledge base and common ideals before zeroing down to the issue at hand. (3) Ask people to be task focused and "self-manage" their work, and by so doing, reduce dependency, conflict, and task avoidance. All these conditions must be present, says Weisbord. Mapping the whole system *cannot be done by the views of management* or of any specific stakeholder group. *It has to be done by all the people in the system*. Neither can the task be accomplished by focusing on a restricted field of inquiry.

9.2.1.3.d Warfield's Design Principles (1990). Warfield's set of principles further inform design thinking. These principles interpret a set of articulated laws of generic design, which in turn are linked with the foundations of the science of generic design. Here, I present some of the principles that can guide our thinking about evolutionary design inquiry. *Variety:* Design inquiry should be carried out in groups whose members are selected to ensure that they articulate the variety and diversity in the design situation. *Alternatives:* The design situation should be represented in dimensions that enable the consideration of a range of alternative design solution options. *Interdependence and integration:* The dimensions of the design situation and the design solution are interdependent and should be explored integratively in order to avoid design errors. *Iteration:* The design process should be iterative in producing design solutions and in assessing solution alternatives. *Ordering choices:* In developing solution alternatives, the design group should consider the sequence in which choices should be made as they evaluate choices. This enables them to make the most salient choices first and the least salient last. *Display findings:* The state of the unfolding design should be continuously made visible by displaying it. This enhances productivity and minimizes reliance on memory. *The design environment* should be carefully designed and equipped to provide maximum support to the inquiry. *Defined processes and roles:* The design program should include formal processes and specially assigned roles by which we pay continuous attention to the content, the context, and the process of design. *Criteria:* Participants in the design inquiry should develop their own self-reflective, self-assessing, and self-governing criteria that will guide and steer their participation.

9.2.1.3.e Christakis' "CogniScope." A long time co-worker of Warfield's, Cristakis applied a thoroughly tested cognitive technology that integrates software as a groupware in what he calls the CogniScope systems. The system has been used in more than 200 applications and documented in more than 400 scholarly papers. CohniScope is a disciplined inquiry process that enables a community of stakeholders to create a design and an action plan based on the design (Christakis and Conaway, 1995, 1996). In the program, conversation among the stakeholder community generates and clarifies a large number of ideas, upon which they generate an interactively created design solution and action plan, which are co-owned by them because they have been created by them.

A designing community employs the CogniScope system to perform three principal activities. It: (1) *generates and clarifies ideas* in response to carefully and properly framed triggering questions; (2) *produces "idea patterns"* that result from an exploration of relationships among ideas in the context of carefully framed generic questions; and (3) *evaluates idea patterns* and action packages based on agreed-upon criteria. In the course of these activities, the various consensus generating group techniques are used to suggest, clarify, and structure ideas; explore and interpret influence patterns; and *work out solution alternatives*. In the course of the design process the three principal activities are carried out in four interrelated stages, described as follow.

Stage 1. At this stage, the designing community defines the design situation by responding to the question: "What should we do?" about the design situation in question. Usually well over a hundred ideas are generated. Such a large number of ideas cannot be handled without the support of software technology offered by the CogniScope. The technology enables the community to organize ideas quickly and effectively into meaningful relational patters. For example, one step at this stage is the structuring of the ideas for exploring "influence relationships" that reveal that the accomplishment of one idea will help in the accomplishment of others. By using an "inference logic system," designers discover these influence relationships in one-tenth of the time it would have taken without the support of the Cogniscope program.

Stage 2: Creating Alternatives. Once the design situation is defined, the designing community begins to explore design alternatives, by responding to the question: What to do and how? Through group work and focused dialogue, the "Hows" are superimposed on the "Whats" and displayed on newsprint on the walls of the designing facility. The designing community will now see whether the "Hows" are indeed addressing the fundamental "Whats."

Stage 3: Choosing the Preferred Alternatives. Within the host of "How" ideas there is an almost limitless number of possible solution alternatives from which to choose the preferred alternative. With the use of the CogniScope program, the designing community can select a limited number of alternatives to be fully deliberated. The deliberation is followed by the formulation and consideration of criteria to be used to select the preferred alternative.

Stage 4: Developing an Action Plan. Once the preferred design solution alternative is selected, the solution idea components are sequenced. This process responds to the question of "When" and "How" will we do the "Whats."

The CogniScope system literature notes that in addition to the inquiry power and proven results of the program, the system produces intangible gains that may be of even greater value than the product of the design effort. This added benefit is manifested in three forms. (1) The system *nurtures organizational learning.* (2) It *fosters cooperation* among stakeholders. (3) It ensures the likelihood of *effective implementation* in that the design will be installed and operated by those who created it; *they own it.* The further development of CogniScope system, introduced earlier, promises the possibility of its use in the public arena in working in the evolutionary design space, in creating evolutionary design solutions of the various component systems of the Fourth Generation HSS.

Activity 29

In the preceding section, I introduced a limited set of design ideas and principles, articulated by various design scholars. It seems that even this limited set can be considered as potential bases for formulating a design approach to conscious evolution.

Your first task is to review Section 9.2.1.3 and make a list of design ideas that you find to be relevant to evolutionary inquiry.

Your second task is to reflect upon your findings, transform them into the context of evolutionary inquiry, and formulate an internally consistent set of design ideas that might guide our work in the evolutionary design space. Enter your formulation in your workbook.

9.2.2 Ideas of Scholars of the "New Science"

Going beyond what we usually consider the domain of sources of systems and design thinking, we can be informed by new systems and design relevant insights that have emerged in the thinking of scientists of the new sciences, such as Bohm, Prigogine, and Davis. It would take a separate book to explore the relevance of emerging scientific ideas to evolutionary inquiry—and such a book should be written in order to further evolutionary thinking. Here, I can project only a fleeting impression of how some of the emerging ideas of the new science might shape evolutionary thinking.

9.2.2.1 Bohm's Wholeness and Order

Bohm, a quantum physicist and philosopher of science, has presented a sweeping theory of wholeness and order. Developed in the 1920s by Bohr and his followers, quantum theory demolished the then prevailing concepts about reality. It blurred the distinctions between cause and effect and object and subject and brought forth a strong holistic element in a redefined world view. Quantum theory forms a pillar in new physics and it provides the most convincing scientific evidence about the essential role that consciousness plays in the nature of reality.

Bohm (1983) suggested that quantum theory dropped the notion of analyzing the world into autonomous parts and placed the emphasis on undivided wholeess. The *world is* not a collection of autonomous and coupled things but rather *a network of relations*. A *division into subjects and objects, inner and outer world, body and soul is no longer acceptable*. Bohm and Peat (1987) define "implicit order" as the enfolding order of reality from which the explicit order of specific phenomena unfolds. His scheme of orders embraces "*generative order*." It is a deep and inward order out of which the manifest form of things can emerge creatively. (In Chapter 6, we explored the relevance of generative order to evolutionary inquiry.)

Bohm suggests that generative order is relevant not only to science (e.g., the science of chaos) but to all areas of experience. I find the notion of generative order to be directly relevant to evolutionary thinking. Bohm sees the manifestation of generative order in the work of the artist. The artist, as the designer, begins with an overall vision, a general idea, a feeling that already contains the essence of the final work in an enfolded way. He captures the overall form in an sketch (the image of the future in evolutionary design) from which, as the painting progresses, details are created gradually, each time building on the whole as the work unfolds and elaborates the whole. "The artist is always working from the generative source of the idea and allowing the work to unfold into ever more detailed form" (p. 158).

In the same way, the evolutionary designer community gradually creates the design solution from the "*generative*" *image of the future*. The painting expresses a visual, outward perception, but it cannot be separated from the inner perception of the artist's *values, beliefs, aspirations, his whole life and knowledge*; it is inseparable from his emotional and intellectual relationship to his theme. So it is with the evolutionary design. Our inner perceptions, inner self, and social self are integrated with our outward perception of the future. It seems that Bohm's example of the generative creation of art is a good metaphor for evolutionary design. It is. But there is much more to it. A thorough interpretation of *Science, Order, and Creativity* (Bohm and Pent, 1987) to evolutionary design would enable us to generate some new and powerful insights into what design is and how it works in the evolutionary design space.

9.2.2.2 Prigogine's Theory of Change

The Nobel Prize winner Prigogine's (1984) ideas present us with a comprehensive theory of change that is directly relevant to the way we can think about evolutionary inquiry. He suggests that instead of being orderly and stable, reality is bubbling with constant change, process, and disorder. The subsystems of a system are continually "fluctuating" and may move far from equilibrium. Driven by positive feedback, at times these fluctuations may become so powerful that they "shatter" the whole system—as the Third Generation HSS has been shattered. At that moment the system is at a "bifurcation point," when it can either disintegrate into "chaos" or leap onto a higher level "order" by self-organization, a phenomenon that Prigogine calls "dissipative structure." We have seen this phenomenon at every "quantum jump " of human and societal evolution. (He called it dissipative because the higher order complexity requires more energy to sustain the emerged system.)

In open systems, such as social systems, this energy comes from the environment. In nonhuman systems the direction taken at the bifurcation point is up to chance; it is inherently impossible to determine. It is different in social systems, however. The various cultures of our societies are immensely complex systems. They are highly sensitive to fluctuation and potentially involve an enormous number of bifurcations. These, then, could *lead our systems on an evolutionary path of reorganization at ever higher levels of complexity*. "From a human point of view, all this is quite optimistic" (p. xx). For us this is truly hopeful, in that even small fluctuations may grow and change the system. "As a result, individual activity is not doomed to insignificance" (p. 313). Yes, we can give direction to the evolution of our system by purposeful design. The concept of such "guided evolution" is the overall theme of this work. We carry the burden, the responsibility, and it is our privilege to guide our evolution and be responsible for it.

9.2.2.3 Paul Davis' Cosmic Blueprint

Paul Davis (1989), a theoretical physicist, rejects the ideas of the universe being at the mercy of randomness or a world slavishly conforming to mechanical forces. The *universe is creative, progressive, and innovative* in character. Its organizational aspects are collective and cooperative, and its perspective is synthetic and holistic. Our goal is not to understand what things are made of but how they function as integrated wholes. *Human society*, says Davis, *"has evolved to the stage where it is shaped and directed by conscious decisions"* (p. 194). Our self-organizing power creates ever greater organizational complexity. "As we consider systems of greater and greater complexity, the concept of a class of identical systems becomes less and less relevant because an important quality of very complex systems is their uniqueness. In their uniqueness social systems "possess logical and structural relationships of their own that transcend the properties of individual human beings" (p. 194). They have their own dynamic behavior and generate their own consciousness and meaning that are their collective cultural attributes. In the evolutionary path of self-organization "the possibility arises that a new threshold of complexity may be crossed, unleashing still a higher organizational level, with new qualities and laws of its own" (p. 196). As a result collective activity might emerge that is now beyond our ability to conceptualize.

Creation is an ongoing process, constantly bringing forth new structures, processes, and potentialities. It is not deterministic. It has "freedom of choice," which also means uncertainty. It is free to create itself as it goes along. The fact, says Davis, that "the universe has organized its own self-awareness—is for me powerful evidence that there is 'something going on' behind it all. The impression of design is overwhelming" (p. 203).

Bringing Davis' self-creating cosmic image down to the scale of self-creating coherence of conscious evolution, we can be both humbled and inspired by the possibility of participating in the grand scheme of creation. Our question becomes: How can we capture the emerging cosmic blueprint and how can we mirror it in our designs?

Activity 30

This activity invites you to reflect upon the ideas of the three scholars introduced in the preceding, identify any ideas that are relevant to evolutionary inquiry, and contemplate their use in the evolutionary design space. Write down your findings. Your second task is to integrate the ideas you noted here into the two sets you already formulated in Activities 29 and 30. You have now a rich set of design principles/ideas that can guide you in developing evolutionary design methodology as we continue our work in elaborating self-guided evolution of society.

Reflections

I am truly inspired by the new horizons that opened up for us as we begin to explore the implication of what systems and design inquiry and the emerging new scientific thinking have for engaging in conscious evolution. Exploring these implications offers us a unique challenge and opportunity to enrich our conceptualization and understanding of conscious evolution and how it might work in the evolutionary design space.

9.3 *Conscious, Self-Guided Evolution: Definition and Exploration*

Conscious evolution provides a *sense of direction for cultural and societal processes* by illuminating those processes with *guiding images*. And the faster we go, as we do at our current evolutionary stage, the further we have to look for signs and images to guide our journey (Jantsch, 1975). The envisioning of such an image was defined by Kenneth Boulding (1978) as a *unified view of evolution* that connects all reality from cosmic/physical through biological/ecological/sociobiological to psychological and social systems. It seeks to understand the evolutionary dynamics through which systems evolve, and *grasp the principles underlying the unfolding of evolution* over space and time.

The core idea of evolutionary guidance, says Jantsch (1981), is that evolution is not the result of one-sided adaptation and a desperate quest for survival, but is an expression of *self-transcendence, the creative reaching out beyond the system's own boundaries*. "We humans *are the integral agents of evolution*, we spearhead it on our planet and perhaps in our entire solar system. *We are evolution* and we are, to the extent of our power, responsible for it" (p. 4).

Weisbord (1992) says that a new paradigm is sweeping the post-industrial world. The paradigm is characterized by the ideas of *learning, empowering, democratizing, partnering*; it wants to bridge the gaps of culture, class, race, gender, and hierarchy. The search is now on to find approaches that are equal to these values. The purpose of such an approach should include: (1) the *discovery of common grounds* and the *imagining of ideal futures*, (2) the *involvement of all sectors of the society*; (3) the provision of task-focused discovery, and (4) the broadening of *global perspectives*, the extension of our horizons, and the promotion of self-management that leads to committed action; and (5) the *building* of *democratic values* and higher quality of life.

In summary, evolutionary guidance implies arrangements and operations that are built into various human activity systems at all levels of society by which these *systems are empowered to give direction to their own evolution* and move toward the realization of their *evolutionary vision*.

Evolutionary guidance is a dynamic process of enabling human systems to give direction their own evolution and develop in these systems the organizational capacity and human capability to: (1) *nurture the physical, mental, emotional, and spiritual development and self-realization of both individuals* and *collectively all members* in the system; (2) extend the boundaries of the possibilities for *freedom and justice, economic and social well-being, and political participation*; (3) *increase cooperation* and *integration among societal systems* and manage conflicts in a nonviolent manner; and (4) engage in the *design of societal systems that can guide their own evolution by* purposeful *design*. By attaining these purposes, we can recreate and empower our social systems as *evolutionary guidance systems*. This recreation requires a fundamental *reorganization of our inner map of* reality away from fear, distrust, and hostility. It requires a change in the way we perceive ourselves and our relationship with others. In this way, eventually, we can create a shared image of the global human future and, at the same time, maintain and respect the diversity of our many cultures and societal systems.

An image of humanity, as Markley and Harman (1982) noted, is a gestalt perception of humankind, both the individual and the collective; in relation to self, others, society, and the cosmos. As a new evolutionary stage emerges, the use of old images creates more problems than it solves. On the other hand, when a new image leads sociocultural evolution, as it did when the Third Generation of HSS emerged during the Renaissance period, it can exert what Polak (1973) called a "magnetic pull" toward the future. As a society moves toward the realization of that image, the congruence increases between the image and the evolving system.

Today we are still extrapolating from the old image of an industrial society. Thus, we are locked into conflicts and struggle of races, religions, and nation-states. The old image is not working for us. It causes widespread frustration, alienation, and social upheavals. It may lead us to the brink of self-destruction. We desperately need a new image. This image shall spring forth from the emerging global consciousness and shared evolutionary vision. Its realization will be enabled by the acquisition of evolutionary competence.

9.4 Who Should Be the Designers?

In the *Conference of the Birds*, the Persian poet Fariduddin Attar tells the story of the assembly of the birds. The assembly was called together by the wise Hoopoe bird, who convinced the assembly that they have to find the King of the Birds and ask his advice if they wish to live their lives to the fullest. But the wise Hoopoe warned them of the many dangers of the journey that leads to the castle of the King. Still, a large delegation of birds embarked on the search. During the journey many fell victim to deathly perils. At the end, only a few reached the castle. At the gate the Chamberlain first refused to hear their plea, but they persisted and finally were allowed into the throne room. To their amazement, no one was sitting on the throne. After great hesitation the birds approached the throne one by one and sat on it. First they became confused. But after a while they were astonished as they looked into each others's eye and realized that they themselves where, as a collective, the KING. The throne was theirs and they together shared the responsibility for their kingdom.

In this section we explore the issue: Who is—or who should be—the designer of sociocultural evolution? The story of the birds represents a philosophical metaphor for answering the question. In sociocultural design we can recognize several approaches. I explore these and elaborate the one that is the only viable approach to evolutionary design.

9.4.1 Four Generations of Design Approaches

We can give account of four generations of designers. The first is design by decree. The second is top-down assisted by a design expert. In the third, a design expert drives the design with some involvement of people in the system. The fourth generation of designers design their own system.

9.4.1.1 The First Generation

For long, the question of who should design was not even asked. It was the "natural order" of things that design was handed down from the top either by legislation or by the leader of the organization. The approach was *"design by decree* or design *by dictate."* People in the system or in the organization were expected and coerced to carry out the dictate. As the story goes, a worker went to the supervisor saying: "I think I have a design idea," to which the answer was: "We did not hire you to think." This design-by-decree approach could be called the pre-design-age approach. At the societal level, all dictatorships—and we have had many of these in this century—imposed the design of life, rules, and norms from the overall level of the society down to the family.

9.4.1.2 The Second Generation

Half a century ago, with the emergence of systems science we entered the design age. In the span of five decades, we have witnessed the emergence of three additional generations of design approaches. The *"designing for others"* approach is dominated by the *expert designer*. He is brought in by top management to propose a design solution. He conducts a problem analysis, engineers a solution, and presents it to the decision maker. He is then paid (at times to never return). The solution is implemented by applying coercion. And it works as much as coercion does in social systems. The expert designer approach has been heavily influenced by systems engineering, operations research, and systems analysis methods. Th expert designer approach can be characterized by what John Warfield often calls "throwing the blueprint over the wall and having someone else build it." Its use was based on the belief that human systems can be manipulated, *the expert knows best, and people better do what the expert says.* Expert-driven design dominated the scene during the 1950s and 1960s. It is unfortunate that at many places the expert consultant/designer still reigns.

Rittel (1984) characterized the first generation approach this way: the designer is invited in by a client and studies and analyzes the problem. Then, he withdraws and, following a step-by-step approach, works out a solution. Then, he comes back to the client and offers the solution to him. But often he runs into an implementation problem because the client does not believe him. And the client is well advised not to believe the expert, because at every step in developing the solution the expert made "ought-to-be judgments" that the client may or may not share, and cannot read from the finished product, offered as the solution.

9.4.1.3 The Third Generation

The next two generations are discontinuous with the first generation approach. Their emergence was guided by a gradual and increasing recognition of the open, complex, indeterminate, and self-organizing nature of human systems and an understanding of their value-laden, purposeful, and purpose-seeking characteristics. Their emergence was influenced by the rejection of the use of hard-systems-thinking-based systems analysis/engineering practice in social systems and the coming to the scene of the soft-systems, and later the critical systems thinking approaches and methods of the 1970s and 1980s. The initial stage of this emerging new thinking guided the development of the third generation approach.

The third generation approach is labeled as *"the expert/consultant designing with users."* In this approach the decision-makers bring the design expert as a consultant into the system, where he or she stays for a while and works from time to time with selected groups of people who represent a cross section of the system. They engage in a practical, task-oriented discourse, led by the expert. Depending on the designer's and the decision-makers' inclination, the "designing with" approach can be regarded as *more or less participative*. If it is more, then it may generate a certain degree of commitment to the design solution. If it less, it calls for compliance and coercion.

9.4.1.4 The Fourth Generation: The User-Designer Approach

This approach based on the assumption (Rittel, 1984) that the *expertise is distributed over all participants*, and nobody has any justification in claiming his knowledge to be superior to anybody else's. The consequence of this assumption is the attempt to develop maximum participation to activate as much knowledge as possible. This is a nonsentimental argument for participation. There are many sentimental and political arguments in favor of participation, but this is a logical one.

The fourth generation of design approach is standing on the shoulder of the third and is continuous with it. I have called this approach the *"designing within the system"* or the *"user-designer"* approach. It is based on the belief (Banathy, 1993a) that although the future is influenced by the past and the present, it is not determined by what has been or what is. It remains open to our individual and collective purposeful intervention, accomplished by design. Human activity systems, organized at various levels of society, from the family to the global system of humanity, can give direction to their own evolution and can shape their own future by engaging in purposeful design. Even more, this approach asserts that *designing our future is our responsibility*; we can and we should take charge of shaping it. We find support of the fourth generation approach from several design scholars.

Nadler and Hibino (1990) present a detailed reasoning for the fourth generation approach which they call the "people design principle." The principle is based on the premise that the concerns and ideas of people in the system should be the basic fabric of design. In design people should work "from the center (themselves) out rather than from the outside (others) in" (p. 220.) The great need for creative and innovative design solutions is matched by the need to install the solution, which requires the active involvement of those who operate the system. They have a *commitment to the solution* because they developed it, they know how it was developed, and because they own it. We all want to be involved in decisions affecting our lives. An we accept and feel good about implementing a solution we help to devise.

Ackoff (1981) suggests that when it comes to considering what a system ought to be, *all* people in the system can make an important contribution. *Their ideas, aspirations, dreams and preferences are all relevant.* The fullest participation provides people with an opportunity to think deeply about the system and to share their ideas with others. This encourages the exploration and development of new ideas and facilitates personal and collective development.

Weisbord (1992) sets forth a set of core values about participation in design. He suggests that ordinary people are extraordinary sources of design. They aspire to create their own future and want opportunities to engage their heads and hearts in the design of their future. He proposes *egalitarian participation; everyone is equal.*

The legitimacy of systems design (Churchman, 1971) rests upon the consideration of the many different perspectives that people in the system have and set forth in the course of the design. Therefore the design process receives its legitimacy from ensuring the maximum participation of stakeholders.

The "designing within" approach is based on the assumption that human activity systems must be designed by those who are served by the system, who serve the system, and who are affected by it. When it comes to the design of social and societal systems of all kinds, it is the users, the people in the system who are the experts. *Nobody has the right to design social systems for someone else.* It is unethical to do so. Design cannot be legislated, it should not be bought from the expert, and it should not be copied from the design of others. If the privilege of and responsibility for design is "given away," others will take charge of designing our lives and our systems. They will shape our future.

9.4.2 Key Markers of the User-Designer Approach

During the 1970s I guided an R&D program aimed at developing an educational system that enhanced and nurtured the authentic culture of Native American youth. Our staff was primarily Native American. Going on site to a reservation, one of the tribal representatives told us kindly that if we came to tell them what to do, we better go home. If we work with them for a while and advise them what to do we might as well go. But if we stay with them, live with them, and learn with them we are welcomed.

The story has two ramifications. At a personal level, from that moment on I did not accept any consultancy role. Then, later on, one of our senior researchers went to a small fishing village in Costa Rica, where he lived with them, learned with them, and worked with them for a year on a project called "Participatory Design for Social Empowerment: a Journey of Trust" (Kavanaugh, 1989).

In responding to the question: Who should be the designers, four answers have been offered: "design by decree," "designing for," "designing with," and "designing within: the user-designer." The justification for the user-designers is that the right to design rests with people in the system. This position meets the test of the criteria of: authenticity, sustainability, responsiveness, uniqueness, personal development and organizational learning, and the ethics of design.

9.4.2.1 Authenticity

The design of a social system is authentic only if it is built on the individual and collective values, aspirations, and ideas of those who serve the system and who are served and affected by it. The design should reflect their collective vision and it should make use of their collective intelligence. A design cannot produce an authentic and intelligent organization if it reflects only the intelligence of an expert (second generation), if it is limited to use of the intelligence of only the top echelon or some selected groups in the organization (third generation). It is authentic only if it makes use of the individual and collective intelligence of the people in the system (fourth generation).

9.4.2.2 Sustainability

A system is sustainable only if its design is accomplished and put in place by the creative, collective, and unconstrained participation and contribution by all people in the system. Such *contribution ensures sustainability* several ways. It enables people to understand thoroughly their system, since they designed it. They know first hand what their role is in the system (since they shaped the roles) and know what they have to learn to play their roles. Participation enables the creation of consensus among those who work together. It creates genuine respect for each other and develops fellowship. It ensures that people will take part more effectively and at a *deep level of commitment in the implementation* of the design since the design represents their individual and collective values, ideas, and decisions. Participation is empowering and the design is empowered by it.

9.4.2.3 Uniqueness

The uniqueness of the individual, the uniqueness of cultures, and the uniqueness of situations have been generally acknowledged. These types of uniqueness have been considered within the contexts of such disciplined inquires as psychology, biology, and anthropology. But the uniqueness of human activity systems always frustrates those who approach human systems and organizations from the perspectives of statistical approaches; standards, averaging, aggregating, and other systematic methods, which are unable to consider uniqueness.

In designing sociocultural systems, we must *take into account a whole range of unique characteristics*, such as the uniqueness of the systemic context; the nature of the system to be designed; the individual and collective readiness and design capability of the people involved; the resources available; the design situation; the values and world views of the designing community; and time, space, and complexity factors.

The kind of design described in the preceding brings forth the guarantee of success, of producing a viable system that ensures much benefit to the people who inhabit the system. But the application of the user-designer approach reaps benefit to the designer community that is even more than I described here.

9.4.3 Special Benefits of the User-Designer Approach

In design research and scholarship we often remark that the most important product of the involvement of stakeholders in design is the unique *opportunity for learning and personal development*. Some of us are prepared to say that such learning may be even more important than the product of design. Few people would question that the ability to design is one of the highest personal and collective capability in an age of ever ongoing change. The second and equally important benefit derived from the collective involvement of all stakeholders is that it provides the best possible *opportunity for organizational learning*. By engaging in design, an organization engages in "double-loop learning" (Argyris and Schon, 1978), as people in the system learn how to envision the purposes, perspectives, values, functions, and modes of operation of their organization and develop insights based on which they can change or (re) design their system, and make continuous contributions to its life.

Reflections

At the onset of this section the question was asked: Who should be the designers? A review of the evolution of design approaches and an exploration of their justification have helped us to understand that the only viable approach to the design of social systems is the one that empowers the people in the system to become competent user-designers.

It has taken us more than forty years to reach the understanding that the design of social and societal systems is the right and responsibility of those who serve the system, and who are served and affected by it. Like the perilous journey of the delegation of the birds, our quest for an authentic, sustainable, and viable approach to design has been marked with many failures and disappointments. Disheartened by these, we often left the shaping of our future and the deign of our systems in the hands of "kings"; in the hands of authorities, futurists, and experts. But now we know that the age of social engineering is over. We can reclaim the "throne." It is rightfully ours.

For me the most remarkable aspect of the emergence of the various design generations is the speed of their development Within three decades we moved from the expert designing systems to people designing their own. It is even more remarkable if we parallel this development with the emergence of conscious evolution. *By the time evolutionary consciousness led us to conscious evolution, the self-directed user-designer approach has emerged as the only viable and logical approach to sociocultural design.* Things are falling into place.

Activity 31
First review the text and select and organize sets of core ideas that you deem to be relevant to the question: Who should be the designers of social systems? Then, in view of the core ideas and the generations of design approaches, speculate about means and methods by which you would enable the community of designers in the systems of your interest to get ready to engage in design. Enter your findings in your workbook.

Summary

The development of this chapter is grounded in the understanding that while up to now the evolutionary forces that have been working in the evolutionary design space were beyond our control, now that we have acquired evolutionary consciousness, we have the opportunity to enter the evolutionary design space and engage in self-guided conscious evolution. The questions we are now confronted are: What is this guiding role? How should we play it? And: Who should be involved? These are the questions for which I developed answers in this chapter. The answers have paved the road of our journey toward self-guided conscious evolution.

10

An Epistemology of Conscious, Self-Guided Evolution

The key task of this work is to *propose an approach to conscious, self-guided evolution.* The task is absolutely *not* to propose a particular evolutionary future, but rather a possible approach to creating our evolutionary future. Developing an evolutionary epistemology requires the exploration of evolutionary knowledge, the acquisition of evolutionary thinking, and the attainment of evolutionary consciousness. The material presented in the previous chapters aimed to create evolutionary knowledge. In addition, completing the activities in those chapters enabled you to acquire evolutionary thinking and gain evolutionary consciousness. The work we have accomplished heretofore has also generated an acceptance of the idea of transcending the existing evolutionary state and engaging in conscious, self-guided evolution.

It is important to note that the epistemology proposed here is *a possible approach* and not *the* approach to conscious evolution. I have developed this epistemology for a reason. There are many voices calling for conscious evolution but few articulate an approach of how to go about it. The approach introduced here is informed by my long-term research and development and applications in social systems design, coupled with my work in evolutionary inquiry. The approach is animated by the public philosophy of self-determination and individual and collective commitment to taking responsibility for our own future.

The evolutionary epistemology proposed here has three major strategies: *transcending* the existing evolutionary state, *envisioning an image* of a desired future state, and *transforming our systems by bringing the image to life by design.* Using the language of evolutionary inquiry, I reword the three strategies as follows:

- Rather than trying to solve the problems of the current evolutionary state and "*extend*" the life cycle of the bankrupt Third Generation, we should *transcend* it.

- Rather than improving and *revising* what we now have, we should *envision a new image* of our evolutionary future.
- Rather than *reforming* the evolutionary state of the Third Generation, we should *transform* it by bringing the evolutionary image to life by design.

The present chapter is organized around the first two strategies. The third strategy is the subject of Chapter 11.

10.1 Transcending the Existing State

In the three preceding chapters, we developed knowledge-based evolutionary consciousness. In particular, we became conscious of the process and products of sociocultural evolution. We gained an understanding of *how evolution has worked* in the evolutionary design space through the various generations of *Homo Sapiens Sapiens* (HSS). We understood the workings of the life cycles of these generations through their stages of emergence, maturation, stagnation, decline, and demise. We have elaborated these stages in the life cycle of the Third Generation and recognized our current many-faceted *evolutionary predicament* grounded in the devolution, decline, and demise of that generation. We arrived at the realization that it would be *futile to try to hold on* to the consciousness and the ways and means of the workings of the evolutionary markers of the Third Generation. Any attempt to continue the present state would lead to increasing social disruptions and hostility, decline of quality of life, increasing chaos, continuing environmental degradation, and possibly the demise of humanity.

Such anticipated conditions leave us with no other viable choice than transcending *the current state and taking a quantum leap* into the evolutionary design space of the Fourth Generation of HSS. This is easy to say but it is most difficult—probably *the* most difficult—task to undertake. Transcending is most difficult because many believe that "solving" our current problems, improving the existing evolutionary state, or reforming our current systems and institutions is the way to go. We witness today these attempts of *defending what is*, and denying the need to change. Evolution is change. The evolution of a new generation of HSS is *big change*. And big change often frightens people.

Now, let us look at some *change-avoiding strategies* that prolong the agony of our current evolutionary predicament. Confronting these, dissolving their arguments against change is a prerequisite to transcendence.

10.1.1 *Strategies of Avoiding Big Change*

There are several change avoidance strategies. One is an outright attempt *to deny the need for change*. The other *is to focus on solving the problems* of our evolutionary predicament, and—by so doing—trying to extend the life cycle of the Third Generation. The third strategy leads us to engage in a campaign of trying to *improve and fix what we have now*: to improve and fix our systems and institutions grounded in the world view of the industrial machine age.

10.1.1.1 Defending the Status Quo

When it comes to the choice of saying as we are or initiating change, many people rush into finding all the reasons why we should not change. The defenders of the status quo can be very inventive in procrastinating and avoiding change. We should recall the four attitudes of dealing with change explored in Chapter 5. We should also remind ourselves what happened with our cousins, the Neanderthals, who abhorred change.

10.1.1.2 The Problem-Solving Strategy

This approach attempts to do away with our current problems. But eliminating what is not wanted does not give us what is desired.

Conventional "rational" problem solving is driven and dictated by the representation of the problem itself. This representation is the result of an in-depth analysis and structuring of the problem. *The focus is on the problem*. Thus, solutions, brought to the problem, stay within the boundaries of the problem itself. It is assumed that the more time we spend on formulating the problem the more likely we will find the solution. A thorough formulation is expected to help us to select the method that best corresponds to the problem. The formulation of the problem is expected to lead us to the formulation of the solution. (The opposite is true: We can understand the problem only in light of the solution.) The problem-solving approach described here is appropriate in dealing with well-defined and *well-structured problems*.

But the problems we face at our current evolutionary predicament are anything but well-defined. They are *ill-defined* and *ill-structured*. Rittel and Webber (1984) call design problems "wicked" in comparison with "tame" problems that are the subject of rational problem solving. Ackoff (1981) holds that designers of sociocultural systems are always confronted with "messy" situations. In contrast with rational problem solving, today we cannot stay within the bounds of our current problem situations. Our *focus must be on the desired future evolutionary state*. We are to create images of the future and leave the baggage of the present behind.

Focusing on the system in which the problem situation is embedded locks us into the current system. But solutions lie outside of the existing system. If solutions could be offered within the existing system, there would be no need to enter into the design space. That is why we have to transcend the existing problem-laden evolutionary state.

10.1.1.3 Improving the Current State

Improvement focuses on *how can we make better what we have now*, and how can we make it more effective and more efficient. Improvement examines a current activity, process, product, or service and determines what is wrong with it, so that we can find substitutes for defects or deficiencies. Often we find out that such *substitution can be much worse than what we intended to correct*.

The second major problem with improvement is that usually it is *focused on the specific parts of the problem situation*. Thus, we take corrective measures problem by problem. This kind of change effort is piecemeal, fragmented, and disconnected. But we should know that the excellence of a part or a system component does not prevent the bankruptcy of the whole. A partial consistency of a structure does not guarantee the consistency of the whole (Sallstrom, 1992). Operating in the evolutionary design space, we seek to create evolutionary systems that manifest internal consistency and external viability as a whole. Parts get their meaning from their role in, and their contribution to, the whole.

10.1.2 Approaches to Transcending the Existing State

The insights we gained from our exploration described in the preceding indicate that once we realize that the only viable alternative is to engage in conscious evolution, we should leave the existing state behind, transcend it, and leap out from its boundaries. However, this often creates anxiety, a feeling of uncertainty, and a fear of a future unknown.

I suggest that the ease or difficulty of transcending the existing state depends a great deal upon our *attitudes toward change*. We reviewed four modalities of attitudes toward change in Chapter 6. We found that those who have a "reactive" (re-activating the past) or "inactive" attitude toward change will have the highest degree and intensity of anxiety about change and often denial of the need to change. Having a "preactive" attitude, on the other hand, would lead people to rush into change whenever certain "trends" indicate or "forecasters" predict a specific future. But people with a proactive/interactive attitude will be ready to *transcend their existing state* and seek to engage in purposeful conscious evolutionary inquiry.

We can differentiate two major strategies for transcendence. *Strategy type "A"* is applicable to people who are not convinced that we should engage in conscious evolution or fear the "risk" of "big change." The task of this strategy is to attempt to reduce and dissolve the anxiety to transcend. People need reassurance that engaging in evolutionary inquiry is the only viable choice. *Strategy type "B,"* on the other hand, is for those who are "ready to go," who are prepared to engage in conscious evolution. They hold that it is a waste of time to analyze what is wrong now. They are ready to transcend the existing state and move into the evolutionary design space. Next, I review these strategics. These strategies are essential components of evolutionary epistemology.

10.1.2.1 Strategy "A"

Unfortunately, I found that our current repertoire of type "A" strategies—applicable to evolutionary inquiry—is very limited. We have a key task ahead of us to enrich the repertoire of the "A" type strategy. I have found the transcendence strategy developed by Ackoff (1981) and Weisbord's (1992) Future Search Conference approach useful to mitigate the fear of change. But most significantly, the application of generative dialogue seems to be a most effective approach to the "A" type strategy.

10.1.2.1.a Reference Projection. Ackoff (1981) calls his transcendence strategy *reference projection.* It extrapolates the performance characteristics of the present state from its past into the future. We ask: *What would happen if we would not change?* A thoughtful exploration of this question usually shocks people into a realization that a "big change" must happen. A consideration of the question leads to readiness to engage in conscious, self-guided evolution. Before I continue describing the other type "A" strategies, an activity is in order.

Activity 32
Develop a scenario in response to the question Ackoff is asking: What would happen if we would not change the ways and means of our societal life and the systems and institutions of the Third Generation HSS? Enter your scenario in your workbook.

Conscious evolution engages our families, groups, neighborhoods, systems in which we live and work, our communities, and societies in evolutionary inquiry in the envisioning of images of a desired future and the design of systems that bring these images to life. These social entities are the human activity entities of evolutionary inquiry. But they can engage in such evolutionary inquiry only if they are creating collective willingness to transcend the existing state and establish a common ground. The next two approaches to type "A" strategy aim to create such common ground.

10.1.2.1.b Creating a Common Ground (Weisbord, 1992). In Chapter 9 we introduced Weisbord's design thinking. He holds that working with various evolutionary entities, we should recognize that people in these entities have their own values, convictions, and ways of thinking. This is what diversity is about. In evolutionary inquiry, rather than avoiding, pasting over, or tiptoing around diversity, we should confront it, use it, and value it. Weisbord suggests that "we should put our energy into staking out the *widest possible common ground* all can stand on without forcing or compromising" (p. 7). Then, from that solid base we can engage in creating our desired future.

10.1.2.1.c Creating a Common Ground by Engaging in Generative Dialogue. The kind of social discourse described here as dialogue (Bohm and Edwards, 1991) leads to an exploration of shared meanings. It opens up into honesty and clarity. Everything that happens in the course of dialogue "is 'grist for the mill' and serves as an opportunity for learning how thoughts and feelings weave together, both collectively and individually" (p. 186). If members of an evolutionary community are able to hold all their assumptions in suspension, they will *generate shared consciousness.* (The root meaning of consciousness is "knowing it all together.") In a dialogue the individual's and the group's "knowing it all together" form a subtle higher unity and *come together in a harmonious way*. In the dialogue event people are able to be honest and straight with each other, they level with each other, and they share content freely. They develop a common mind, a *shared mind*, that can think together in a new and creative way. They awaken their *collective intelligence* and feelings of genuine participation, mutual trust, fellowship, and friendship. They can think and talk together. Shared meaning and understanding flow freely in the group. However, they can do none of this if hierarchy or authority is represented in the group. The organizing principle of dialogue implies a *change of how the mind works*. In true dialogue a new form of consensual mind emerges, generating a rich, creative order between the individual and the group as a more powerful force than is the individual mind alone. People who learn the potential power of such a dialogue will be able to transfer the spirit of dialogue into their activities and social relationships and into the systems and communities in which they live and work. Being able to attain such a state of transference creates the condition for transcendence. The learning and using of dialogue may create a new culture in the various social entities of evolutionary inquiry communities. (I will elaborate the meaning and use of dialogue as the preferred mode of communication in evolutionary inquiry in Chapter 12.)

10.1.2.2 Type "B" Strategies to Transcendence

The use of several type "B" strategies to transcendence is applicable to people and groups who are *"ready to go"* to enter into the evolutionary design space. They are ready to transcend the existing state rather than engage in an analysis of problems or explore the past and present. These "ready to go strategies" are introduced next.

10.1.2.2.a Nadler and Hibino. Nadler and Hibino (1990) suggest that rather than engaging in an analysis of the current state we should focus on *envisioning our purposes of the future.* Purposes challenge existing assumptions and thinking. In the present case they challenge the ways and means, the systems and institutions of the Third Generation. "The purpose principle gives you a mechanism for seizing the opportunity to transform the problem into productive change" (p. 111). If people accept the problems of the current evolutionary state as presented, it leads them to "obvious solutions" and eliminates the opportunity for breakthrough solutions. Willingness to transcend is an important defense against the analysis-first—and the technology traps—of conventional problem solving or improvement approaches.

10.1.2.2.b Hammer and Champy. Hammer and Champy (1993) have not addressed evolution per se, but focusing on the business community they in fact call for transcendence. They suggest that the structure, management, and performance of American business throughout the twentieth century have been shaped by principles laid down two centuries ago. It is time, they say, that we retire those principles and start over by developing a new set. This "starting over" and the development of new principles and practices does not mean analyzing current problems and trying to fix them. It does not mean tinkering with our existing systems and making incremental changes or "jury-rigging" them so that they can work better. It means asking this question: "If we were re-creating this company today, given what we know and given current technology, what would it look like?" They call for "tossing over old systems and starting over" (p. 31).

The world in which our systems operate today has changed beyond the limits of their capacity to adjust. The principles on which they are organized were superbly suited to the conditions of an earlier era. Their basic organizing principles are sadly obsolete. Hammer and Champy are today the most vocal advocates in the corporate world of the type "B" strategy of transcending and leaping out from the existing system. For them this strategy means not only leaving behind the past and present but also unlearning the principles and techniques that brought success to our systems in the past. These principles and techniques do not work anymore.

10.1.2.2.c Banathy's Proposal. Banathy's Proposal for a type "B" strategy (1996) is based on a recognition that when a new stage emerges in the evolution of a society, such as when the post-industrial knowledge age emerged in the second part of this century, the continued use of old "cognitive maps" that have guided our thinking and actions create increasingly more problems. Many, if not most, of our social systems still operate based on the *outdated cognitive maps of the industrial society* of the bygone machine age. These old maps lost their viability. As I said in one of my columns on systems design (1991), regardless of how much money and effort we put into it, we cannot improve or restructure or fix a horse-and-buggy so that it becomes a spacecraft.

Today our horse-and-buggy style systems, born in the last century, operate in a continuous crisis mode. They surely face continuing decline and eventual termination unless they: (1) understand the new realities of our transforming era and grasp their implications for evolutionary inquiry, (2) are prepared to leave their old ways of thinking and doing behind and learn new ways, (3) transcend their existing system, and (4) envision a new image of their future. Once we accomplish (1) and (2) and understand the new realities and their implications for our individual and collective lives, and if we are prepared to learn new ways of thinking, we are *empowering ourselves to transcend the present evolutionary state.* Transcending is probably the most troublesome aspect of evolutionary design. In addition to the anxiety and ambiguity that transcending the system generates, the crucial issue that we face is: Do we have the willingness and evolutionary competence to leave the past behind and enter into the territory of the not yet known? Transcending the familiar, the "what is now and what is known" requires that we develop such individual and collective willingness and competence.

10.1.2.2.d Dealing with Change. Bridges (1991) suggests that "change is the name of the game today, and organizations that can't deal with it effectively aren't likely to be around long" (p. ix). The effective way of *dealing with change requires transcendence and transformation*. Unless transformation occurs, change will not work. And the starting point of transcendence is the ending that you will have to make by leaving the old situation behind. We have to let go of the old reality, the old setting, the old thinking.

The first step of transcendence is to "mark the endings." Do not just talk about "the leaving behind" but create activities that dramatize the ending of what has been. Commenting on his strategy for "leaving behind," Bridges says that he is not urging to let it go slowly piece by piece, just the opposite. "Whatever must end must end. Don't drag it out."

Reflections

"One does not discover new lands without consenting to lose sight of the old shore" (André Gide, French novelist). In this section, I have focused on *transcendence*, which is the initial strategy of self-guided evolution. We reviewed conditions that are to be present for us to recognize the need for change. We differentiated change that aims at the improvement of the existing situation from the discontinuous "big change" of evolutionary leap. We explored two types of transcendence strategies.

We have found that there are people who have a high level of anxiety when they are challenged to leave the existing situation behind. They are rather fearful in the face of uncertainty and the unknown that "leaping out" brings with it. Thus, our best approach in such a situation is to be sensitive to the anxieties that emerge as people are faced with big change. The type "A" strategy reviewed here aims to reduce those anxieties and enhance the likelihood of developing an informed judgment that might lead to a commitment to enter the evolutionary design space.

The type "B" strategies help those who are ready now to transcend the current situation and engage in evolutionary inquiry. Restating Bernard Shaw: They dream about an evolutionary future and ask why not?

I close my reflections with an adaptation of Chris Jones's (1984) remark. Entering into the evolutionary design space is a process of unlearning what we know of what exists, leaving behind what we call the "status quo." We need to reach the point where we are able to lose our preconceptions sufficiently to truly understand life and learn new ways we see it.

Activity 33

Task 1: Review the core ideas on transcendence and contemplate the difference between solving problems and improving an existing system and change that invite transcendence. Looking at a system that you are involved in develop a rationale for transcendence. Enter your rationale in your workbook.

Task 2: Choose a system in which people have a rather anxious (even fearful) attitude toward change. Review the strategies introduced as type "A" and select and synthesize those that appear to be most useful in reducing anxiety and leading to a decision to engage in evolutionary inquiry. Explain the reason for your selections in your workbook.

10.2 Envisioning a Guiding Image of Our Evolutionary Future

Evolutionary guidance implies arrangements and operations that are built into the various human activity systems at all levels of the society to enable them to design their own future. Created by our families, our neighborhoods, by the systems in which we live and work, our communities and societies, and ultimately our global system, these arrangements and operations will guide the work of these systems in the evolutionary design space.

In this section, I introduce various strategies and methods that might be useful in exploring and envisioning our evolutionary future and creating its first image. The design journey that leads from the "leaping out" from the existing system to the final outcome of a comprehensive description of the future system is a long, adventurous journey. It begins with a fleeting vision of the "what should be" and it proceeds from there through a *series of "pictures" of the desired future.* These pictures become ever more elaborated, ever more specified and differentiated, and increasingly clearer and detailed through the process of the evolutionary design inquiry.

The first leg of this journey is the envisioning of an image of the future system. It calls upon our *creative intelligence, our unrestricted imagination, and our unbounded thinking* and becomes possible only if we free ourselves from the here and now; if we leap out from and transcend the present state which would only lock us into what now exists and would constrain and restrict our creative envisioning and imaging. In the present section, we explore the major activities of envisioning and creating the first image of the system.

10.2.1 Envisioning the Future

> Where there is no vision, the people perish.
>
> *Proverbs XXIX. 18*

For the term "vision" we find a two-pronged definition in Webster (1979): "the act or power of imagination" and "the act and power of seeing." In evolutionary inquiry, *we use the power of imagination and the power of envisioning the future* in the course of our "vision quest." Our vision quest commences as we leap out from the existing system and capture a vision of an *ideal future*. The image of our evolutionary future must be an ideal image so that it can pull us—like a magnet—in the course of our evolutionary journey.

10.2.1.1 A Vision of an Ideal Society

First, we seek a vision of the ideal society that would embed the system in which we live and work. Our quest is guided by several questions, such as: What kind of *society do we wish to create*? What is our *vision of an ideal society* that will give inspiration and meaning to our quest? What is that larger vision that will have the power to guide us toward the desired future state?

10.2.1.2 Valuing the Ideal

If an *ideal society* were to exist right now, what aspects/characteristics of that society would you *value the most*? In our design we often state this question as follows: You have dreamed last night that you live in an ideal society. You awake and find that now you live in that ideal society. What in that society do you value the most?

10.2.1.3 What Can We Do about It?

Once our quest is rewarded with an image of the society we wish to have, only then do we ask the question: What is *our vision of a system* through which we can make a contribution toward creating that society?

10.2.1.4 Evolutionary Epistemology

Evolutionary inquiry is *disciplined inquiry* which is guided by a clearly articulated epistemology. Our present task is to construct such an epistemology. But evolutionary inquiry must also be an *informed inquiry* based on knowledge. Therefore, an *evolutionary inquiry community*—be it a family, a neighborhood, an organization of any kind, a real or virtual community—must avail itself to knowledge sources that inform the community about the kind of questions proposed in the preceding paragraphs.

The outcome of our vision quest most likely will be a highly inspirational formulation, which expresses at the most general level the evolutionary future we seek. The questions introduced earlier should be considered "triggering" questions that guide the conversations of our evolutionary inquiry community in its vision quest. Well-conceived *triggering questions*, well defined *methods of conversation*, clearly *articulated epistemology*, and relevant *knowledge bases* provide a structured, disciplined, and informed approach to vision quest and image creation. The conversation at this stage should lead to the formulation of a shared vision, a shared picture of the future that presents the first *shared collective identity, a collective sense of destiny in the designing community.*

I describe here a method, adapted from Moore (1987), and develop it as one possible method of vision building. In collective evolutionary inquiry, consensus building methods are used throughout the entire process. An example of a vision building scenario follows.

Upon establishing an evolutionary inquiry community, the group assigns the role of coordinator to one of its members. Then, they begin to formulate the kind of triggering questions introduced in the preceding. Next, members of the group have a contemplating period and begin to write down their response to the triggering question. Then, the items are read and clarifying questions are asked. The items are recorded on newsprints attached to the walls of the room. There is now an opportunity to integrate (pull) some of the visioning ideas and cluster them into idea systems that are complementary and internally consistent. Then, the ideas might be prioritized. If several groups take part, the groups might report back to the entire inquiry community. The newsprints, developed by the small groups, are attached to the wall and the larger community begins to work with the visioned ideas to create a synthesis of the visions. Findings are recorded. This process and others are introduced in some detail in Chapter 12 in the section, "Conversation."

Activity 34

To have a grasp on the process of envisioning, you are asked to arrange for an activity that would enable you and others to experience vision quest in a real-life context. The group you identify can be your own family, people in your neighborhood, any group with shared interests, friends from an organization you belong to, a study group, and so forth. You should arrange an appropriate place and time that will enable the group to spend several future building inquiry sessions. The group should follow the process described earlier in exploring the three inquiry items: ideal society, valuing, and our contribution to the ideal. The sessions should be distributed in time, allowing participants to work with selected knowledge sources in between the sessions. Enter your findings and your assessment of the sessions in your workbook.

10.2.2 Creating the Image of Our Evolutionary Future

I develop this section in three parts. The meaning of societal image is explored first. Next, an approach to the creation of an evolutionary image is introduced. Then, these two processes of evolutionary epistemology will lead to the introduction of the notion of evolutionary guidance systems.

10.2.2.1 The Notion and Role of Image in Evolutionary Inquiry

> The image not only makes society, society continually remakes the image.
>
> *Kenneth E. Boulding*

Changes in societal images are of particular importance today. Our society is now undergoing a transformation that is more profound, more intensive, and more dynamic than any time since the emergence of the Cro-Magnons. This transformation is to be guided by an evolutionary image that is being shaped in the evolutionary design space of the emerging Fourth Generation HSS. This image is going to be vastly different from the outdated image of the industrial era of the Third Generation. The persisting use of the outdated image is *creating a dangerous evolutionary gap* between what is and what should be.

Whenever we wish to study the notion of image, many of us go to Boulding's classic book, *The Image* (Boulding, 1956). Boulding's first proposition is that *behavior depends on the image*. The behavior of members of an evolutionary designers community depends on the image they have about the future. That image will guide them in their evolutionary design.

Bouding's second proposition is that our experiences provide us with *messages that produce changes in our image*. A majority of the messages do not bring about change. There are some, however, that call for adjustments or that lead to a clarification of what had been vague before. This type of change attempts to adjust or fix the current evolutionary state. However, when a message hits the nucleus of the image, "the whole thing *changes in a quite radical way*" (p. 8). Images are resistant to change. Thus we often reject a message that challenges our prevailing image. If the message persists, we begin to have doubts. "Then one day we receive a message which *overthrows our previous image* and we revise it completely" (p. 9). The goal of this work is to convey such a message.

Our images are rich and complex. Boulding classified their dimensions as spatial, temporal, relational, and personal (which is the picture of the individual in the midst of the universe of his environment). Modifiers of the image are conscious/unconscious/subconscious, certain/uncertain, clear/vague, having or not having outside reality. The image can be private or publicly shared.

Our present concern is primarily the *public image, the evolutionary image that makes the future society*, the image that at the same time is shaped by the evolving and transforming society. It is important for us to understand the dynamics of this *mutual shaping*, inasmuch as we are to create a new evolutionary image. Earlier we named this dynamic recursive shaping *coevolution*.

The basic bond of any society, culture, subculture, or organization is an image, the essential characteristics of which are shared by the individuals participating in the group. In designing our evolutionary future, it will be up to designer communities everywhere to create the societal image, understand what it means to them, and how an evolutionary guidance system can be designed that both *unfolds that image and shapes it*.

10.2.2.2 Sources of Creating the Image

When a new image leads sociocultural evolution, it *provides direction for change* and gives meaning and lends reinforcement to each movement toward the realization of the image. We are now in *search of a new evolutionary image* to direct us toward the conscious evolution of our evolutionary systems. In creating the image, we look for sources that can guide us in creating the image. There are two kinds of sources: *invariant* that have guided humanity through the ages and sources that yield knowledge and insight that reveal important *variants*.

10.2.2.2.a The Invariant. Huxley (1945) called the invariant the "perennial philosophy." The invariant elements can be found in the traditional lore of various cultures as well as in advanced philosophy and in the primitive and the most advanced forms of religion. The central characteristics of this image include the notions of *harmony, balance*, and *wholeness*; the universe of *consciousness*: individual, collective, cosmic; the awareness of and the quest for the realization of the *higher self*; the limitlessness of *human potential*, motivation toward development and *creativity*; and being directed by higher consciousness that enables our participation in conscious evolution.

10.2.2.2.b The Variant. The variant sources of the image are the *sets of core values* and beliefs that the members of the evolutionary designer community define as ones they share. The *variant is unique to the designers*. They develop these as they ask what values they would cherish and hold important in an ideal future society. An example follows by introducing a set of values that were nominated as sources of creating an ideal image of education (Banathy, 1991).

- There are two *absolute values: the individual and the global system of humanity*. (Arrangements of human activity systems between these two are sociocultural inventions that should serve both. Systems of learning and human development are such arrangements.)
- The most valuable resources on earth are the *uniqueness and the unique potential* residing in the individual, in the family, and in our various societal systems.
- Among the highest order values of human rights are *the freedom and the right to learn*.
- The attainment of *inner quality of life* and its enrichment should be central in life's aspirations.
- High value is placed on *ethical, moral, and spiritual development and cultural diversity*.
- The increasing significance of *maintaining interpersonal and social relationships* as a key value in societal life, coupled with the value placed on *cooperation*.
- Among all the societal functions provided to individuals and collectively to the society, the *nurturing of learning and human development* is of the highest value.

10.2.2.3 The Shaping of an Evolutionary Image

The findings, generated in the course of the vision quest, become an *essential knowledge base* that is used in the course of creating the image of the future evolutionary system. This "essential knowledge" emerges from the entwinement and integration of *the vision* that was formulated, *the core values* that the designers collectively articulated, and *the knowledge base* that emerges from working with knowledge sources. This "essential knowledge" is the basis upon which we can create the image of our evolutionary future. This image provides us with a "broad-stroked picture" or a "macro-view" of the future evolutionary system.

10.2.2.3.a The image creating strategy unfolds by attending to the following tasks: (1) *Organize the knowledge*—generated in the course of the previous activities—*into an internally consistent system of core ideas and core values that represents and elaborates the various interacting dimensions of the human experience. (2) Label each domain with an evolutionary marker. (3) Organize the markers into a system of markers that represent the first image of the Fourth Generation HHS.*

The power of the image depends upon the thoroughness and completeness of the work we have accomplished—and continue to accomplish—in the evolutionary design space. The more complete and detailed the envisioning process, the more core values and core ideas we define and agree upon, the more thoroughly we define experiential domains and organize their markers, the more attention we give to organizing the markers and their interactions into the *image*, the more confidence can we have in the power of the image.

10.2.2.3.b Working with the Markers. Does the term "markers" sound familiar? It was probably one of the most often used words in Parts I and II. In the context of the evolutionary journey as well as in elaborating knowledge about evolution, we used the term "markers" to denote "evolutionary markers." Evolutionary markers, organized in a system of markers of a particular evolutionary generation, comprise the *evolutionary systems model* of a particular evolutionary generation of HSS. In the course of the emergence of a new generation of HSS, the human experience is reorganized at a *higher level of complexity* by the emergence of a *higher level consciousness*. This new consciousness then organizes the various dimensions of the human experience into an internally consistent sociocultural system, in which the markers interact, coevolve, and mutually influence and shape each other. This emerging and coevolving, maturing evolutionary process leads to a state where the *evolutionary qualities* of balance, harmony, internal consistency, synergy, symmetry, synchronicity, and wholeness are all manifested in the behavior of the new evolutionary generation. We also noted that successive evolutionary generations may have *different and differentiated evolutionary markers*. Markers also may *increase in number* and in individual and collective *complexity*. Furthermore, individual markers may acquire different characteristics. Human and societal *evolution is a continuous process of change*: *change by learning and learning by changing*. This changing/learning is due to *maturation*, to the unfolding of—yet unrealized—*evolutionary potentials*, to *coevolution* with the changing *environment* and—most importantly—by the driving force of *creativity* and the guiding and organizing force of *consciousness*.

We are now ready to move into the evolutionary design space of the Fourth Generation HSS and engage in a future creating process as described in Activity 35. The process of addressing the tasks of this activity constitute a major part of self-organizing, self-guided evolutionary epistemology. By accomplishing the three-pronged activity, we will create *the first systemic normative representation* of the Fourth Evolutionary Generation HSS in the specific context in which we live. This is a conscious process of creating a specific image for a specific evolutionary inquiry community.

Activity 35
In creating the image of a desired future evolutionary state your task is now to work with the findings of your vison quest. It is from these findings that we are to create the image of the desired future evolutionary state. We accomplish this by clustering our items that seem to belong into a category. Then, we give names to the clusters. We next nominate these clusters as our evolutionary markers of the Fourth Generation. In accomplishing this task, follow the consensus building process. The outcome of this task is the first high-resolution systems model of the Fourth Generation. Construct a figure that shows the systemic organization of the evolutionary markers. Enter your findings in your workbook.

Reflections

We have made great strides in creating and using methods of evolutionary epistemology. We developed methods of creating an evolutionary image. These methods comprise the core activity of working in the evolutionary design space. We also created an image of the Fourth Generation, which is a specific image of our desired future system. It is this image around which our evolutionary systems will be designed. The image we develop is a *unique image*. It reflects and represents *our unique individual and collective values, hopes, and expectations, and the uniqueness of the specific situation of the designing community.*

Out of the many unique images created by the various evolutionary design communities could emerge a *"mega-image,"* if and when we establish *common ground* among various designing communities. This will be a long process, but mini-minuscule in the larger time frame of sociocultural evolution. Now, I am ready to introduce the idea of the evolutionary guidance system.

10.3 Evolutionary Guidance Systems

As an introduction to an elaboration of evolutionary guidance systems (EGSs), in this section I retrace the path of moving from the idea of evolutionary guidance to a generic formulation of an EGS. Then, I explore how EGS works (and should work) in the evolutionary design space. I will identify pitfalls in designing EGSs, and present an example of applying an EGS in a specific functional context.

10.3.1 From Evolution to Evolutionary Guidance to Evolutionary Guidance Systems

What follows here is a short review of our learning journey, starting from a brief restatement of evolution, to revisiting the notion of evolutionary guidance, and arriving at the point of exploration of EGSs.

10.3.1.1 A Brief Restatement of Evolution

Evolution means *"unfolding."* It is a process by which successive evolutionary generations unfold creatively in the evolutionary design space of HSS. The unfolding brings forth the various interacting dimensions of the human experience, organized into the *wholeness* of an emerging evolutionary generation.

As we saw in Chapters 3 and 4, the emerged whole can be represented as a systems model of a specific evolutionary generation of HSS. If we develop evolutionary consciousness and engage in conscious evolution, the unfolding process described here will be manifested in all of our human systems, from the family on to the global human community. It has been an unconscious process heretofore. Now it can become conscious, and we can guide it.

10.3.1.2 Evolutionary Guidance

Evolutionary guidance is a *dynamic process* in which human activity systems operating at the various levels of the society engage, in order to give direction to their evolution. Engaging in evolutionary guidance implies building into the various human activity systems arrangements and operations by which those systems can move toward the realization of their envisioned evolutionary image. Conditions of such engagement include: (1) a *commitment* to conscious self-guided evolution, (2) the acquisition of evolutionary *competence* by evolutionary learning, (3) the acquisition of *a knowledge base* relevant to the various dimensions of the Fourth Generation HSS, and (4) the *learning and using of methods* of evolutionary epistemology by which people in the various human activity systems are empowered to give direction to their own evolution individually and collectively. Of the four conditions, we have already elaborated the first two. The rest of the work will address items (3) and (4).

10.3.1.3 Evolutionary Guidance Systems

As we have stepped over the threshold of conscious evolution, we are in search of *means and methods* by which we can consciously and purposefully guide the unfolding of the evolution of the Fourth Generation. What follows here is a definition and presentation of such *ways and means.* The means is *an intentional system,* which can be built into the various human activity systems to enable these system to guide their entry and their work into their own evolutionary design space. I call this intentional system the *evolutionary guidance system (EGS).* Later, I propose *methods* by which EGSs can be designed and operationalized.

The EGS described next is *just an example.* It has been offered as a *generic* EGS that can be adapted for use in several functional contexts. It has already served that function. Thus, *the EGS presented here is not prescriptive.* It is descriptive and contemplative. It may assist others to *develop* their own EGS. Before I introduce the generic EGS, I propose a set of functions addressed by EGSs.

10.3.2 Key Functions of Evolutionary Guidance Systems

It will be beneficial at this point to state the key functions of EGSs. A statement of functions will assist in the unfolding elaboration of evolutionary epistemology, the work in the evolutionary design space of EGSs, and it will also serve as criteria for judging the usefulness of EGSs.

10.3.2.1 The EGS Is Ideal Seeking

Based on an ideal image of a future evolutionary state, we design an EGS that has the function of guiding the work of people in the various human activity systems who are engaged in creating their own evolutionary future. It is essential that the image we create is an *ideal image.* We place the ideal out there on *the horizon, way out in the future.* The ideal becomes a magnet that pulls us toward it. The ideal inspires us. It is the ideal to which we can make a commitment.

10.3.2.2 Turning the Arrow Around

We do not create our evolutionary future by working out from the problem space of the current evolutionary state by trying to fix problems or improve the current state. We transcend what is and jump far ahead in the future seeking an ideal evolutionary state. As stated previously, we place the ideal image on the horizon, and then *work back from it.* We turn the arrow around. We ask: Given our definition of the ideal system and the constraints of our present situation, how much of the ideal can we put into operation right now? We answer this question by *"putting up for experimentation and testing"* various alternatives and select and develop the most powerful, the most promising, the most feasible alternative. We call the most feasible alternative the evolutionary system (ES). The ES is a representation of the EGS that can be implemented here and now. The ES becomes the operating system of the evolutionary designing community as it moves toward the ideal system of the EGS. Figure 10.1 depicts what we have explored in the preceding.

10.3.2.3 The Moving Horizon: Evolutionary Design Never Ends

The next implication of the function of EGS is that moving toward the ideal means that *design never ends*. As we experience living in the evolutionary system we have designed as a "here and now" representation of the ideal system, we are becoming aware of *how* our system moves us toward the ideal. We also become aware of what I have called in my earlier work the design *law of the moving horizon*. Namely, as we move toward the horizon, the horizon moves ahead of us. We develop ever higher levels of collective consciousness that give birth to ever higher quality values, aspirations, and images. Animated by such consciousness and inspired by emerging values and aspirations, we recreate our ideal image, which then leads us to continue our design. In fact, we *never leave the evolutionary*

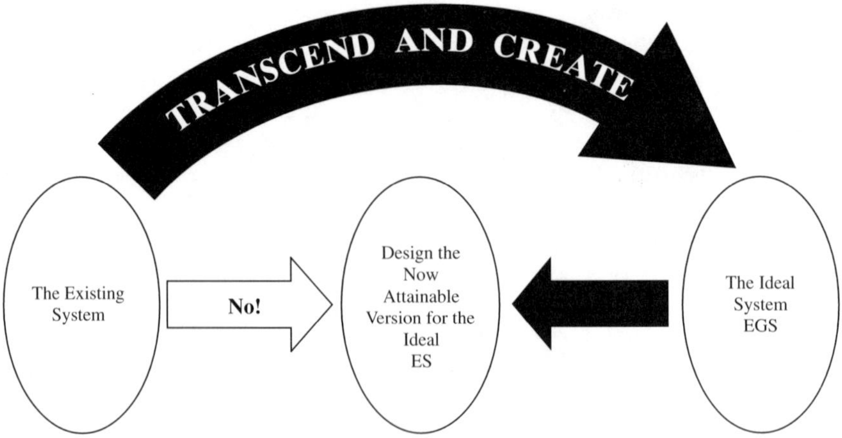

FIGURE 10.1 Turning the Arrow Around

design space. As Erich Jantsch remarked: *We are evolution and we are responsible for it.*

10.3.2.4 We Learn as We Evolve and Evolve because We Learn

As we travel on the road of our evolutionary design journey, we *learn from what we do* and we are conscious of the way we evolve. We increase our capability to become *ever more competent in design* as well as *ever more competent in living in the evolutionary system we designed*. We learn increasingly more about evolution itself, which enables us to guide our evolution. We learn to make new distinctions, by which we increase complexity. This, then, will lead to the *reorganization* of our system at a higher level of complexity. *Evolutionary learning never ends.*

10.3.2.5 Unfolding New Potential

Species evolve by enfolding evolutionary potential. New potential unfolds as a species emerges from the evolutionary design space. In the course of the life of an evolving evolutionary generation, *new potential is built upon unfolded potential*. Evolutionary potential is not fixed. It can learn. Evolution always offers new possibilities. While up to now the unfolding of potential was brought forth by evolutionary forces operating in the evolutionary design space, now it *becomes our task to unfold* and *create evolutionary potential*.

10.3.2.6 Realizing Evolutionary Qualities

Evolution aspires to perfection. Our journey toward understanding how sociocultural evolution has worked helped us to recognize the *qualities* that evolution aims to achieve. We suggested three categories of qualities: *core qualities, enabling qualities, and meta-qualities. Core qualities* include balance, harmony, internal consistency, synergy, and symmetry. Self-referencing, self-organizing, coevolution, and cooperation are *enabling qualities*. Collective consciousness, wholeness, and creativity are *meta-qualities*. These qualities are becoming measures of the health of the evolutionary systems we design.

10.3.2.7 Breaking the Vicious Cycle

The final point I wish to make is that the accomplishment of the tasks elaborated in the preceding enables us to *break the vicious cycle* that has operated in previous evolutionary generations. Guided by the ideal, and continuously designing our lives toward the ideal, we can avoid the *evolutionary traps* of change denial; the status quo of habitual orders of customs, laws, and institutions; rigidity and inflexibility—all leading to decline, decay, and the eventual destruction of an evolutionary generation. Guided by the ideal image, *we can avoid the fall into the evolutionary traps, provided* we continue to move toward an ever more desired and hopeful ideal evolutionary state.

10.3.3 A Generic Evolutionary Guidance System

If guided evolution is possible—as I suggest it is—we face three critical questions: (1) What systems enhance the creative purposeful unfolding of human evolution—from the family on to the global human community—along the multidimensions of human experience? (2) What dimensions represent the wholeness of human experience? and (3) How can we go about designing these systems?

In the preceding subsections, I have developed an answer to the first question, as I have explored a detailed definition of EGS and reviewed its functions. What comes now is providing an answer to the second question: What are the existential dimensions of an EGS? What are the dimensions that account for the wholeness of the human experience? I introduce these dimensions in the next section. Before I proceed, you are asked to complete an activity.

Activity 36
It becomes your task now to define the evolutionary dimensions that represent the wholeness of human experience. These dimensions will become the markers of your EGS. The knowledge base of this activity is the findings you produced in working with the earlier activities of the present chapter. Introduce your findings in your workbook.

10.3.3.1 The Evolutionary Dimensions of a Generic Evolutionary Guidance System

Here I introduce a generic image of an EGS as an arrangement of a set of interacting dimensions of the wholeness of human experience. The dimensions comprise a set that is contemplative. By no means is it prescriptive, it serves only as an example. The set I introduce was presented in previous publications (Banathy, 1987a, 1996, 1998). As a generic set, it is context free. It is not constructed in the functional context of a particular human activity system. The construction of the set has been inspired by several thinkers of the public philosophy arena, starting with the Greek philosophers who identified truth, plenty, goodness, and beauty as necessary and collectively sufficient pursuits for the development of man. The set is contemplative as it invites thinking about what dimensions should constitute an EGS. The generic set has the following dimensions:

- A *social action dimension* aimed at ensuring *social justice, the attainment of a civil society,* and the continuous *increase of cooperation* and *integration* of our societal systems at ever higher levels of complexity.
- A *socio economic dimension* with a focus on *economic justice,* the *nurturing of social capital and social economy,* and the practice of *integrated and indigenous development.*
- A *moral dimension* of our lives that is manifested in actions guided by *self-realization, social, and ecological ethics.*
- A *wellness dimension,* the *nurturing of the physical, mental, emotional, and spiritual health and well-being* of the individual and the society.
- A *learning and human development dimension, aiming at the full development of individual, social, and societal potential,* coupled with the creation of evolutionary learning communities and an evolutionary learning society.
- A *scientific dimension,* which is ethically based, mobilized for the enhancement of societal evolution, promoting human and social betterment, and is in the service of the common good.
- *A technological dimension,* which is placed under the guidance of the values and aspirations envisioned in formulating the evolutionary image and that serves the quality of life of all.
- An *aesthetic dimension* in the pursuit of beauty and *cultural and spiritual* values, the various forms of *creativity* in all aspects of life, all aiming at the enrichment of our inner quality of life.
- A *communication dimension* that enables high-quality human and social interaction. It applies methods of communication that enhance authentic participation and collective decision making and extend the time and space boundaries of the human experience into ever broader scope, creating systems for local to global communication.
- A *societal/polity dimension* of self-determination and self-organization, of genuine participation in governance. This is a dimension that nurtures peace making and social reconciliation, one that promotes continuous action for human betterment and the *improvement of the human condition.*

The definition of the various dimensions of the set reflects my own public philosophy, my own values and beliefs as well as the collective consciousness I share with groups I am working with. (This statement underlines the notion of the *uniqueness* of each and every design situation.) The set proposed is *a possible set*. In any particular EGS, a substantive definition of the dimensions will be animated and influenced by the collective consciousness of the particular human activity system that engages in evolutionary inquiry.

At this point, several observations are in order. First, there is *no claim that the dimensions set is complete*. However, it is sufficiently inclusive to serve as an example in creating EGSs. Second, when the set was used as an example for applications in various contexts, invariably *it has been added to or modified* to fit the context and the intentions of the groups that applied it. Third, each of the dimensions calls for the *development of a knowledge base* that is used in the full development of the dimension. Once developed, the statements represent a "common ground" for the people who create their EGSs. Fourth, the set is just a set, waiting to be organized into an EGS. Finally, I have yet to see a human activity system that would demonstrate the conscious and purposeful application of the dimensions described here.

10.3.3.2 The Systemic Components of EGSs

An EGS has three major components: a set of *experiential dimensions, a set of generic evolutionary perspectives,* and a *set of organizing perspectives*. The set of dimensions attempts to define the evolutionary markers of an EGS. The purposeful, knowledge-based development of these dimensions will provide a powerful agenda for the self-directed evolution of our human systems. The second component of an EGS is a set of *generic evolutionary perspectives* that guide our work in the evolutionary design space. The third component is comprised of a set of *organizing perspectives that are specific to the EGSs* we design.

Figure 10.2 portrays an image of organizing the experiential dimensions into a systemic arrangement of an EGS. In the center of the figure are a set of *evolutionary perspectives* and a set of *organizing perspectives* that guide the design of the EGS. The two sets have an integrating, systems creating effect. Observe at the bottom left-hand corner of Figure 10.2 the term "interactive" dimension. It refers to the interacting process that integrates the various experiential dimensions of the EGS into an image of the future desired evolutionary system.

10.3.3.3 Evolutionary Perspectives

The evolutionary perspectives that operate in the evolutionary design space are *generic* to all EGSs. These are *collective consciousness, wholeness, and creativity.*

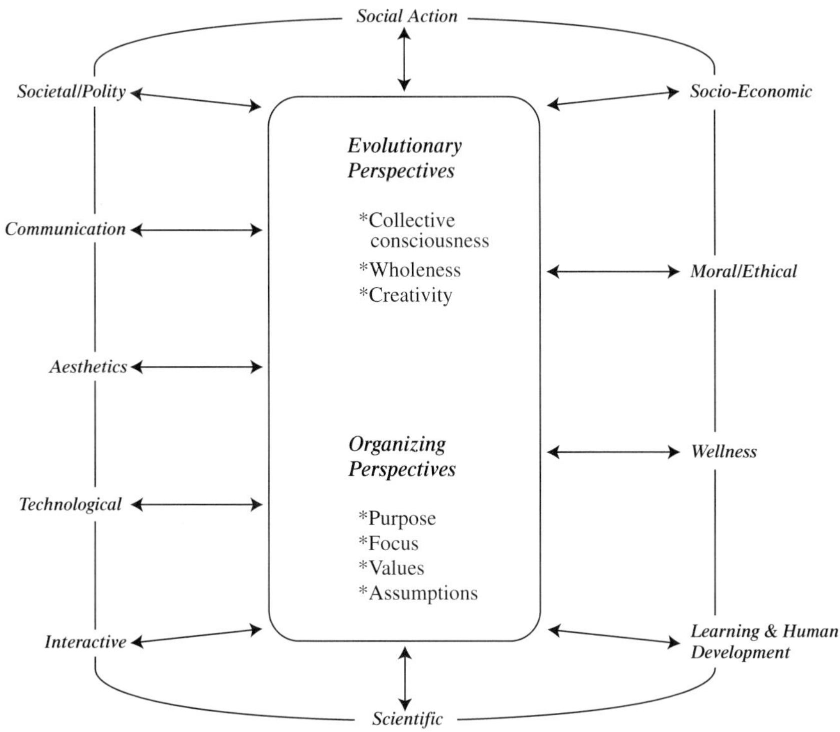

FIGURE 10.2 The Generic Image of Evolutionary Guidance Systems

10.3.3.3.a Collective Consciousness. Collective consciousness is the force that triggers the emergence of an evolutionary generation. It is infused in all of the dimensions of the EGS. It creates a shared view of the world as well as generates and maintains internal consistency. Without the development of collective consciousness, an evolutionary experimentation will fall apart. With it, it will become whole and will prosper.

10.3.3.3.b Wholeness. Wholeness is a meta-quality of an evolutionary system. The attainment of wholeness is a core purpose of emergence. The whole organizes the parts, the various experiential dimensions—the evolutionary markers—into an evolutionary system.

10.3.3.3.c Creativity. Creativity is the core evolutionary force that operates in the evolutionary design space. Creativity brings forth novelty. It creates design solution ideas and formulates from design ideas design alternatives. It is creativity that enables us to engage in conscious, self-guided evolution.

10.3.3.4 Organizing Perspectives

Organizing perspectives of EGSs are *specific* to any human activity system that enters the path of conscious evolution. Perspectives include the following.

10.3.3.4.a Purpose. Earlier, we identified a set of *generic purposes* of evolutionary guidance. Beyond these, there are purposes that are specific to the human activity system which engages in its own evolutionary inquiry. These purposes are unique properties of the evolutionary designer community; thus, the EGS designed shall reflect this uniqueness. Purpose is the source of power to act. Only shared purpose generates the individual and collective commitment needed to engage in evolutionary inquiry.

10.3.3.4.b Focus. Focus is specific to the functional contextual nature of a system for which we design the EGS. For example, a health prevention/maintenance system shall have *the dimensions of wellness and education* as the focus of design. The other *dimensions* are built around it and are integrated with it. This is the way of building wholeness.

10.3.3.4.c Underlying Values. Underlying values are of two kinds. Values that are generic evolutionary values were introduced earlier as evolutionary qualities, such as harmony, balance, synergy, symmetry, and wholeness. Another set is comprised of values that people in the system collectively identify and seek to realize. This set will guide decision-making in the design of specific EGSs. In designing the EGS, values are transformed into observable qualities of the EGS and become the basis of assessing the adequacy of the EGS.

10.3.3.4.d Underlying Assumptions. Underlying assumptions constitute the philosophical bases of evolutionary inquiry. For example, each and every individual has the right to be involved in decisions affecting his or her life. It is not only a right but also a responsibility. In the present context it means that no one has the right to design EGSs for someone else.

10.3.4 Pitfalls to Avoid in Designing EGSs

A pitfall is a hidden mistake or a lack of clarity that may undermine or even destroy the inquiry. In evolutionary inquiry we can recognize two kinds of pitfalls: underconceptualization and shortcomings.

10.3.4.1 Underconceptualization

In considering the design of the EGS we should be particularly concerned about avoiding *underconceptualization*. An obvious underconcepualization is compromising the ideal and not seeking and defining the most promising, the truly ideal evolutionary image. Not seeking the ideal degrades the inquiry into a short-term focused design activity. A *"dimension poor"* EGS is another and rather frequent manifestation of underconceptualization. Most of our social systems suffer from this. For example, most families organize their lives by limiting attention to the consideration of the economic dimension, health, and education. Systems of science and technology seldom go beyond their own domains with some attention to economic and political issues. Most so-called health service systems focus on sickness and disease control rather than on health promotion and health education. They are concerned primarily with the economics of their services and often neglect the other dimensions. Our business and production systems—even our governance systems—are driven by economics, mostly disregarding social and economic justice, the moral and ethical, the spiritual, and the aesthetic dimensions. As mentioned earlier, I have yet to find a human activity system of any kind that is built from the conscious and purposeful integration of the experiential dimensions of the human/social experience. Ergo, most of our social systems are dimension poor and underconceptualized—and we suffer for it.

10.3.4.2 Shortcomings

In addition to the kinds of underconceptualizations mentioned earlier, there are several potential shortcomings that threaten an adequate creation of EGSs. These include: (1) A *lack of internal consistency* among the dimensions. This happens, for example, when the definitions of the various dimensions are not grounded in shared knowledge, a shared value base, and/or clearly stated underlying assumptions. (2) The characterizations of dimensions are made at various levels of resolution; thus, they are *uneven, incompatible,* and confusing. (3) Underestimating the significance of the *disciplined nature* of the inquiry, the *need of a knowledge base* in formulating the dimensions, and the *time required* to create the EGS. (4) *Limiting the involvement* of the various "stakeholder" groups, thus sacrificing the authenticity of the inquiry. These are just a few, but they are adequate examples of shortcomings.

Activity 37
Review Sections 10.3.3.2 through 10.3.3.5 and Section 10.3.4. Respond to the following questions: What additional experiential dimension(s) would

you consider as being essential in designing an EGS? Define these new dimensions. How would you redefine the dimensions I introduced? What other evolutionary and organizing perspectives would you take into account in designing an EGS? Reviewing the pitfalls: What other items would you add to the list of underconceptualizations and shortcomings? Enter your findings in your workbook.

10.3.5 What If? An Example of an Evolutionary Guidance System

What would happen if the idea of EGSs becomes reality? What would happen if our human activity systems would engage in purposeful evolution along all the dimensions proposed in the preceding sections?

Let me speculate about the "what if" in the context of the most basic human activity system, the family. I am envisioning an evolutionary guidance system that would enable the family to shape its future and design and develop itself along the lines of the dimensions described previously. The example might generate further thinking, exploration, and conversation about the idea of creating EGSs.

The family today is basically a socioeconomic unit. Its primary concern focuses on the economic necessities of existence, health, and safety, and often education and spiritual development, but in most cases little attention is paid to the other dimensions. The question is therefore asked: What kind of families would we have if the family purposefully developed an agenda for its own life and its evolution along all the dimensions of evolutionary guidance proposed here? I will attempt to answer this question by outlining briefly a possible representation of the ten dimensions in family EGSs.

10.3.5.1 The Social Action Dimension

In the social action dimension, we might envision the creation of a family agenda for the development of social consciousness, and the realization of this consciousness in the family and in contributions made by members of the family to promote social consciousness in the community. The social action dimension can be manifested in cooperation, mutual respect, help, and the practice of social justice within the family and its promotion in the systems in which members of the family live and work.

10.3.5.2 The Economic Dimension

In this dimension, beyond attention to economic necessities, an EGS would guide the establishment of economic justice within the family. The family would actively seek to develop social capital within itself and promote it outside of the family. An agenda would be developed for the wise and value-based use of the family's economic resources.

10.3.5.3 The Moral and Ethical Dimension

In any society, the family is called upon to establish the foundations of ethical and moral consciousness and behavior in members of the family and collectively in the family as a whole. This dimension calls for the development of an agenda that will include: the establishment of ethical standards to live by; the purposeful shaping of moral character and a moral world view; and the nurturing of self-realization ethics that guide members of the family and the family collectively to develop their individual and family potential. The realization of social ethics would promote respect for each other within the family for all other individuals and cultures, and the development of genuine concern for others. By practicing ecological ethics, the family learns to live responsively and in harmony with nature.

10.3.5.4 Nurturing Wellness

The nurturing of physical, mental, emotional, and spiritual wellness has always been a salient concern to the family. The evolutionary challenge here is to develop an agenda for the family and its members for: (1) the development of physical fitness as individual and collective activity; (2) mental wellness, which actively promotes individual and collective self-reflection and creativity and attention to continuous cognitive development; and (3) emotional wellness, the foundation of which is love, compassion, altruism, caring, helping, and sharing. For the development of this dimension the family offers the most ideal social context. An evolutionary purpose is the constitution of the family as a spiritual unit, and as such, the nurturing of faith and beliefs in all its members.

10.3.5.5 Learning and Human Development

In the industrial society educational functions were mostly assumed by schooling and, with it, the school became identified as the "legitimate" institutionalized form of education. Even today, at best we ask families to support and cooperate with the school. We do not assign families the role of becoming a force for learning and human development and becoming a full partner in education. Within the evolutionary perspective, developed here, the family assumes responsibility for nurturing the full development of its members and it takes responsibility for its own collective learning and development. A new task is added: The development of evolutionary competence (as described earlier) that will enable the family to direct its own evolution and shape its future. If we reconceptualize the educational function as described here, we shall talk about equal partnership between the family, the school, and many other societal institutions that can become learning territories, and may constitute the total ecological habitat of learning and human development.

10.3.5.6 The Scientific Dimension

The scientific dimension can have a very special role in the family as an evolutionary unit of the society. Recognizing that the now emerging society is a knowledge-based society and that science is the breeding ground of knowledge, the family can offer the first context in which to develop interest in—and excitement about—science and about discovery and creativity. It can offer opportunities to "do" science at a scale that is appropriate to the family.

10.3.5.7 The Technological Dimension

This dimension offers an interesting agenda for the family. Earlier, I identified the critical evolutionary gap between our highly advanced technological intelligence and our sociocultural intelligence and wisdom. The recognition of this gap as a topic for family discussion, and the bridging of this gap by collective family learning and action, can set the basis for developing the kind of human wisdom needed to gain the upper hand, to guide choices, and to make wiser decisions about the use of technology, as well as to promote appropriate technology. The use of technological devices in the home and in the community offers a very practical real-life context of learning and making informed choices.

10.3.5.8 The Aesthetic Dimension

The aesthetic dimension offers the learning and nurturing of the pursuit of beauty: what it is and how it can be created and appreciated. The cultivation of this dimension nurtures the enjoyment and creation of arts, literature, poetry, and music by members of the family as well as a shared, collective activity by the whole family. Aesthetics can be manifested by creating beauty in and around the home, by the enrichment of our inner beauty and the inner quality of our lives. Aesthetics can offer a very specific and satisfying evolutionary agenda in the life of the family, both collectively as a unit and individually for all its members.

10.3.5.9 The Communication Dimension

The nurturing of genuine, open, and honest communication is a core requirement in the life and evolution of the family. The family council is a fine functional context for developing and practicing genuine communication. The method of consensus building described earlier is a most appropriate approach to communication in the family. Family members can learn this easily within the family so that they can become experts in it as they enter the larger social arena.

10.3.5.10 The Polity/Governance Dimension

This dimension of family life can become the domain within which to coordinate, guide, and govern the evolutionary development of all the other dimensions. The public philosophy that guides this dimension is that each person has the right to participate directly in decisions affecting his or her life. This kind of philosophy should become a shared family philosophy. When brought to life in the family, this philosophy nurtures genuine participative democracy, the peaceful and negotiated resolution of conflicts, the integration of the family as a social unit, participative management and decision-making in the family, and the nurturing and encouragement of the use of these functions in the community and in larger societal contexts. Developing these characteristics, members of the family will be able to exercise creative participation in social systems, in the community, in public service, and in public decision-making.

Activity 38
Develop the ten dimensions introduced in the preceding as an evolutionary agenda with your family. If circumstances would not make this feasible, work with a group to which you belong. Enter your findings in your workbook.

Reflections

Extending the "what if" speculation, just imagine what would happen if all of our societal systems, our communities, our public and private institutions, and our entire governance system from local to global would redesign themselves as evolutionary guidance systems. Such an engagement would enable us to take charge of our future and take part in shaping decisions that affect our lives. It would mean a new way of life for all of us individually and collectively. It would mean the establishment and functioning of all of our systems as evolutionary guidance systems.

The "what if" idea has seen light in a variety of contexts. In a special issue of *World Futures* (1993) we reported on the design of EGSs in a variety systems, including the design of an EGS for the retirement of a couple (Frantz and Miller, 1993), the design of an EGS for a nonprofit professional/scientific association (Bach, 1993), the use of EGS for evolutionary systems management of organizations (Wailand, 1993), and a mediation approach that fosters evolutionary consciousness and competence (Pastorino, 1993). In a different publication, Biatch (1995), Blais (1995), and Dils (1995) gave their account of the use of EGSs in systems of their interest. A recent project supported by the CDC applied the EGS approach to the design of healthy communities (Martin, Banathy, and Ward, 1999).

Summary

The key proposition advanced here is that societal systems are purposeful systems in which purposeful design can guide evolution. With the emergence of the process-oriented, self-organization paradigm of human systems, evolution became the integral aspect of self-organization, in which design becomes the core process of evolution. Thus, in the evolution of societal systems, design is the central activity and competence in design is a commodity of the highest value.

In this chapter, we explored the front part of evolutionary epistemology, namely, transcending the current evolutionary state, creating an evolutionary image, and based on it, defining an EGS. The front part of the epistemology clarifies our intent to engage in evolutionary inquiry. I am reminded of Kenneth Boulding's admonition: "Intentions are fairly easy to perceive, but often do not come about. Design is hard to perceive, but it is design not intention that creates the future" (1995, p. 212). So the main task of conscious evolution is the design of the EGSs that will guide us to bring the intended evolutionary future to life.

11

The Design of Evolutionary Guidance Systems

Design, in the context of evolutionary inquiry, is a creative, decision-oriented, disciplined inquiry that aims to formulate expectations and aspirations for our evolutionary future. We create a value-based ideal image of the evolutionary system we wish to design. Then, based on the image, we define and design our evolutionary guidance system (EGS). As we engage in the design of our EGS, we formulate alternative representations of the future system, devise criteria by which to evaluate those alternatives, select and describe the most promising alternative, and prepare a plan for the development of the selected design.

In the previous chapter, I indicated that in evolutionary design inquiry, we are involved in two design operations. First, we design an EGS, which is an ideal representation of the evolutionary future toward which we intend to move. Then, given the ideal, working back from the ideal, we design the evolutionary system (ES) which brings to life the now attainable representation of the ideal. But there is also a third system, *the design inquiry system* (DIS), which the evolutionary designing community creates to engage in the design of the EGS and the ES.

In the present chapter, I introduce a generic architecture for the design of all three systems. Then, I adapt the generic architecture for the purpose of designing the DIS. The use of the architecture for the design of the ideal seeking EGS follows. Then, I present the approach to the design of the ES. We develop the ES by working back from the ideal EGS and creating a now attainable ES. In the last part, I explore the issue of how to get ready for evolutionary design.

11.1 A Generic Architecture for the Design of Evolutionary Systems

Conscious evolution leads us to enter into the evolutionary design space. This space can be conceptualized as a system of several interacting component spaces. The relational arrangement of these component spaces presents us with a generic architecture of evolutionary design. We can identify five component spaces (Banathy, 1996): *the space in which we define what the inquiry is about, the design information and knowledge space, the design solution space, the design experimentation and valuation space, and the design outcome modeling space.* Figure 11.1 depicts these spaces and their relational arrangement.

The design architecture depicted here is *generic*. It can be applied to the design of all the three evolutionary systems—the DIS, the EGS, and the ES. A brief definition of the component spaces of the generic architecture follows.

11.1.1 The Definition Space

In this space we define what the design inquiry is about. For example, in the present case, the definition of the EGS will include a description of the ideal image of the EGS. In designing the ES, the definition will focus on the description of the EGS.

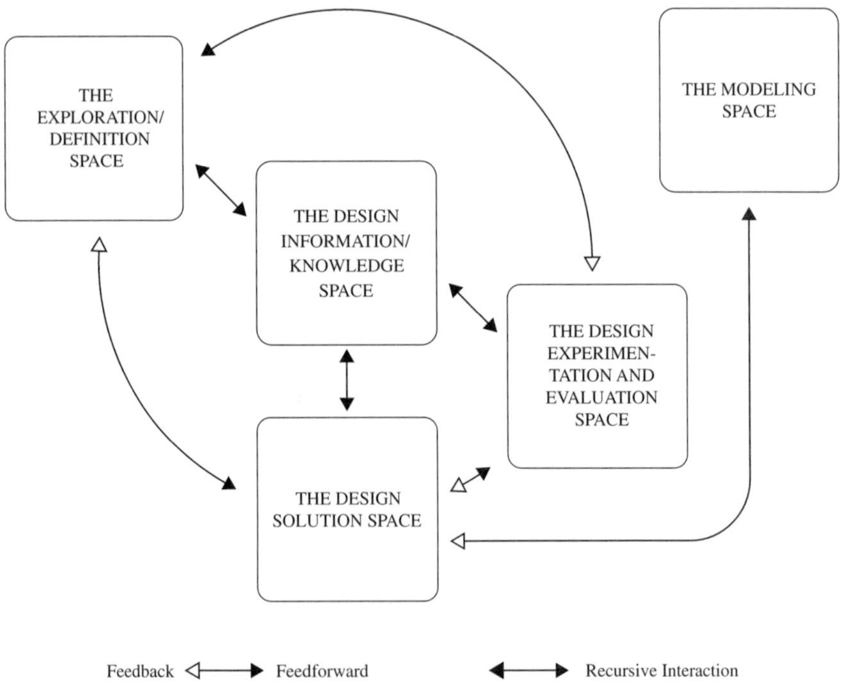

FIGURE 11.1 The Generic Design Architecture

11.1.2 The Knowledge Space

This space is dedicated to containing all the information that serves as the knowledge base of the particular design inquiry. This is an open system in that during the design process we shall need knowledge that cannot be determined a priori.

11.1.3 The Design Solution Space

In this space design alternatives are created and considered, and substantive decisions are made that shape the evolving design solution.

11.1.4 The Evaluation and Experimentation Space

Here, we create the criteria by which to evaluate whatever we created in the design solution space. For example, the key source of formulating the criteria for assessing the design solutions of EGS is the ideal image created in the definition space.

11.1.5 The Modeling Space

In this space we display the systems model of the design solution as well as the model of the systemic environment with which the system we designed interacts and from which it receives support.

11.2 Designing the Design Inquiry System

The first task of the evolutionary designing community is to design the design inquiry process that they would apply in the design of the EGS and the ES. A differentiation is in order. While designing an evolutionary system we ask what that system should be. In designing the design inquiry we ask how we should design that system. In designing the design inquiry, designers are to review design approaches, models, methodologies, methods, and tools and select or develop those that are most appropriate to: (1) the systemic context of the design situation, (2) the desired future system, (3) the design competence and characteristics of the designing community, (4) the perceived complexity of the design task, and (5) the knowledge and resources base available to support the design effort.

The process of designing the design inquiry requires a framework, an architecture, within which designers create the design inquiry. In this section, first I identify the components of the design activity, such as approaches, models, and methods about which decisions are to be made, followed by the introduction of the inquiry architecture within which these components are considered, selected, and displayed as a model of the design inquiry program.

11.2.1 Components of the Design Activity

Components of the design inquiry include design approaches and design models, methodologies and methods, and design tools. I briefly described these here.

11.2.1.1 Design Approaches

Approach is a most general term. Examples include the ideal systems approach, soft systems approach to design, a critical systems approach, breakthrough thinking approach, and critical heuristics of social systems design. An approach is a system of internally consistent principles of design inquiry, for example, the ten principles of breakthrough design approach, defined by Nadler and Hibino (1990).

11.2.1.2 Design Models

A model of design inquiry is an epistemological or process model, a descriptive representation of a system of internally consistent, theory-based design methodologies, with their component methods and tools. Examples include the design models of Ackoff (1981), Banathy (1979), Checkland (1981), Christakis and Conaway (1995), Flood and Jackson (1991), Jones (1984), Nadler (1981), and Warfield (1990).

11.2.1.3 Design Methodology

A design methodology is an internally consistent system of methods selected to be used in a design situation. It could be subsumed by a design model, and incorporates several design methods and tools. Examples include transcendence (transcending the existing system), creating the image of the future system, and transforming the image into a design of the future system. Other examples are: creating a rich picture of the problem situation, exploring relevant systems, designing a conceptual model, and a methodology for means/methods analysis.

11.2.1.4 Design Method

A design method addresses a particular design task and is a constituent of a design methodology. For example, creating an image of the future system might involve methods for envisioning the future system, establishing the boundaries of the inquiry, considering major options, and formulating core ideas and values.

11.2.1.5 Design Tools

Tools are specific techniques applied to address the most detailed aspects of design inquiry. Nadler and Hibino (1990) describe more than twenty tools. Warfield and Cardenas (1994) presented tools such as idea-writing, nominal group technique, delphi, option fields analysis, option profile, tradeoff analysis, and so forth.

11.2.2 Architecture for Designing the Design Inquiry System

The architecture for designing the design inquiry system is adapted from a generic inquiry architecture (Banathy, 1993b, 1996). The architecture used here is a specific case of the generic, offered as a framework for designing the design inquiry program. The outcome or the product of this inquiry is a description or model of the design program, depicted in Figure 11.2.

As pictured in the figure, the inquiry proceeds from the Definition Space, in which designers set the stage for the design of the design inquiry program. From this space the process moves into the Design Solution Space, in which the various design approaches, strategies, methods, and tools are considered and selected. These are tested and validated in the Design Program Evaluation Space. The validated design inquiry program is described in the space marked as the Model of the Design Program. The Organized Knowledge Space, in the middle of the figure, is where all the information and knowledge relevant to design of the design inquiry program are deposited. The figure shows arrows that connect the various spaces. Multiple arrows indicate a spiraling process guided by the type of selections that are to be made in the design solution space. The two directional arrowheads indicate recursiveness. Activities in the spaces are described next.

11.2.2.1 The Definition Space

In this space we display design information relevant to the systemic environment of the design inquiry, the nature of the desired future system, the

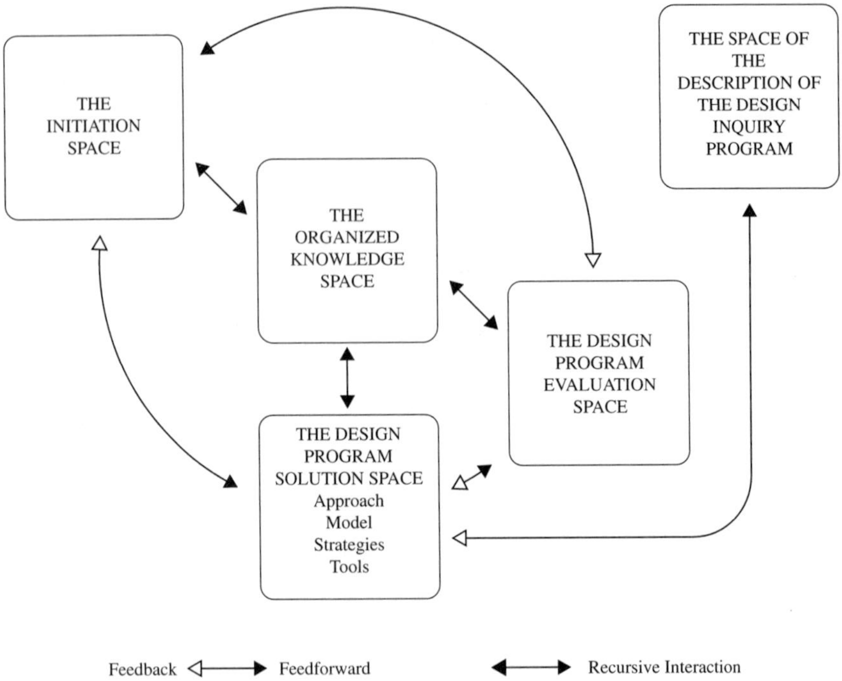

FIGURE 11.2 Designing the Design Inquiry

design competence and the sociocultural and socioeconomic characteristics of the designing community, the perceived complexity of the design task, and the time constraints (if any) of the design of the future system.

The second major task to accomplish while working in the definition space is the formulation of Organizing Perspectives that will guide our thinking in designing the design inquiry program. These perspectives make explicit the values and qualities that the evolutionary designing community wants to realize through the design activity. (These are described in Chapter 12.)

11.2.2.2 The Organized Knowledge Space

This space is occupied by the following types of knowledge: (1) Knowledge about design in general and specifically about design approaches, models, methodologies, methods, and tools. (2) Knowledge about systems and design thinking. (3) Knowledge about the evolutionary designing community and its environment. (4) Knowledge about ways that the system we design can be modeled. (5) Information about organizational capacity and human capability available to the design effort. (6) Information about resources available to support the inquiry effort. In this space, knowledge and information will be continuously acquired and displayed as design work calls for them in the design solution and evaluation spaces. Thus, this space of the architecture is open and evolving.

11.2.2.3 The Design Solution Space

In this space the designers consider available options in terms of design. Activities are never considered in isolation but always in their interrelationship and interaction. What is implied here is that if the designers select an ideal design approach—as is the case in the present situation—the models, methodologies, methods, and tools should be relevant to the ideal approach and should be internally compatible with and consistent with each other. As the designers proceed toward the selection and development of a design inquiry program, they may find they need additional information or knowledge; thus they call for these from the design knowledge space.

11.2.2.4 The Design Program Evaluation Space

Here we define the criteria by which to evaluate the various alternatives that emerge from the design inquiry solution space. Designers should develop both external and internal criteria for the testing of design alternatives. *External criteria* might include probing into the general validity of the approach or model examined, for example, articulated theoretical base, evidence of testing and successful use in social systems design, and internal consistency with other design choices made. *Internal criteria* assess a "goodness of fit" with the overall design effort and with the kind of system designers aspire to build.

The following example represents a design of a design inquiry program developed for the institutional design of an R&D Laboratory (Banathy, 1987b). The criteria had four categories: (1) General assessment of use, such as making a judgment if the alternative identified is realistic and feasible as well as effective to use in the institutional context of the laboratory. (2) The inquiry power of the alternative. Does it allow going beyond the existing boundaries of the system and its systemic environment? Does it offer flexibility of use? Does it allow continuous review and modification? Is it sensitive to evolution and emergence? (3) Is there a systemic match, for example, is the alternative considered appropriate to use in the context of a system such as the laboratory? Can the design inquiry alternative fit in with and transfer into the general inquiry approach of the laboratory, or does it require a change in it? (4) Does the alternative considered nurture unrestricted exploration and the use of creativity? Is it sensitive and supportive toward new initiatives? The formulation of the criteria should be ongoing during the design of the inquiry program. Its guidance should be derived from the definition, the organized knowledge, and the design solution spaces.

11.2.2.5 The Displaying/Modeling Space

This space is the "resolution space" of the inquiry, in which designers display the description of the most promising design inquiry program they devised and that they intend to use in designing the desired future system.

11.2.3 The Dynamics of the Inquiry

The design of the design inquiry program is not accomplished in a step-by-step linear fashion. It is carried out through the dynamics in recurring spirals and recursive, mutually influencing interaction among the design activities and design spaces. The dynamics operate as we explore and reexplore the various spaces and integrate information, knowledge, experience, and evaluation relevant to the emerging inquiry alternatives of the design inquiry program. More specifically, as we proceed with the design inquiry, and as solution alternatives emerge in the design solution space, we continuously revisit the definition space to gain more insights about the issues our design inquiry should address. The same is true with the knowledge space. As we work with design solutions, we shall draw upon increasingly, and in a more focused way, on the sources of design knowledge. We shall often find that we need new knowledge to inform and enlighten our creation and selection of solution alternatives. We also repeatedly move into the design testing space to evaluate (based on the criteria we identified) and reality test the emerging inquiry alternatives. This testing shapes the model of the inquiry program; it validates or questions it. This process is the main source of attaining confidence that there is a "goodness of fit" between the inquiry program and the criteria we established. The knowledge space also offers sources of design testing. In the knowledge space we deposit information about design testing and evaluation means and methods. The design inquiry that intersects all these spaces eventually converges as it moves into the modeling space, where it displays the product of the inquiry as a representation or a model of the design inquiry program.

Another aspect of the dynamics of the inquiry is manifested in the dynamics of divergence–convergence, operating in the design solution space. It relates to the search, creation, and selection of inquiry alternatives. Initially, we seek to consider a number of alternatives; thus we operate in a divergent mode. An assessment of the alternatives against stated criteria eventually will lead us to converge and to the selection of the most promising design inquiry program.

Activity 39

You are *not* expected to develop a comprehensive design inquiry program. To create your own understanding of how the design of a design inquiry program might work, however, it is suggested that you speculate about the task of developing such an inquiry, and outline briefly how you would go about the design such a program. Enter your findings in your workbook.

Reflections

The design of a design inquiry program, portrayed in this section, is an integral part of the process of getting ready for evolutionary design. This process of "getting ready" has received only scarce attention in the design literature. In the first, second, and third generation design mode, it was the design expert or consultant who, when invited, was already ready to move in with his or her "toolbox." The very moment a position is taken that it is the prerogative of people to design their own system, we must embrace the idea that people in the system reconstitute themselves as the designing community. They need to learn to engage in the design of a design inquiry program that best fits the design situation.

11.3 Designing the Evolutionary Guidance System

Here I adapt the generic architecture to the design of the EGS. First, I define the component spaces and their function, followed by a description of how the design work is accomplished in the various design spaces. Then, we consider the dynamics of design inquiry and of divergence and convergence.

11.3.1 A Description of the Design Spaces

Below, I introduce a description of the five design spaces in which the design of the EGS will operate.

11.3.1.1 The Exploration/Image-Creation Space

This space is the genesis space of evolutionary design. We enter into this space once we have committed ourselves to *transcend* the existing evolutionary state and engage in conscious evolution. Working in this design space we generate a *common ground*, formulate our collective *value system*, envision the *ideal image* of its evolutionary future, and define our EGS.

11.3.1.2 The Design Information and Knowledge Space

In this space we display: (1) a description of our *shared values*, the *ideal image*, and the *definition* of the EGS we generated in the Exploration Image-Creation Space; (2) a description of the *knowledge base* of the dimensions of the EGS, (3) a description of the situational and the environmental *context* in which the designing community operates, and (4) the *design knowledge* base, which contains information about the various design methods and tools. Our work in this space is not a "one-shot deal." We constantly need new knowledge to support the creation and (e)valuation of emerging alternative design solutions.

11.3.1.3 The Design Solution Space

As the designation of this space indicates, it is this space in which the core design activity takes place. *All the other design spaces are built around it*. It receives the input to design from the Exploration Image-Creation Space. Its knowledge base comes from the Information/Knowledge Space. The outcomes of the design solution are assessed in the Experimentation/Valuation Space, and the final product is displayed in the design outcome Modeling Space. In the Design Solution Space, we (1) formulate the purpose of the EGS that responds to the image we created, (2) design the system of functions that need to be carried out to attain the desired outcomes, (3) design the EGS as a human activity system that has the organizational capacity and human capability to carry out the functions, and (4) design the systemic environment that will support the EGS.

11.3.1.4 Experimentation/Valuation Space

In this space we make judgments about the *adequacy and relevance of the unfolding design*. We experiment with the design solution alternatives and make a judgment about their potential future-building value. We are to formulate a set of criteria in this space. The criteria have two main sources. The first is a synthesis of the core values, the image, and the evolutionary and organizing perspectives of the EGS we generated in the Experimentation/Image-Creation Space. This synthesis will guide our decision-making in the course of the design. The second source of assessment probes into the adequacy of response of the emerging design to the stated purpose, formulated in the Design Solution Space.

11.3.1.5 The Modeling Space

In this space we display the model of the EGS and the model of its systemic environment. Design work to be accomplished in this space will be described in Section 11.3.5.

11.3.2 Design Dynamics

We can speak of two kinds of dynamics: one that operates among the five design spaces and one that works as we make design decisions in the divergent–convergent mode.

11.3.2.1 Dynamics Operating among the Design Spaces

The dynamics of design operating in the various spaces of the design architecture are depicted in Figure 11.3.

The figure portrays the flow of the process of design through the various spaces of the design inquiry. The lines that connect the spaces stand for the various spirals of the inquiry process. Each spiral is a complex of several component spirals that represent design alternatives that we construct and evaluate. The arrows indicate the recursiveness of the process. The blackened arrows show the direction of spiraling and the white arrows show the direction of the feedback process.

In the course of spiraling through the design spaces, *we create and evaluate various potential solution alternatives* of our EGS. The spirals represent the flow of design inquiry as it goes through the substantive domains of systems design. These domains are the formulation of the core definition, the definition of specifications, the selection of functions, the design of EGS, and its systemic environment. As we spiral through the spaces, the interaction is never one directional, but recursive and mutually influencing/shaping. Feedback and feedforward are always ongoing two-pronged operations.

11.3.2.2 The Dynamics of Divergence and Convergence

The second type of dynamics is the interplay of divergence–convergence, as shown in Figure 11.4. This dynamic has two manifestations in the course of our design inquiry.

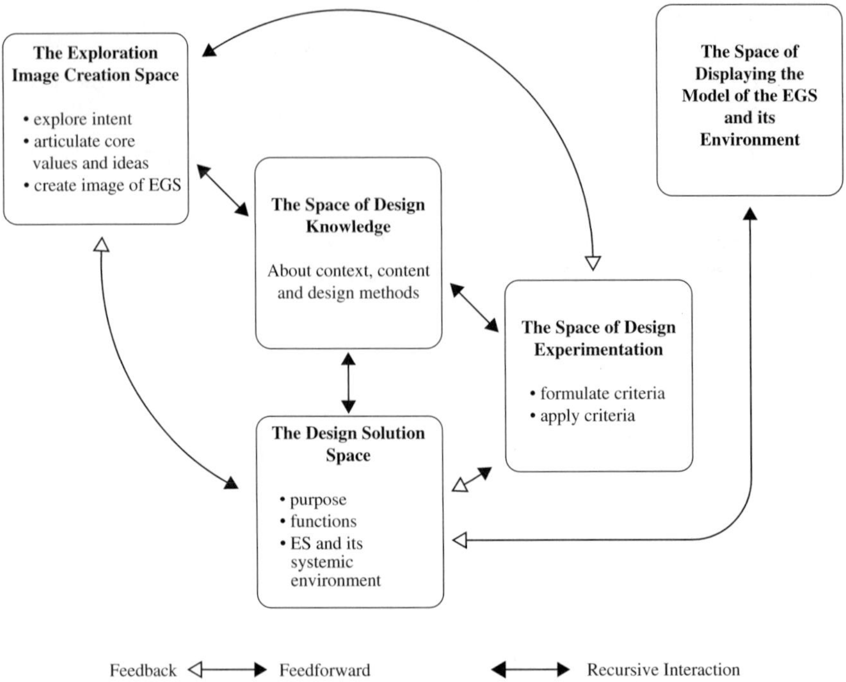

FIGURE 11.3 The Design Architecture for Designing the Evolutionary Guidance System

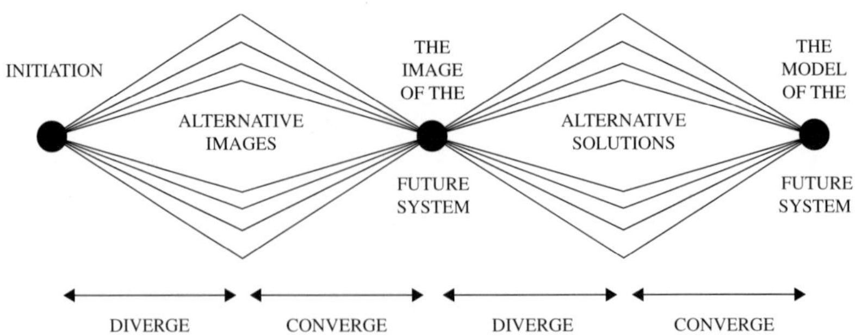

FIGURE 11.4 Dynamics of Divergence and Convergence

11.3.2.2.a Dynamics in the Exploration and Image-Creation Space. In this space, we first operate in the divergent mode as we consider a large set of core values, core ideas, experiential dimensions, and evolutionary and organizing perspectives. Then, we converge, as we make choices that we collectively agree upon and *create an image of the future system.*

11.3.2.2.b Divergence–Convergence in the Design Solution Space. For each of the substantive design domains (core definition, functions, enabling systems, systemic environment) we first diverge, as we create a number of alternatives for each, and then we converge as we evaluate the alternatives and *select the most promising and most desirable alternative.*

11.3.3 The Workings of Design within the Architecture

Returning to Figure 11.3, we can see that it portrays the flow of the design process as we move through the various design spaces of the evolutionary inquiry. The figure also shows how the five spaces are related. Our journey through the spaces is accomplished in five spirals.

11.3.3.1 Spiral 1: Formulating the Purposes of the EGS

During spiral 1, we formulate the purposes of the EGS. The purpose is to interpret the image we created in the Exploration Image-Creation Space. We explore several questions: What is our ideal EGS about? More specifically: What is our aspiration in serving humanity, the larger society, our community, and most importantly members of our evolutionary designing community individually and collectively? What is the shared commitment that can give everyone in our evolutionary community clear direction and guidance? A synthesis of answers provides us with a comprehensive definition and the purpose of the future system. Keep in mind that we are to develop several alternative purposes.

Design Dynamics. The purpose spiral is connected with the knowledge base that informs and provides content for the decisions we make. We spiral through the experimentation/valuation space, where we assess the alternative purposes based on the criteria we displayed in this space. This valuation will guide us in converging on and selecting the purposes. Then, we spiral back to the exploration space, where we might reformulate our design ideas and the image in view of the purposes.

11.3.3.2 Spiral 2: Selecting Functions

During spiral 2 we ask such questions as: What are the key functions that have to be carried out that enable our EGS to guide our ES toward the ideal as defined in the purpose statement? How do these functions interact to constitute a systems of functions? What are the subfunctions and how do they integrate into subsystems of the key functions? What are the component functions of the subfunctions? How can we organize the subfunctions in subsystems of functions? By answering these questions we unfold the system's complex of functions that constitute the first systems model of the EGS. This system model of functions is a system of verbs. Unfortunately, in most organizational work, once the purposes are stated, people move on to establish the structure of the organization. In systems design there is an iron law: *Form follows function.*

Design Dynamics. The functions spiral connects to the knowledge base and is assessed (in the Experimentation/Valuation Space) by asking: Do the functions we identified have the power and the complexity to respond to the purpose as stated? Next the dynamics lead back to the Exploration/Image-Creation Space, where we ask if there is anything to be changed or reconsidered based on the decisions we made about the functions.

11.3.3.3 Spiral 3: Designing EGS as a Human Activity System

The design work to be accomplished in the third spiral responds to the overall questions: Given the functions as identified earlier, what organizational capacity and what human capability are needed to carry out the functions? Who should do what? More specifically: (1) What design will enable the EGS to guide the functions, energize and inspire people in the system, ensure the availability of needed information and resources, and provide for continuous evolutionary and organizational learning? (2) Next, we ask: What systems components, and what kind of people in those components, have the required capabilities to carry out specific functions? How should we integrate the selected components in relational arrangements? What resources should be allocated to whom?

Design Dynamics. The EGS and its components draw on the knowledge base and are assessed in the Experimentation and Valuation Space against the functions and the purposes of the EGS, as well as against the values that were collectively agreed upon in the Exploration/Image-Creation Space. We should be reminded that several alternative EGS design configurations will be created as the spiral goes through several iterations as we experiment with and assess EGS solution alternatives. Moving with a spiral, initially we are in a divergent mode as we create a number of alternatives. But gradually we converge, as we make a value judgment about the alternatives and select the most desirable solution.

11.3.3.4 The Fourth Spiral: Defining the Environment of the EGS

The fourth spiral leads us back into the Design Solution Space. The work we accomplished at the preceding spiral of the Design Solution Space provided us with the design of the EGS as a human activity system. Now we ask: What is the environmental system of an EGS? We can speak of two kinds of environment. One is the *general environment*. This environment is everything that lies beyond the boundaries of the EGS. That portion of the general environment that is in constant interaction with the EGS, that has the information and other resources needed to support and to coevolve with the EGS is called the *systemic environment*. It is this systemic environment in which the EGS is embedded that we need to define and eventually design in a coevolutionary interaction.

Design Dynamics. In defining the requisite systemic environment, we need to connect and interact with the various design spaces. We start out by exploring and defining the information and resource needs of the EGS we have designed. To do this we need to connect with the Information/Knowledge Space and spiral to the Experimentation/Valuation Space where we are to test the adequacy of the various alternatives of the systemic environment we defined. Circling back to the Exploration Image-Creation space we shall pay specific attention to the ability of the systemic environment to support the various experiential domains of the EGS.

Activity 40
It is not expected that you can design an EGS in the course of working with this book. But to gain a "feeling" for getting involved with evolutionary design inquiry, you are asked to continue your work with the designing community you worked with earlier in creating the image of EGS. Given the image your group defined, I ask you to guide the group in what can be called a design simulation. As you know, each spiral is a "bundle" of spirals that produce the various alternatives called for in a particular design space. In this simulation we ask your designing community to spiral through the various design spaces *just once* and produce just one alternative and describe it only briefly. Enter your findings in your workbook.

Reflections

A description of the process of evolutionary systems design, introduced in this section, might be viewed as a journey that took us through the various spaces that comprise the *territory of evolutionary design*. I mapped out the design territory by introducing a design architecture that provided a comprehensive framework for the design of an EGS. The application of the methodology of social systems design to evolutionary inquiry is a *novel undertaking*. The approach described here is *one possible alternative* of evolutionary epistemology. In the realm of systems design it would be a serious underconceptualiztion if we would be satisfied with having only one alternative. Thus, the evolutionary community is called upon to focus on evolutionary epistemology and explore different approaches to the evolutionary design inquiry. We should *create and test approaches by which we can shape our individual and collective futures.*

11.3.4 Modeling the Design Solution

Following a definition of the concept of systems model, I describe how to build systems models and introduce systems models that portray EGSs. The term model, as used here, is a mental or conceptual representation or description of a future system. We can display the outcome of our design, and model the design solution by constructing three System Models: a Systems-Environment Model, a Functions/Structure Model, and a Process/Behavioral Model (Banathy, 1992). The *models jointly provide a comprehensive characterization of the evolutionary system.* Models of social systems are built from systems concepts and principles that represent the context, the content, and the process of a particular system. The relational arrangement of these concepts and principles can be organized into systems models. Two examples highlight the preceding statements. (1) Input, transformation, and output are familiar systems concepts. Their relational arrangement gives us the following systems principle: input is transformed into output. (2) Feedback and adjustment are also systems concepts. Their relational arrangement gives us two principles: (a) feedback informs us about the adequacy of output and (b) based on the feedback information adjustments might be introduced in the system. The three systems principles formed in the preceding can be arranged into an image of a very low resolution general model of systems, depicted in Figure 11.5.

By observing various types of human activity systems and studying their behavior, we recognize characteristics that are common to them. Once we have identified and described a set of systems concepts that are common to social systems, and observed and discovered certain

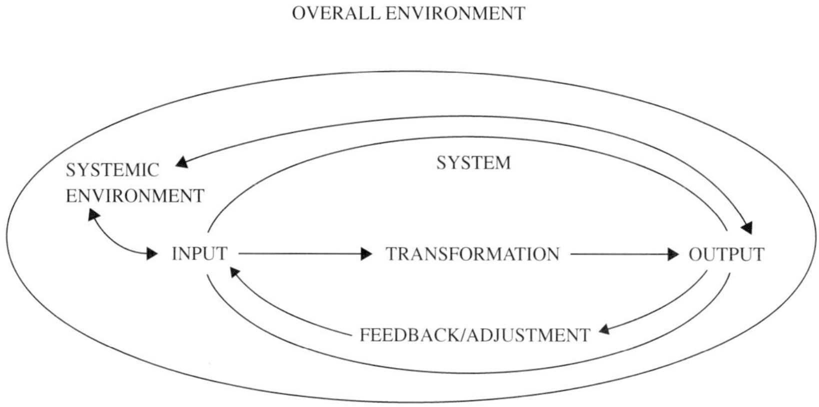

FIGURE 11.5 A Generic Model of Systems

relationships among them, we can construct systems principles. A systems principle emerges from an interaction and integration of related systems concepts. Next, we can organize related principles into certain conceptual schemes, called systems models.

Following the line of reasoning described earlier, I constructed three systems models (Banathy, 1973, 1992). The first model defines social systems in relationship to their environment. This is called the Systems-Environment Model. The second focuses on what the system is. It is called the Functions/Structure Model. The third portrays how the systems operates through time. It is called the Process/Behavioral Model. Models are frames of reference that we can use to examine and talk about the system that the models represent. We work with models all the time. When we exchange ideas about something, we usually do so by using conceptual models. In a discourse, it is helpful to have a common frame of reference (model) of what we talk about, so that we have some assurance that everybody is on the same wavelength. The three models that I introduce are most beneficial in systems design. If we describe the outcome of our design in terms of models, we can examine our product, explore it, and everybody involved will be able to understand what we have designed. A description of the three models gives us only a broad-stroked picture of modeling evolutionary systems. For a detailed presentation of the models consult my earlier work (Banathy, 1992).

11.3.4.1 The Systems Environment Model

This model is an organized and relational arrangement of systems concepts and principles that enables us to describe and represent relationships and interactions that portray mutual independence between an EGS and its systemic environment. As a "lens," it projects a bird's eye view of the landscape in which the system operates. The model provides:

- A detailed description of the geopolitical, the socio cultural, and socioeconomic environments that influence the EGS and establishe its requirements. It defines those systems in the immediate (systemic) environment that affects the EGS and that is affected by EGS. It is this environmental system upon which the EGS depends for its support and into which it sends its output.
- A description of the boundaries of the EGS, such as the spatial, geopolitical, economic, social, sociocultural, sociobiological, psychological, temporal, ethical, and technological. We shall also define the nature of boundary judgments to be made and how those judgments are made.
- A definition of all input entities that enter the EGS as well as the output entities that the EGS sends to its environment. The systemic environment is a key input source.

11.3.4.2 The Functions/Structure Model

This model provides us a lens through which we can focus our attention on what the EGS is and what it does. The source of describing this model comes primarily from the solutions we generated in the course of working through the spirals in the design solution space. This model is a snapshot of the EGS, projecting a still picture image at a given moment. In presenting this model, we describe: (1) the purpose of the EGS, (2) the functions that the EGS carries out to attain the purpose, (3) the organizational arrangement of the EGS and its components as they are engaged in attending to functions, and (4) the relational integration of the components into the structure of the EGS.

11.3.4.3 The Process/Behavioral Model

Tihs model provides the evolutionary designing community with a lens that focuses on what the EGS does through time. It projects a motion-picture image of the EGS and it represents how the EGS behaves as an ideal pursuing, changing, living system in the context of its environment. The model describes how the system:

- Processes input and more specifically: how it identifies, receives, screens, assesses incoming information and resources; and how it sends the input on for transformation.

- Transforms input to output, more specifically: how it carries out transformation production, transformation facilitation, and transformation guidance.
- Processes output; how it develops and applies the output model, how it facilitates and guides the output process and dispatches the output into the environment.
- Carries out the assessment of the output by collecting evidence of the relevance and adequacy of the output; how it analyzes and interprets evidence, and if indicated, how it constructs models of adjustment and how it introduces adjustments. This process also describes operations that should be activated in case there is a need for changing the EGS.

This description of the three models gives us only an orientation for the workings of the three models in the evolutionary design space.

Activity 41
Your designing community is asked to engage in another simulation. Using the information you developed in completing the design simulation of your EGS, you are asked to describe your EGS with the use of the three models. Enter your description into your workbook.

Reflections

People who begin to work with the three models often ask: Which model is a true representation of a sociocultural system? Which is the most important? The answer is: No single model or even a combination of two is sufficient in portraying sociocultural systems. Each has its own function. The Systems-Environment Model maps the space in which the system lives. It portrays the relationships and the coevolutionary interactions with the systemic environment. The Functions/Structure Model depicts what the system is at a given moment in time and how it is structured. The Process/Behavioral Model tells as how the system operates and lives through time. Only if considered jointly, as if superimposed upon each other, do *these models tell us the real story of an EGS*. Only if we integrate them do they reveal their true nature.

11.4 Designing the Evolutionary System

In this section I will first make a distinction between the evolutionary guidance system (EGS) and the evolutionary system (ES). Then, I will introduce the process by which to design the ES. In conclusion, I will explore the functional contexts in which ESs are created.

11.4.1 Making a Distinction between the EGS and the ES

Earlier, I differentiated the ideal EGS from its temporal, at a given moment in time manifestation: the Evolutionary System. Here I further develop the difference between the two by depicting and describing the difference between the two systems. (see Figure 11.6.)

The figure helps us to see the overall evolutionary design space within which we observe the operation of several aspects of evolutionary inquiry.

- On the figure we can see depicted the *major strategies of evolutionary epistemology*. The overarching arrow shows the strategy of transcending the existing state, then creating the ideal EGS and bringing the ideal to life by working back from the EGS and designing the ES.
- We differentiate the purpose of the EGS from the purpose of the ES. The purpose of the *EGS is to guide us* toward an ideal state, to exert a magnetic pull on our ES. The purpose of the *ES is to be guided by the EGS*, to respond creatively to its magnetic force and at any moment in time to strive to manifest the properties of the EGS. Therefore, in designing the ES, we place the ideal model of the EGS in the definition space. (See Figure 11.7.)

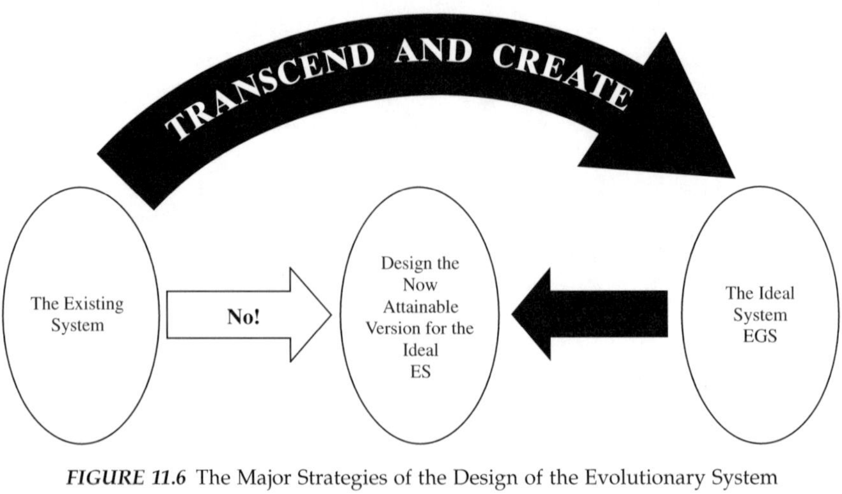

FIGURE 11.6 The Major Strategies of the Design of the Evolutionary System

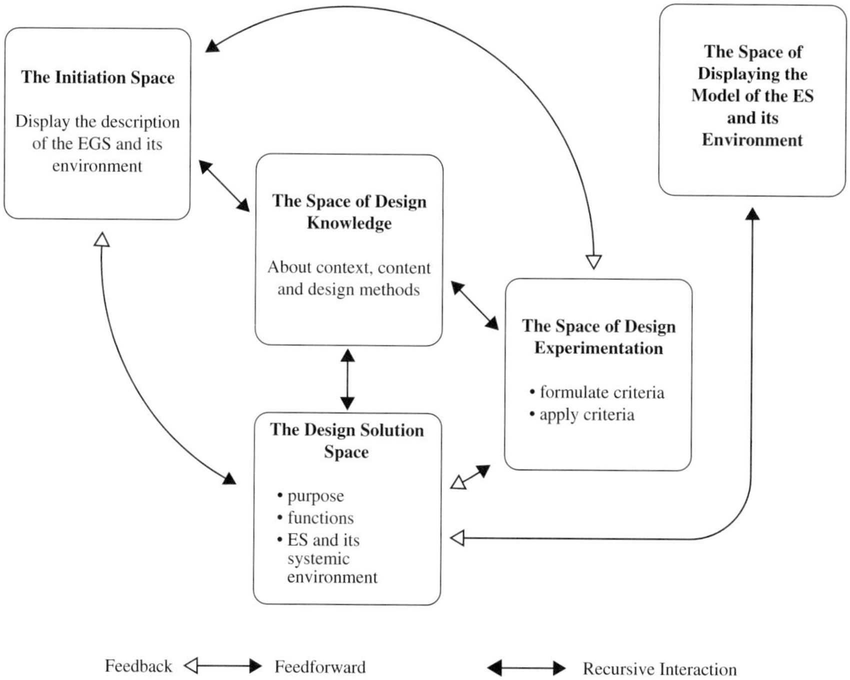

FIGURE 11.7 The Design Architecture for Designing the Evolutionary System

- We can also see a "prohibition" in operation. The arrow on the left hand side tells us that we *should not try to stay within the boundaries* of the existing evolutionary state by trying to work out from it by attempting to fix the present state.

11.4.1.1 The Evolutionary Guidance System

An EGS is designed to guide us toward an *ideal evolutionary future*. We describe the ideal and place it on the horizon. It becomes a magnet that pulls us toward it. It is our guiding star toward which we travel. The *law of the moving horizon* operates in our pursuit of the ideal. As we move toward the horizon, the horizon moves ahead of us. The EGS will also change. New insight, continuous evolutionary learning, changing environmental conditions, unfolded evolutionary potential—all these forces impinge upon our evolutionary design communities so that they will *continuously re-create* their EGSs.

11.4.1.2 The Evolutionary System: Definition and Organizing Perspectives

The definition of the ES will show the contrast between it and the EGS. The organizing perspectives help us to move out from the definition to the design of an ES.

11.4.1.2.a Definition. An ES is a temporal, here-and-now manifestation of an EGS. It is designed by saying: The ideal EGS is out there in the distant future. It guides us toward a desired evolutionary future. But we have to organize ourselves somehow in a way that we can create a system that can be put in place today. We have to design it so that it will have the potential to move us ever closer to the ideal EGS. The EGS will guide us on our journey toward the desired future evolutionary state. This guidance will enable us to avoid the *"life-cycle trap"* of past evolutionary generations. Those generations—once they reached maturity—became fixed and rigid, maintained the status quo, and were unwilling to change. They went downhill to "devolution" by declining and decaying. The *ES constantly evolves* and changes as it travels on the path of its evolutionary journey guided by the EGS.

11.4.1.2.b Organizing Perspectives help us to organize our thinking about the ES and its design. These perspective include the following:

- We are to *create an ES that reflects the EGS* in all its aspects. It should be animated by the same collective consciousness; attend to the same functions; and should have the same systems properties, the same structure, the same experiential dimensions, and the same evolutionary and organizing perspectives as the EGS. However, it is less than the ideal manifestation of the EGS, which can be operationalized here and now. Its purpose is to respond to the magnetic pull, to the guiding force of the EGS, and move the ES toward the EGS.
- The *ES is designed today and for today.* It is a "here and now system." It is a temporal system. Therefore, it is not constraint free, while the EGS is free from any constraints. Working with the constraints of the here and now, we try to remove as many of the them as we can, while accepting and working with those we cannot remove.
- The ES is an *ideal seeking system.* We seek to realize as much of the ideal—as represented by the EGS—as we possibly can.
- The ES is a *learning system.* It is involved in evolutionary learning so that it develops the evolutionary competence needed to be an

evolutionary system. Furthermore, it also learns from the evolutionary experience by assessing what works and what does not.

- The ES is an *evolutionary potential creating system*. In the course of its evolutionary journey, it unfolds evolutionary potential that is still enfolded and, as a learning system, it creates new evolutionary potential. As we observed in the lives of previous evolutionary generations, evolutionary potential builds upon evolutionary potential. Evolution is a potential building venture.

11.4.2 The Design of the Evolutionary System

Earlier, I introduced a generic design architecture and used it to portray an approach to the design of EGSs. In contemplating the design of ESs, I propose the use of the same design architecture, but with a few variations from the EGS design approach.

The ES can be designed only once we have designed an ideal EGS. The EGS is designed for an *imagined ideal shared future*. Such an imagined ideal future can and should be shared by many of us, living in many systems from local to global. Given the image of such an ideal future, we can engage in the design of specific, temporal ESs that live and work in different functional contexts, under different constraints. While an EGS can be designed as generic to many ESs, an *ES* is *designed as a unique system* in the myriad of our evolutionary communities.

An evolutionary epistemology should have the power to offer approaches and methods for the design of EGSs, and then it should offer approaches and methods for the design of ESs. In the first part of this chapter, we explored an approach to the design of an EGS. In what follows, I first offer an approach to the design of ESs, then I propose functional context alternatives within which the approach can be applied.

11.4.2.1 The Design of ES Works within the Feasibility Space of the Evolutionary Design Space

Evolutionary design inquiry is three dimensional. The evolutionary design architecture is two dimensional. The third dimension is seen as three extending spaces in the evolutionary design space, shown in Figure 11.8.

The arch of the Ideal Design Space denotes the *ideal design space* in which we create the ideal EGS. The design of the EGS is set in the *future*. The EGS is not attainable at the present. The arch of the Feasibility Space marks the design space of ES. It denotes a *feasible design space*, in which we design the ES, which is attainable in the *present*. I could add a third

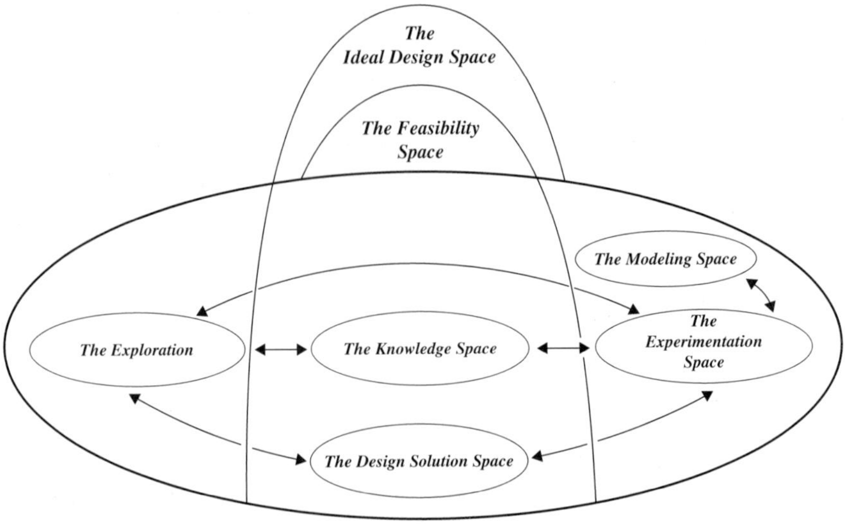

FIGURE 11.8 The Ideal Design Space and the Feasibility Space

arch, which I call the fantasy or science fiction space. Fools design in this phase. Still, at times designers can play the fool and enter into this space. Occasionally they might get some insight, some strange design ideas that might prove to be the best. But most importantly, they might stretch the power of their imagination and creativity.

11.4.2.2 Modification in the Definition Space

In designing the EGS, we called the definition space of the design inquiry *Exploration/Image-Creation*. Work in this space leads us to the definition the image of the EGS. In designing the ES, at the front-end of the design, we display the outcome, the three models of the EGS. Then, the design proceeds by working out from the models of the EGS. We tell ourselves that the purpose of our designing the ES is to bring to life, to realize as much of the EGS as possible.

11.4.2.3 Modification to Allow for Constraints

A key characteristic of an EGS is that it is *constraint free*. It should be created as the most ideal, most desirable evolutionary system we can possibly imagine and aspire to attain. On the other hand, a designing community that creates an ES has to consider constraints in order to bring forth an ES that can operate now. Faced with constraints, designers of the ES have to make judgments as to how to deal with constraints. As part of their design they might seek ways to remove some, to mitigate some, and accept and work with some.

11.4.2.4 The Issue of Distance

Distance means the distance between the EGS and the ES. Two considerations operate here. First we are to consider how far in the future, how far on the horizon do we place the EGS? There is a dilemma here. The further out the EGS is, the less powerful its magnetic force may become. For example, the technology of communication has played a crucial role in the evolution of our species. Major shifts from sign communication to the symbolic communication of speech, to writing, to print, to electronic communication, and the Internet have brought forth the major, discontinuous shifts of emerging evolutionary generations. There are people and groups today who consider *telepathy* as a potential viable mode of communication in our evolutionary future. Still, I would hesitate to include telepathy as a viable mode of communication. Thus, in designing EGSs, judgments have to be made as to how far to go. We encounter the same problem on the other end. If we design a conservative, less than ideal EGS, it may act as a very weak magnet, losing its pulling power.

Furthermore, designers of ES may struggle with the issue of *how hospitable or how hostile is the environment* within which they design the ES. We have ample examples to suggest that many visionary design attempts are strangled by a bureaucratic, status quo protecting environments that are hostile to change. For example, many visionary attempts to create new educational designs are stifled by the educational establishment, even though the state of education is the primary concern for most people today, and rightly so.

11.4.2.5 Variations in Functional Contexts

There are a great number of possible functional context configurations within which designing communities can engage in designing their future. There are many types of system that might engage in their conscious, self-guided evolution. The family is the basic unit of conscious evolution. Others are neighborhoods; a group of people with common interests; a particular community that defines itself as such; systems in which we live and work, such as educational, non-profit, public service, and business communities; professional organizations; the polity; and government.

All the above system types can be considered to constitute an evolutionary design community. In pursuing the design of their EGSs and ESs, these communities have options. They can do their design on their own, or they can *link up with others*. The most likely and most desirable link-up would happen when they consider the design of EGSs that can be generic to several designing communities. At a base level, families within a kinship might link up. (Such a link-up could result in benefits beyond the design of an EGS.) A broad-based designing community could be established by linking up several designing communities. In fact *the more the better*. Such a link-up could be reasonably expected to happen within a community in which there already exist subcommunities. Virtual communities, such as communities within a profession, various educational institutions, and several public agencies could form another type of evolutionary design alliance. When it comes to the design of ESs, even though a particular ES is specific to a designing community—large or small—bringing them together would create an evolutionary learning community that would enable them to learn from each other by exchanging ideas and information about both how they are designing and what the outcomes of their design are.

Reflections

Evolutionary design inquiry is a novel domain of social systems design. Evolutionary epistemology and the kind of issues we have considered in Chapter 10 and in the present chapter present novel territories and novel approaches to social systems design. In this section, we have added another novelty by considering the variations in the design modes of EGSs and ESs, as well as variations in the functional context configurations, and variations in the design modes of evolutionary epistemology. There is much to be learned in this realm. Evolutionary design inquiry is in its early years. It invites the attention and work of design scholars as well as all of us as evolutionary design practitioners.

11.5 Building a Design Culture

All of what I described in the preceding can happen only if the evolutionary designing community becomes competent in sociocultural design. To engage in self-guided, conscious evolution evolutionary design, we must learn the whats and hows of evolutionary systems design. And this can be accomplished only by the acquisition of a design culture.

The evolutionary epistemology introduced in the preceding was applied to the design of DISs, EGSs, and ESs. To enable evolutionary designing communities to apply the proposed evolutionary epistemology, they have to develop competence in evolutionary design. (In Chapter 9, we already explored the development of evolutionary competence. Now that we explored evolutionary epistemology, we have to extend evolutionary learning requirements to include competence in evolutionary design.) But there is another aspect of this issue. Inasmuch as conscious evolution of the Fourth Generation is a task that challenges all of us, we must consider nothing less than the acquisition of competence in evolutionary design by all of us. The challenge that confronts us is nothing less than building a design culture in the society. Without it, conscious evolution cannot come to life.

11.5.1 Design Culture: What It Is and What It Means

Design culture is a learned pattern of behavior. Its acquisition enables the collective creation of novel phenomenon. It integrates: (1) the design's own distinct ways of thinking and knowing; (2) design concepts and principles that constitute design inquiry; (3) methods and means by which creativity is applied in such actions as envisioning, imaging, inventing, assessing, and creating design solutions; (4) the use of the "language" of design, including "modeling" which is the design's specific form of expression and representation; and (5) conversation and consensus building as the special social communication behavior of collective design. Finally, in the context of evolutionary inquiry, design culture is manifested in action that aims at the creation of ESs that realize the aspirations and expectations of the evolutionary designing community.

11.5.2 The Concept of the Design Culture

The understanding that design culture is an essential component of the wholeness of human experience has emerged only recently. Popper (1974) made a distinction between three worlds. The first world is the world of things, the second is the world of subjective experience, and the third is the world of our creations of all kinds which have their own autonomous laws. The disciplined inquiry by which the three worlds are explored is: science for world I, the humanities for world II, and design for world III. Worlds I and II are the domains of descriptive science, while design is prescriptive. Table 11.1 presents these three domains of disciplined inquiry.

Design culture is now recognized as the third culture which complements the cultures of science and the humanities (Cross and Warfield, 1987). A lack of any one of the three cultures leads to a grave loss of substance and value in the quality of human experience. But today, design

TABLE 11.1
The Three Cultures

	Science	Humanitites	Design
FOCUS	The natural world Problem finding Describe "what is"	The human experience Understand the human experience and Portray it	The man-made world Solution finding What "should be"
PRIMARY METHODS	Experimentation Pattern recognition Analysis Classification Deduction	Analogy Metaphor Criticism Valuation Induction	Modeling Pattern formation Synthesis Conjecture Abduction
WHAT IS VALUED	Objectivity Rationality Neutrality Concern for "truth"	Subjectivity Imagination Commitment Concern for "justice"	Practicality Creativity Empathy Concern for "goodness of fit"

culture is not yet part of the general human experience. While our schools provide literacy and competence in the cultures of the sciences and the humanities, they provide design learning only to a few professions. Today we do not provide general education in design.

11.5.3 Rationale for Building a Design Culture

Design literacy and design competence are the two sides of the design coin. We have by now recognized that in an age of explosive growth of all kinds of design, we are at the mercy of those who design for us. We individually, and collectively as a society, are uninformed design illiterates. When bad designs are thrown at us the best we can do is complain. (And we have certainly no capability to participate competently in the design of systems in which we live.) Faced with this debilitating predicament, design literacy comes to our rescue. It will enable us to understand what design is, what design does, how it does it, and what the impacts of designs are on our quality of life. Design literacy can create informed users of products and systems in the creation of which technical expertise is required. But design literacy is only one side of the coin of design culture. The other side is design competence. When it comes to sociocultural systems—the systems we inhabit—and evolutionary systems, it is we the people in the systems who are the experts. Only we have the right—and carry the responsibility—to design our systems. No one, but no one, has the right to design human activity systems for someone else. The age of social engineering is over. But we can exercise self-directed evolutionary design only if we acquire a design culture.

11.5.4 Building a Design Culture

During times of relative stability, characteristics of previous eras, piecemeal adjustments were able to bring our systems in line with the slow rate of change in the societal environment. But in a time of accelerating and dynamic changes and transformations—characteristics of our current era—when a new stage is unfolding in societal evolution—piecemeal adjustments, fixing, or improvements of the old designs of the current evolutionary state will create more problems than they can solve. The emerging "new realities" of the massive societal changes of the post-industrial information/knowledge age require continuous design activity at all levels of the society. They require us to engage in the evolutionary design of the Fourth Generation HSS.

We have a choice. We can relegate design decisions to others who "represent" us and make decisions for us. Or we can empower ourselves by: (1) acquiring design literacy and using it to make informed judgments and choices of designs, (2) developing design competence, and (3) assuming responsibility for conscious evolution by designing our own evolutionary systems. Furthermore, the building of a design culture enables us to create a truly participative democracy about which we talk so much today, but which—in a true sense—is not yet part of the human experience.

Activity 42
Review the preceding section. Make note of the core ideas of design culture. Create an agenda for the development of design culture for the group you have worked with previously on the activities of this chapter. Introduce your findings in your workbook. Keep in mind that the development of a design culture is a prerequisite to engaging in evolutionary design.

Summary

In the first part of the present chapter I introduced a generic architecture for the design of evolutionary systems and suggested the use of systems models for portraying the outcome of design. In the second part I explored the use of the generic architecture for the design of design inquiry, the EGS, and the ES. In the third part, I proposed that if our evolutionary designing communities aspire to design EGSs and ESs, they must develop a design culture.

We have reached the point in our work when we can say with some confidence that evolutionary epistemology has become transparent for us. In the next chapter we shall explore the application of what we have learned about conscious evolution and its working in the evolutionary design space.

12

The Agoras of the Twenty-First Century

> Never doubt that a small group of thoughtful,
> committed citizens can change the world:
> Indeed, it's the only thing that ever does.
>
> *Margaret Mead*

We have arrived at the last station of our evolutionary learning journey. We stop for a moment and reflect on the experience of the journey and ask "looking ahead" questions. The end of our present journey marks the beginning of our never ending self-guided evolutionary journey. We have learned enough, and have developed enough confidence to apply what we have learned in working with evolutionary designing communities as well as continue our evolutionary learning from our ongoing evolutionary work. We are ready to join our communities, help them to learn what we have learned, and enter with them into the evolutionary design space.

We know that throughout history the truly fundamental changes in societies and in the life of humankind have not happened by the decree of rulers and potentates, or by laws constructed by governments. If anything, their efforts focused on maintaining their power by preventing change, or making small incremental adjustments if they absolutely had to. As Margaret Mead remarked, only small groups of thoughtful and committed people can change the world. Such has been the case in all the major faiths, spiritual movements, and the idea and value systems of humankind. So it will be now as the Fourth Generation of *Homo sapiens sapiens* (HSS) emerges through a major evolutionary movement.

The evolutionary quantum jump, the *big change, will happen in our myriad of communities*, living and acting all over the evolutionary landscape. They will become the forces of conscious evolution. We have already left the darkness of the Third Generation and entered a new evolutionary landscape that becomes visible in the emerging sunrise of the Fourth Generation. We can already see that people in many communities are now ready to reclaim their basic right to take part directly in decisions affecting their lives, take responsibility for shaping their future, and guide the evolution of their communities.

Starting with the family, groups of all kinds—neighborhoods, the systems in which we live and work, communities of committed citizens everywhere, smaller and larger collectives that share interest and purpose—are ready to enter into the evolutionary design space. Some of them are now engaged in conscious evolution, and will in ever larger numbers engage in building the civil societies of the Fourth Generation. Embarking on their evolutionary journey, they will map out their evolutionary path as they search for and formulate an ideal evolutionary image that will guide them in the design of their evolutionary guidance system (EGS). I call these evolutionary designing communities the *Agoras of the twenty-first century.*

The citizens of Athens gave us great gifts. The idea of democracy, practiced on the "public sphere" of the Agora, is the greatest of those. This great gift, now a heritage, enriches our lives as we embark on our evolutionary journey and aim to *forge a way of life that will truly represent the democratic idea.* We shall build evolutionary learning and design communities and a society in which thousands and thousands of Agoras will bloom on the evolutionary landscape. These communities will keep our lives purposefully creative and focused and personally and collectively meaningful and rewarding. The New Agoras, the evolutionary design spheres and communities of conscious evolution, will offer us functional contexts of self-guided evolution that are tangible and can be brought to action.

The present chapter is developed in three sections. First, following an exploration of the story of Agora, I propose ways and means by which the myriad self-organizing and self-guiding evolutionary designing communities might engage in the evolutionary design space and design their EGSs and ESs. I will contemplate the various organizational forms these communities might take, how they might work, link up with each other and create an *"evolutionary web,"* and constitute an *evolutionary design movement* in the twenty-first century.

In the second section, I suggest a system of qualities that we might seek to realize in our design inquiry and in our evolutionary designer communities, as well as in the two evolutionary systems: the evolutionary guidance systems and the evolutionary system. This system of qualities becomes the benchmark of assessing our work in the evolutionary design space.

In the third section, I propose a set of characteristics that the Agoras of the twenty-first century should manifest in their work in the evolutionary design space. I see them as designing communities who live by evolutionary ethics, who are in search of the ideal, who are seeking consensus in creating their EGSs. I see these New Agoras as building healthy and authentic communities, in which they bring their creative force alive. Above all, I see them making the democratic idea their value and belief system and making it work in their communities, in their institutions, and in governing themselves.

12.1 *The New Agoras*

First, I review briefly the Agora experience of the classical Greek period. Next, I explore the Agora concept as an approach to establishing a variety of forms and arrangements of our evolutionary designing communities. Then, I contemplate ways and means of how these communities might work.

12.1.1 The Agora Experience of the Athenians

The Agoras of classical Greece were the places of assembly in the city-states, usually located at the city's central plaza. It was during the sixth to the fourth centuries B.C., during the height of the Greek classical period, that democracy was established and flourished in the city-state of Athene. It was practiced in the "public sphere" of the Agora, a place of about 26 acres, declared as an open public space. It was surrounded with plane trees that provided shade and gave the Agora a country ambiance (Burn and Burn, 1980). The citizens of Athene held forty statutory Assemblies a year on the Agora, and others when the need arose. These meetings provided the citizenry the opportunity to take part in deliberations and make decisions about issues that affected their lives and the life of their community. The procedure of the Agora was governed by the democratic constitution, which was established in 507 B.C. at Athene by Cleisthenes. It provided for the election of magistrates by lot and established the supremacy of the Assembly. Later, Pericles completed two constitutional arrangements by adding a third, a provision for public service. This arrangement brought the everyday citizen into a more active role in the service of the common good than at any time before or since. Athenian democracy thus represented the widest possible diffusion of political power among its citizens, the widest ever practiced in our history.

12.1.2 The "Agora Concept" for the Twenty-First Century

The Athenian Agora lived as a most inspiring manifestation of society's life during the Second Generation of HSS. It was a shining moment in the history of the human experience. Hannah Arent (1956) holds that true democracy was lived only once, namely, as it was manifested in the life experiences of the Athenians. She suggests that it became possible only because the Athenians established the Agora, where they were able to make collective decisions about issues affecting their lives and their community. *The Agora experience was unique in the history of humankind.* It became lost in the darkening and declining centuries of the life cycle of the Second Generation.

Now, we can bring the Agora experience back to life again. We have arrived at an evolutionary stage when we have acquired evolutionary consciousness, and, as a consequence, we have become responsible for engaging in guiding our own evolution. To exercise this responsibility, we are in search of *public spheres, New Agoras*, where we can not only (re)establish true democracy, but also bring it alive as a shared culture (a democratic culture) and (re)constitute a method and procedure by which our institutions could serve us and establish arrangements by which we can govern ourselves (establish a cultural democracy). Most significantly, we need to hold up democracy as both a guiding idea and a process by which to work in the evolutionary design space.

Today there are no designated Agoras, no "public spheres" available to us where we can engage in evolutionary inquiry. Csikszentmihalyi (1993) comments on this lamentable state:

> It is incredible that in our society we are spending trillions of dollars on armaments, space exploration, supercolliders, and inefficient social-service bureaucracies, yet we have no budget and no program to enhance the match between our dreams and the institutions that are supposed to make them real. At the very least, on would expect every community to have a beautiful place—an amphitheater in a park, a lofty hall—where people could meet to discuss their public concern and where decisions affecting their representatives could be made. Such town meetings would be cheap even if free caviar and vintage champagne were served, compared to the sums now wasted in programs nobody benefits from. (p. 271)

The image of a "public sphere" evoked by the author is one of many possible Agora arrangements.

12.1.3 The Agora Experiences of Our Evolutionary Designing Communities

The Agora experiences of our evolutionary designing communities could take a variety of forms and arrangements. Each designing community will be unique, working in a unique environment, having unique members, with unique aspirations, resources, and capabilities. We can expect, however, that from their engagement in self-guided evolution will emerge a newly formed, shared collective consciousness that will guide their work in the evolutionary design space. Here, I describe a few possible organizational forms and working arrangements for our new Agoras. They can be adapted or transformed to become the kind of Agoras described earlier. They can be created as evolutionary design spaces in forms yet unimagined. They can apply the evolutionary epistemology developed in this work, and most likely create others. What these Agora evolutionary inquiry systems could become and could accomplish is subject *only* to the limits of our commitments, our imagination, and creativity. What is most likely to happen is that all of them will put the Internet or other emerging technologies in their service.

12.1.3.1 A Fellowship of the Future

In the course of our learning journey, we have visited several times with Mihaly Csikszentmihalyi (1993). He proposes the establishment of "Cells of the Future" by forging fellowship among small groups of people who share an interest in evolutionary inquiry. We have already heard him say that what we have to seek as the core quality in evolutionary design is *harmony*. He sees communities of kindred spirits emerging who have a shared aspiration about the future, and who are dedicated to moving us toward attaining greater harmony. What would such a community of individuals be like? What would they do? How would they engage in evolutionary inquiry? Csikszenmihalyi contemplates these questions for us.

12.1.3.1.a The Evolutionary Cell

> An ideal social unit for accomplishing a task is a group, small enough to allow intense face-to-face interaction. One in which members participate voluntarily, and in which each person can contribute to a common goal by doing what he or she knows best. A "cell" of this type is likely to be a complex social unit, and one that allows the greatest amount of flow for its members. (pp. 285–286)

Today, there are not many opportunities to belong to the type of group described here. The entities we belong are usually large, involuntary, and top-down driven. The voluntary groups we might belong today usually focus on the task of the here and now, regardless of how altruistic they might be.

12.1.3.1.b Purpose and Tasks. The purpose of evolutionary cells is to collect information and develop and share a knowledge base that serves the evolutionary inquiry and take action that advances the cause of evolution. Csikszentmihalyi sets forth four tasks for evolutionary cells. One calls for the assembly of information about the economic state of the neighborhood or community. Another is to understand and portray the landscape of the political and governing forces, their activities, and make judgments about how they affect the well-being of the community. The third task is to determine the skills and the dispositions of members of the cell and organize and coordinate their work. The fourth task is to integrate the knowledge and the substantive work of the cell, understand their systemic relationships, and work on ways to advance harmony within the cell as well as in the community.

A "fellowship of the future," created by the evolutionary cells, is one possible form and arrangement for establishing an Agora that would advance evolutionary inquiry. Assessing it from the point of view of the evolutionary epistemology developed in Part III, it seems that the evolutionary cells would create conditions that would make it possible to transform the cells into evolutionary designing communities. This would come about once they would create a collectively shared value base and an evolutionary image of the future and engage in the design of EGSs.

12.1.3.2 The Cogniscope Approach

In Chapter 9 we reported on the Cogniscope approach (Christakis and Conaway, 1995). Cogniscope is a thoroughly tested cognitive technology that integrates software and groupware. It requires a special "Agora environment," a high-tech environment arranged and equipped to support the work of the designing community. The program is designed to enable the generation and clarification of a large number of ideas about the issue of interest. The ideas judged by the group to be the most powerful are selected and organized into systems of ideas. Based on these the group generates an interactively created design solution and action plan. As a consensus building methodology, Cogniscope could be well considered by evolutionary designing communities as an approach to carry out the design of EGSs and ESs.

The adaptation of Cogniscope technology to evolutionary inquiry could be achieved by using it to accomplish the following: generate a shared value system, create an evolutionary image, based on it design an ideal image of EGSs, and working back from it, design an ES as the operating system of the evolutionary designing community. (Cogniscope has been applied to evolutionary design at one of the annual Design Conversations of the International Systems Institute.)

12.1.3.3 Future Search

The *Future Search, All Design Approach* (Weisbord, 1992) brings people together to achieve in a collaborative mode breakthrough innovations and the creation of a shared vision. Its main operating principle is that the world is moving from experts designing for people to people designing for themselves by including everybody to design their own systems. Involving everybody is the best strategy if we want to achieve long-term commitment, meaning, and community. The approach is based on the following assumptions: The world is knowable by ordinary people and they can organize their knowledge collectively. They are capable of creating their future by design. They want to have the opportunity to engage their heads and hearts in determining their future. The nature of their participation is egalitarian and they value diversity. The Future Search approach leads its user community to create its vision of a desired future. As such, therefore, the approach can well serve as the front-end part of designing evolutionary systems. The approach can be applied in creating the first image of an EGS. Then, the community would use the design architecture described in Chapter 11 and create the EGS. Finally, working back from the EGS, the community can design its own ES.

12.1.3.4 The Conversation Community: Agora Stewardship Development

Having attended systems conferences for many years, by the late 1970s some of us in the international systems movement began to question the usefulness of the traditional conference format as a forum for systems learning, knowledge development, and systems design. Searching for a more meaningful and productive format, we initiated research conversations as a method of choice of collective communication in disciplined inquiry. (The nature of conversations as disciplined inquiry is explored in the last section of this chapter.)

A group that engages in such conversation is called a *"conversation community."* Conversations focus on an issue of shared interest. Organized by the International Systems Institute, we conducted some thirty conversations in several countries. These conversations have addressed the question: How can we apply the insights we have gained from systems and design thinking to the improvement of the human condition? Over the years, the methodology of the conversations became ever more productive. During the last ten years or so, the focus has been on research and development of methods and approaches to the design of sociocultural systems, evolutionary learning, on the creation of evolutionary learning communities, and the design of EGSs.

Several members of our Conversation Community initiated evolutionary inquiry back in their homes. I briefly describe one that was initiated and reported by one of our Conversation Fellows (Sapiro, 1999). The North End Agora of Boise, Idaho, was designed by residents of the neighborhood. Its design began in early 1997 and its implementation began in late 1998. This project engaged the neighborhood in the design of an EGS. It took a group of residents a year and a half to create the first design. What makes this effort so potentially powerful is that it represents the convergence of three "keys": the first is the context, the neighborhood. The second key is the subject of the design: communication and community-building. The third key is the process used to create and continually re-create the system: ideal systems design. In this essay, Sapiro tells the story of how the Agora was created, what lessons may be learned from the experience, and what may lie ahead.

In addition to getting involved directly in the design of EGS, members of the conversation communities aim at *building competence* in order to assume multiple roles in evolutionary inquiry, namely: (1) to provide resources and opportunities to others for the development of evolutionary competence, (2) to coordinate the work of an evolutionary designing community, and (3) to engage in research that advances evolutionary design approaches.

I suggest that the kind of program described here can be designated as *stewardship development for conscious evolution*, where stewardship means assuming shared responsibility for the advancement of conscious evolution. I hold that the kind of stewardship development described here could become a critical component of the evolutionary inquiry movement. Beyond the development of resources for evolutionary learning, through geographically distributed groups of evolutionary stewards, they could enhance the building of a network of evolutionary designing communities, a network of Agoras.

Activity 43

Assume that you are a member of a stewardship group that has responsibility for coordinating an evolutionary stewardship development program in your region. *What would you do to enable the establishment of evolutionary designing communities? How would your stewardship group serve them?* Take into account the potential use of the Internet. (Wasn't the Internet created for this purpose?—although its creators didn't know it.) Enter your story in your workbook.

12.2 *The Qualities We Seek to Realize in Our Evolutionary Systems*

Designers of a particular system of their interest always aim at realizing certain qualities in their creation, whether the nature of the systems they design is technical, sociotechnical, socioeconomic, or sociocultural. For example, in the design of a technical product, design qualities sought might include fidelity to the intended purpose, reliability, safety, ease of use, durability, economy, affordability, aesthetics, and so forth. If anywhere, it is in the design of evolutionary systems that the explicit articulation of qualities is of crucial importance.

We address the "quality" question three times: first, as we design the design inquiry system of the evolutionary community; next, as we design the ideal system of the EGS, and then, as we design the operating system: the ES. A system having certain qualities begets a system having similar qualities which in turn begets a system of similar qualities. *A system of qualities constitutes the DNA of an evolutionary system.*

12.2.1 Qualities of the Designing System

Qualities are organized into five categories: (1) qualities that the design inquiry itself manifests, (2) individual and collective qualities of people in the system, (3) qualities manifested in the designing community as a whole, (4) evolutionary qualities, and (5) systemic qualities that designers wish to realize in the system they design.

12.2.1.1 Qualities of the Design Inquiry

In design inquiry various qualities are sought; one is the use of the highest quality intellectual technology of design. We also seek that the design be authentic and sustainable, user-friendly, and aesthetic. Furthermore, it should enable the seeking of the ideal and the consideration of the uniqueness of the design situation.

12.2.1.1.a Seeking the Highest Technical Quality of the Design Inquiry. We seek the most up-to-date and tested design model, methodology, methods, and tools that evolutionary designing community could use.

12.2.1.1.b The Qualities of Authenticity and Sustainability. Design inquiry is authentic if, and only if, it is conducted by the evolutionary designing community with the full participation of its members. The inquiry should provide opportunities for the incorporation of their individual and collective values and ideas in the design of the evolutionary systems. Genuine participation will ensure the sustainability of the evolutionary systems, because people who participate in the design will take part more effectively and with a greater commitment when the time comes to implement and operate the design. People in the systems feel that they "own" the design, because they created it. Therefore, the design inquiry we use should ensure both authenticity and sustainability.

12.2.1.1.c The Quality of Ideal Seeking. In designing evolutionary systems the design inquiry should enable us to seek an image of the evolutionary system that is the most ideal we can create, that reflects our highest aspirations and expectations.

12.2.1.1.d User-Friendly Language. To enable the full participation of members of the evolutionary designing community, the technical language of systems and design technology used should be transformed to become meaningful to the designers. It should be compatible with the everyday language use of the designing community.

12.2.1.1.e The Quality of Uniqueness. Each and every design situation, design environment, and designing community is unique. The design inquiry applied should have built into it the capacity to take into consideration and capitalize upon these aspects of uniqueness.

12.2.1.1.f The Quality of Aesthetics. This quality comes into play in two ways. Through their participation, members of the designing community can bring into the design individually and collectively their aesthetic values and ideas so that the design, once implemented, will be aesthetically pleasing. If this quality is explicitly sought and realized in the course of design, then the designers' involvement in the design inquiry will become in itself an aesthetic experience.

12.2.1.2 Nurturing Individual and Collective Qualities

The design inquiry system should nurture the individual and collective potentials and capabilities of members of the evolutionary designing community. This effort should include the purposeful nurturing and involvement of the individual and collective intelligence of members of the evolutionary designing community, the development of their human potential through continuous individual and collective learning, their attainment of competence in evolutionary design, their emancipation from dominance of any form, and the full activation and engagement of their creative potential and ability. Thus, design as a human activity becomes not only system creation but also an individual and collective learning and human development experience.

12.2.1.2.a Individual and Collective Intelligence. These are design qualities of the highest value. It is in the nature of authentic design to call upon the intelligence and talents of each and every member of the evolutionary designing community. The activation of collective intelligence means the application of individual intelligence in a purposeful, focused, and coordinated way. Evolutionary design calls upon a wide scope of different capabilities and the design inquiry should create conditions by which these capabilities are offered and called upon freely, are exercised in a cooperative and collective way, and are directed toward the common purpose of evolutionary design.

12.2.1.2.b Individual and Collective Potential. Individual and collective potential is a quality that design inquiry is to call upon and should nurture. The inquiry should provide resources and opportunities to members of the evolutionary designing community for continuing individual and collective learning and human development.

12.2.1.2.c The Quality of Design Competence. The quality of design competence to a major degree defines the quality of the outcome of design. The attainment of design competence requires a special kind of learning, which enables members of the evolutionary community to acquire a design culture. Design culture empowers members of the designing community to shape their own futures individually and collectively and participate competently in the design of evolutionary systems.

12.2.1.2.d Emancipation Is an Aimed at Quality. In a true participative design mode, there are no chiefs, no designated authorities. Dominance stifles creative involvement. Design flourishes only if there is equity among the participants. Everyone should have the same right and responsibility in making contributions. This may be the most difficult idea to accept by those in authority. Liberation and emancipation from dominance become one of the highest qualities to be realized in evolutionary design.

12.2.1.2.e Creativity. There are qualities without which design cannot happen. Without creativity there is no design. Creativity is central to design. We cannot understand design unless we understand creativity. The quality of creativity can be realized once we understand what it is and how it works. Once we are aware of the internal and external conditions that are to be met to activate and sustain creativity, and once we know the barriers that prevent it as well as the ways to overcome those barriers, then—and *only then*—are we in a position to use the individual and collective creative potential of the evolutionary designing community.

12.2.1.3 The Qualities of the Evolutionary Designing Community

The designing community seeks to attain high ethical qualities, knowing well that only ethical people can design ethical systems. In their ethical stance, they especially seek the quality of being sensitive to the effect of the design on future generations. As a community, they wish to become both responsive and responsible; they will seek the quality of diversity in their own community. They develop among themselves the qualities of unconditional acceptance and respect for each other. Their dominant aspiration is to become a learning system that will seek to establish a shared systemic world view, a quality of the highest order.

12.2.1.3.a An Ethical Community. Only an ethical system can design an ethical system. Individual members, as well as the collective of the evolutionary designing system, should manifest the quality of high moral and ethical standards. It might be useful for the community to establish a *code of ethics* for itself as a guide.

12.2.1.3.b Responsive and Responsible. These qualities are being part of being ethical, but deserve special emphasis. Being responsive means that designers respond to not only their own aspirations and desires but also consider those of the less fortunate, the oppressed, the weak, and very specifically the effect of the design on future generations. The quality of responsibility means that the designing community collectively takes responsibility for the design it creates.

12.2.1.3.c The Quality of Diversity. Diversity is an essential condition of evolution of life. The loss of diversity in life creates monocultures and fragile conditions. Only diversity can generate in the evolutionary designing community the broad multifaceted design intelligence required in conscious evolution. The evolutionary designing community should represent cultural and racial diversity. Exclusion of diversity from design or from a designing community and seeking only the involvement of designers with similar viewpoints builds brittle designs that will not weather the dynamic of a diverse world.

12.2.1.3.d The Quality of Unconditional Acceptance and Respect. This quality is a core requirement of maintaining viable designing communities. It is a requirement that is more than a condition of civility and humanness. In the designing community only such acceptance and respect for each other shall ensure that team members will be willing to offer solution ideas and make creative contributions spontaneously and without any fear of being rebuffed or ridiculed.

12.2.1.3.e The Quality of Being a Learning System. The evolutionary designing community needs to become a learning system. This quality ensures that arrangements, operations, and structures are in place by which to build the required organizational capacity and human capability to engage in continuous organizational learning within the designing community. This quality will ensure continuous coevolution and creative interaction with the environment, and readiness to engage in continuous design and in the ever ongoing renewal of our evolutionary systems.

12.2.1.3.f The Quality of Having a Shared World View. A world view attributes meaning to what we observe and experience. It shapes our behavior and our actions. This is another higher order quality that emerges from an interaction and integration of a systems view of the world, a systems view of the evolutionary systems in which we live, and a design view of change and evolution. These views create and guide the emergence of designing consciousness. They will become explicit in shared core values and ideas in the collective world view of the designing community. They constitute the philosophical and theoretical basis of creating evolutionary design solutions.

12.2.1.4 Evolutionary Qualities

In designing evolutionary systems, the evolutionary designing community seeks to realize evolutionary qualities in their inquiry as well as in the systems they design. Our understanding of what these *qualities* are is grounded in an understanding of how evolution works. These qualities include harmony, balance, internal consistency, synergy, symmetry, self-organization, coevolution, and the meta-quality of wholeness. These qualities constitute a system of qualities. A particular evolutionary state of HSS is heathy and fulfills its potential only if these qualities are manifested in the behavior of a particular evolutionary generation. When these qualities are not manifested in an evolutionary state, wholeness is lost, and the evolutionary generations enter the downward slope of decline and disintegration.

12.2.1.4.a Balance. Balance refers to a desirable qualitative state among the various evolutionary markers. It is a state in which the markers, animated by the new level of collective consciousness, have coevolved, mutually influenced each other, and attained a balanced state. *Coevolution* was the enabling quality that has helped to attain such a balanced state. An out-of-balance state comes about when one or more of the markers dominates others. In the lifecycle of an evolutionary generation an out-of-balance state indicates disorder and a declining stage. For example, during the declining stage of the Third Generation of HSS, the *dominance* of machine-age thinking and the industrial age technology over society, nature, and the health of people created many harmful effects.

12. 2.1.4.b Harmony. Harmony is a quality that is enabled by intensive *cooperation*, internal compatibility, and mutual assistance among the evolutionary markers. The more intensive the cooperation and the more explicit the mutual assistance the higher is the quality of harmony. I consider harmony as a quality of the highest order, the most desirable evolutionary state.

12.2.1.4.c Internal Consistency. Internal consistency is a qualitative state of a sociocultural system that is derived from the manifestation of the qualities of *balance* and *harmony* and—more importantly from the continuous infusion of the meta-quality of *collective consciousness* in all the evolutionary markers. A lack of internal consistency moves the systems toward a crisis of consciousness and a chaotic state.

12.2.1.4.d Synergy. Synergy is another core evolutionary quality. It is manifested in the interaction and the integration of the parts (components) of a system. Synergy is animated by collective consciousness. It is the result of the whole organizing the parts, bringing forth the coevolution of the evolutionary markers.

12.2.1.4.e Symmetry. Symmetry develops from synergy. The depth and dynamism of the material realm of evolution are mirrored by the corresponding depth and dynamism of consciousness, says Elgin (1993). When these are symmetrical, a conscious organism has the capacity to become fully self-referencing and self-organizing. One of the characteristics of the "declining" stage of evolution is a state in which the material side outruns and dominates the consciousness side, resulting in a symmetry break. Symmetry break also signals the decline of an evolutionary generation.

12.2.1.4.f Self-Referencing, Self-Organization, and Self-Renewing. Self-referencing, self-organization, and self-renewing are complementary evolutionary qualities. A balanced, harmonious, internally consistent, synergistic, and symmetrical sociocultural system may be attained by the continuously ongoing process of self-reflection and self-referencing, when the system is always aware of itself and its purpose. It knows how it is doing, knows what it desires, what it aspires to attain, and how to attain it by self-renewal.

12.2.1.4.g Consciousness, Creativity, and Wholeness. The meta-qualities of consciousness, creativity, and wholeness are directly involved in attaining the core qualities. The infusion of consciousness into the evolutionary markers is a necessary condition of the emergence of a new generation of HSS. Shared consciousness will integrate the evolutionary markers so that the new entity will have the evolutionary core qualities defined previously. It is the continuously ongoing *creating* process by which this infusion interpenetrates the evolutionary markers and by which the core qualities are attained in the evolutionary design space. The ultimate meta-quality, *wholeness*, becomes manifested in the life of an evolutionary generation when all the other qualities are attained.

12.2.1.5 The Qualities of Being Systemic

Designers seek to realize systemic qualities in their design. These qualities are grounded in an understanding of systems and design concepts, principles and models, and their manifestations in the behavior of social systems, in their nature, and in their internal and external relationships. These qualities are to be realized in the course of the design inquiry process, in the evolutionary designing system, as well as in the life of the evolutionary systems.

12.2.1.5.a Qualities of Systemic Behavior. These behavioral qualities include wholeness and the indivisibility of the system, the seeking and fulfillment of purposes, attention to positive and negative feedback and to requisite variety, and attaining robustness.

12.2.1.5.b Desirable Qualities in Relationship with the Environment. These include continuous awareness of and interaction with the environment, the seeking of a coevolutionary and cocreative relationship with it, and the relating to larger and larger spheres of the environment to become ever more relevant to the system's contexts and to make the environment ever more relevant to the system. These qualities comprise the external dynamics of the system.

12.2.1.5.c Desirable Qualities of the Internal Dynamics. These include purposeful and guided emergence (rather than chance emergence) that comes about by the purposeful interaction and integration of the evolutionary markers, recursiveness that guides interaction among functions and components, and purposeful attention to the use of feedback and feedforward loops.

Figure 12.1 displays the five domains of qualities we seek to realize in the design inquiry, in the evolutionary designing community, and in the evolutionary systems.

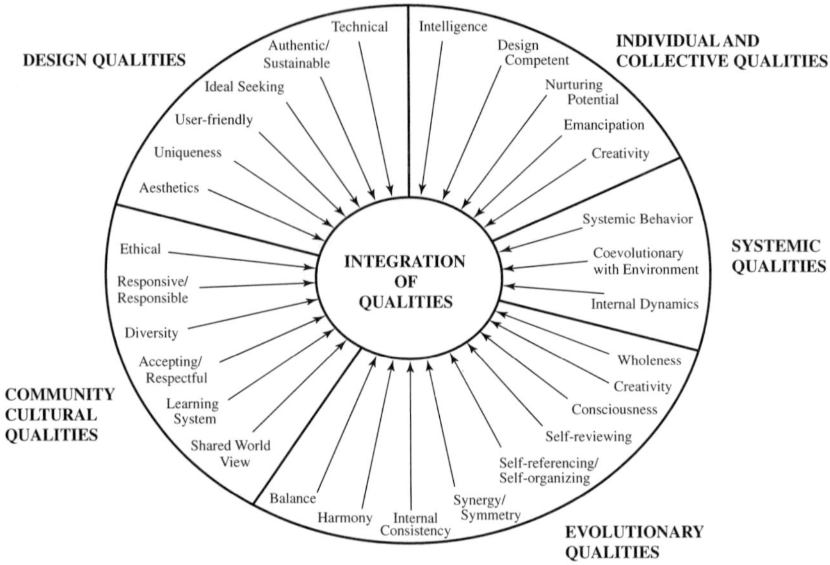

FIGURE 12.1 The Qualities We Seek

Reflections

In the course of evolutionary design, we seek qualities that become the properties of the design inquiry, the members of the designing community, and the designing system itself. The overall quality of the system emerges from the interaction and integration of the qualities we seek to attain. (See Figure 12.1.) That overall quality is more than their sum. Therefore, we are to conceptualize the sets of qualities as a system. First, we look for internal consistency and compatibility among the qualities that we seek in order to bring to life the evolutionary system we wish to create. Then, we are to design and *integrate them into a system* by identifying relationships among the qualities. We should stipulate recursive interactions among qualities that through mutual influence reinforce each other. Thus, qualities become strengthened and the overall quality emerges.

Activity 44

Three tasks might be helpful in constructing your own understanding and interpretation of qualities to be manifested in evolutionary design. First, review the qualities described in this section and propose others that come to mind. Then, test for internal consistency. Ask yourself: Do the qualities support and reinforce each other? Next, select one quality from each of the quality sets and ask: What evidence would I accept as an appropriate manifestation of the realization of that quality? Enter your findings in your workbook.

12.3 *Essential Characteristics of the Agoras of the Twenty-First Century*

One more aspect is left to explore in imagining Agoras for the twenty-first century. It is to contemplate the essential characteristics of the nature of the New Agoras as designing systems. I characterize the nature of New Agoras as *ethical systems*. They are *ideal seeking and creative. They are capable of building consensus* by disciplined conversation. They create themselves as *authentic, healthy, and nurturing communities*. In their lives they *manifest democratic culture* and *build cultural democracy*.

12.3.1 *Evoutionary Ethics and Morality*

In the course of this century we have developed evolutionary consciousness and more recently we grasped the potential of conscious evolution, the potential of giving direction to our own evolution and the evolution of the systems we inhabit, the evolution of our communities and our society by purposeful and deliberate design. A core requirement of evolutionary inquiry is that it should be *guided by evolutionary ethics* and manifest evolutionary morality. I explore the meaning of evolutionary ethics, the multilevel nature of ethics, morality in evolution, ethics as aesthetic conversation, and the product and process of evolutionary ethics.

12.3.1.1 Evolutionary Ethics

Jantsch (1980) holds that *ethical behavior enhances evolution*. Ethics takes on a regulatory function in evolutionary systems. This regulatory function in the human world consists of rules of behavior but also as a distinct inner experience. As an integral aspect of evolution, ethics is experienced directly by way of the dynamics of self-organization and as a creative process. In the life of evolutionary systems, in which possibilities for action are available in such a rich spectrum, it is primarily our intentions, desires, and preferences that are guided by ethics and thus guide our work in the evolutionary design space.

12.3.1.2 Multilevel Ethics

Ethics is multilevel. It includes personal and transpersonal ethics, the ethics of social systems, and ultimately what Jantsch calls evolutionary ethics. He notes, however, that we are still far from formulating and implementing an evolutionary ethics. "What in the Western world we call ethics is a behavioral code at the social level which is primarily geared to ensure the free unfoldment of the individual." (Jantsch, 1980, p. 265). For this reason we talk almost exclusively of the rights of individuals and particular groups but almost never of their responsibilities. As Vickers (1981a) pointed out, *rights are static and defensive,* while *responsibilities imply creative participation* in the design of the human world.

According to Jansch (1980), the ethics "that dominates the Western world is therefore an individual ethic in the disguise of socially committing behavioral code. It is not a multi-level ethics in the true sense. Morality, in contrast, is the direct experience of ethics inherent in the dynamics of evolution. The higher the number of levels and the intensity at which we live, the higher the number of levels and intensity at which our morality becomes effective." (p. 265). *Ethics is a manifestation of consciousness* and being guided by it is a major aim in our design of evolutionary systems. The ethics that an evolutionary design community establishes for itself will guide the behavior of the community and the process of making design decisions.

12.3.1.3 Morality and Evolution

Csikszentmihalyi (1993), discussing morality and evolution, says that "In every human group ever known, notions about what is right and what is wrong have been among the central defining concerns" (p. 139). Moral imperatives have become necessary because "evolution, in liberating humankind from complete dependence on instinct, has also made it possible for us to act with a malice that no organism ruled by instinct alone can possess. Therefore, every social system must develop 'memes' to keep the intergroup harmony that genes no longer can provide. These memes constitute the moral system, and generally, they have been the most successful attempts humans have developed to *give desirable direction to evolution*" (pp. 159 and 160).

For over a century, it has become fashionable in the social sciences to suggest that different cultures develop entirely relative and arbitrary moral systems. "In fact what is so remarkable is how similar the world's major moral systems are in considering 'good' to be the achievement of the kind of harmony within consciousness and between people that we have called negentropy, and which in turn leads to higher levels of complexity" (Csikszentmihalyi, 1993, p. 160). *The great moral systems of the world are congruent*, despite differences in emphasis and variations in the metaphors used to explain why some things are right and others wrong.

All ethical systems propose to direct evolution by channeling thought and action from the past to the future. "The past—represented by the determinism of instinct, the weight of tradition, the desires of the self—is always stronger. The future—represented by the ideals of a life which is freer, more compassionate, more in tune with the reality that transcends our needs—is by necessity weaker, for it is an abstraction, a vision of what might be" (Csikszentmihalyi, 1993, p. 162). The new, the hopeful, and the creative appear to be more ephemeral than what has already been tried. The realist, who deals with the here and now, belittles the "impractical" idealist who invests energy in the stuff of a "blue-sky world." *Without the realist we could not survive. But without the idealist we could not evolve.* The choices made in our continuing evolution are to be guided by a moral system that takes into account the wisdom of tradition, yet is inspired by our vision of the future.

The preceding conversation on morality and evolution brings into focus the individual and collective challenge we face in our evolutionary design communities. As we engage in the conscious evolution of our EGS and our ESs, our inquiry and our design decisions should be guided by evolutionary ethics and morality.

12.3.1.4 Evolutionary Design as Aesthetical Conversation

Ethics, says Churchman, is an ongoing aesthetical conversation (1982). This conversation should never stop. "Ethics is an eternal conversation. The reason that ethical relativism ('different strokes for different people') is so bad is that it stops conversation. Relativists are only sure of one thing, their relativism. They actually think that ethics is a search for absolute values, and since it is, it is a hopeless enterprise. Since *ethics is an eternal conversation*, its conversation retains its aesthetic quality if human values are regarded as neither relative nor absolute" (p. 57).

In evolutionary design the ethical conversation goes on at every decision point and it guides each and every choice we make. At each decision point we ask: What is the ethical base of this choice? In considering design alternatives we ask: Which of the alternatives manifest best the ethical position upon which we build our design? Ultimately, *that ethical position is the guarantor*, the conscience, and the soul of our evolutionary design.

12.3.1.5 The Ethical Imperative: Impact on Future Generations

What are the implications of our design for future generations? Concern for future generations is the key imperative for Churchman (1982). We are obliged, he says, to consider the impact of our design on those who come after us. Our design should be such that it *enhances and expands their options*. Thus, the moral law with respect to future generations is: "We should undertake to design our societies and their environments so that people of the future will be able to design their lives in ways that express their own humanity" (p. 21).

12.3.1.6 The Ethical Product and Process

What is the relationship between the ethical process of evolutionary design and an ethical product of design? We raised this question at one of our annual international conversations on social and evolutionary design. The research team on ethical design proposed that the outcome, the product of *the evolutionary design process, is ethical insofar as the process used to generate the product is ethical*. The group suggested that we can speak of (1) the ethical design of a system; (2) an ethical system that produces the design (an ethical designing system); and (3) an ethical system that is the product of both. Furthermore, the product of the design should include explicitly stated ethical standards (4) by which to assess the ethical quality of the evolutionary system we design, and (5) serve as guidelines for the behavior for members of the ES.

Reflections

Ethics provides us with a mirror in which we can see if our evolutionary design is guided by ethical imperatives. If we have such a mirror, then ethical design conversation replaces potentially aggressive conflict among members of the designing community with informed and value-based exchange of ideas and perspectives. Thus, we can say that the outcome of design is ethical insofar as the process used to generate the outcome is ethical.

Ethical behavior enhances evolution. Ethics is multilevel, including personal and transpersonal ethics, social ethics, and evolutionary ethics. Evolutionary ethics, at the top of the ethical hierarchy, includes all the others. In the course of designing guided evolution what is right and wrong have been among the central defining concerns.

It has become fashionable for moral relativists to claim that human systems are arbitrary products of various cultures. In fact what is remarkable is how similar the various moral systems are. Ethics involves an eternal conversation. Moral relativism stops this conversation.

A key task of evolutionary design communities is to create a common ground by *defining a collective ethical position and applying it in making design choices*. This position should include consideration of the impact of our design on future generations. Our design should be such that it enhances and expands their options.

Activity 45

In view of the conversation on the ethics of evolution, presented in this chapter, ask the group you have worked with on previous activities: At what points of the design inquiry and in what manner would you give considerations to ethics. And: What was the ethical position taken by the group in working with the activities? Describe your findings in your workbook.

12.3.2 In Search of the Ideal: Nothing Less than the Ideal

The previous section highlighted the role of ethics in design. We understood that design inquiry is value based and should be guided by shared ethical and moral imperatives. In this section we introduce the imperative of the ideal. Evolutionary design inquiry should always be guided by the ideal. First, I connect ethics in design with the design of ideal evolutionary systems. Then, the pursuit of the ideal is set forth as an evolutionary imperative. In conclusion, the implications and consequences of ideal seeking in evolutionary design are explored.

12.3.2.1 Ethics and the Ideal System

In *Search of a Way of Life* (1948), Singer developed the notion of ideal in the philosophy of ethics. He proposed that ethics should always be discussed in the context of human ideals. Singer, however, recognized that there is a dialectic between the idealist and the realist. Commenting on Singer's view, Churchman (1982) says that "the realist is a down-to-earth, practical person who tries to solve the practical, hard problems of everyday life. The realist goes to management development programs and expects to find out what to do next Monday to become a better realist. The idealist tries to understand the human saga in terms of ideals and their meaning in the very long run. He sees that there is a constant struggle towards an ideal society. He tries, as best he can, to explain what that ideal might be" (p. 133).

Envisioning a creative interaction between ethics and the ideal, I believe that ethics is aimed at *searching for the "ultimate good,"* which is the ideal. We can also recognize the interaction between *inspiration*—which is coming from the beauty of the ultimate—and the *aspiration* that aims to define and bring it to life. Evolutionary design, inspired by ethics, leads us to create visions and images of an ideal future that we aspire to attain. *Inspiration and aspiration jointly give us the courage and determination to pursue the ideal.*

12.3.2.2 Evolutionary Guidance Systems Are Ideal Pursuing Systems

"We humans are unique in our ability to formulate and pursue ideals, desired states that we can never attain but to which we can always come closer. If we are to pursue the ideal continuously, we must never be willing to settle for anything less. We must never be either permanently discouraged or completely satisfied. We must always be able to generate visions of something more desirable than that and must pursue these visions (Ackoff, 1981, p. 40). In our EGSs we collectively pursue states we know we cannot attain today. We still draw satisfaction from approaching such states. As many wise men observed, there is more satisfaction in pursuing an end than in attaining it.

There is an ancient Hungarian legend of hunters of antiquity who pursued the "miraculous white stag." They followed it through endless obstacles, but they never hunted it down. It was the joy of the chase that inspired them. They had the vision that the miraculous stag would eventually lead them to the promised land. This story is a metaphor for ideal pursuing systems.

The pursuit of the ideal can provide cohesiveness and continuity to life. By formulating and pursuing ideals we put meaning and significance into our lives and by so doing derive satisfaction. The ideal system refers to the perfect, the best and flawless "prime" system that achieves the ideal we seek to attain. Ideal conditions refer to a state and conditions in which systems can be designed in an ideal way. An EGS is a conceptual representation of a system that represents the ideal. It can guide us throughout the evolutionary design inquiry. Thus, the conception and articulation of the ideal is a most practical approach to the design of evolutionary systems.

12.3.2.3 Reasons and Benefits of Engaging in the Pursuit of the Ideal

Reasons and benefits include participation in forging our evolutionary future, addressing aesthetic values, attaining consensus, and engendering commitment to pursue the ideal (Ackoff, 1981).

Pursuing the ideal by engaging the evolutionary design community facilitates *genuine participation*. It provides an opportunity to work with others in the system, to think and learn about their system, to contribute their ideas to the evolutionary design, and thus affect the lives of members of the evolutionary design community. Furthermore, participation in idealized design enables members of the designing community to incorporate their *aesthetic values* into the design and thus improve the quality of their individual and collective lives. Participating in the design of an EGS and an ES is a most rewarding aesthetic experience in itself.

Evolutionary design calls for the generation of *consensus* among those who participate in the inquiry. We can attain consensus because we focus on ultimate values. When agreement is reached on the ideals, differences over means often can be easily resolved. Most importantly, consensus generates *commitment* to bringing the design to life. Those who have a hand in developing the EGS and the ES will have strong commitment to implement those systems. They become the owners of the system. Finally, evolutionary design inquiry engenders, stimulates, and *releases individual and collective creativity.*

12.3.2.4 The Evolutionary Imperative of Pursuing the Ideal

In times of relative stability and slow change, characteristics of the past, piecemeal adjustments, and incremental changes were adequate to bring social systems in line with gradual changes in the societal environment. Under those conditions, the task of social planning was one of making ad hoc improvements in existing systems. However, those so-called "good old days" are passe.

In the later part of this century, we entered an era of rapidly accelerating dynamic changes, discontinuities, and transformations. Today, *incremental changes and fixing the existing system no longer work.* We have also found out that social planning and design that extrapolates from the present and experts' predictions of the future does not work either. This situation leaves us with no other serious option than to develop disciplined inquiry that enables us to define and design our own future.

Evolutionary design in the context of human activity systems is a decision-oriented disciplined inquiry. It is aimed at defining what our future should be and producing a description or a model of an EGS that has the organizational capacity and collective human capability to bring about what we defined as the ideal future evolutionary state. In this work we are concerned with the design of such a future state. *The EGS is an intentionally created system through which we seek to fulfill purposes that are collectively formulated by those who are in the system.*

12.3.2.5 The Implications and the Consequences of Pursuing the Ideal

In this concluding section, I explore the implications and consequences of engaging in the design of EGSs and ESs.

12.3.2.5.a Pursuing the Ideal: Evolutionary Design by the People for the People. Evolutionary design intentionally creates an EGS that fulfills the purposes as envisioned by an evolutionary design community. It is a process by which visions, ideals, ideas, values, and aspirations are shared, collectively agreed upon, and articulated by the community. The community then engages in design to create the EGS, that will manifest their shared vision, that represents their shared ideas, aspirations, values, and ideals. The kind of design described here is not a top-down, design-by-directive approach. It is not the kind that is designed by experts for other people. It is a creative process *designed by the people and for the people who constitute the designing community.*

12.3.2.5.b Shaping the Ideal and Being Shaped by It. Evolutionary design is a journey toward the ideal. Only the ideal is worth the effort that is required to undertake the evolutionary journey. *Creating our evolutionary future is a demanding disciplined inquiry.* But its reward lies in the progress we make toward it. It is the ideal that will shape us. A metaphor highlights this "shaping."

There is the story of a prince, born with unsightly deformities, who commissioned the sculpturing of a statue of a beautiful young athlete. The statue was placed in the royal garden. The young prince sat in front of the statue all day long, wanting to transform himself into the image of the statue. Years of persistence rewarded him. He became a living image of the statue.

Commitment to the ideal means a determination *to create the most inspiring ideal system,* one that will act as a magnet and pull us toward its realization. The ideal system is not some science fiction speculation. For us it is real. It is a system that we want to pursue. The ideal model—like the statue in the royal garden—will always be in front of us. As we focus our eyes on the ideal, it will guide our continuous movement toward it. It is the ideal—like the beautiful statue created for the prince in the fable—that will inspire us to become ideal-like. The ideal and our aspiration to attain it give the inspiration and the courage to pursue our design. *Inspiration and aspiration shape each other as we shape our ideal system.*

12.3.2.5.c Be Prepared for the Long Term. In my twenty years of work at an R&D laboratory, where I guided some fifty projects and programs, I was often confronted with the question of how much time it would take to design or redesign a particular system. My answer was that a genuine and viable design always calls for: (1) a commitment to develop design competence, (2) a readiness to engage in a long-term process, and (3) a willingness to commit adequate resources. It would take at least a year or two. In almost all cases the response was, "We don't have that much time to this design thing. We must demonstrate visible change by the time of the next board meeting." Of course they were not prepared to face the question: What will happen if they do not engage in design?

Today, we are spending billions of dollars trying to fix systems that should no longer exist. The educational arena is one of the best example of this. The greatest hurdle to genuine and viable change is the pervasive mentality of demanding short-range results. *We live in an "instant-quick-fix" culture.* The implications of an ideal seeking evolutionary inquiry is the need to *bring about a cultural change from a short-range to a long-range view and perspective.*

12.3.2.5.d Evolutionary Design Never Ends. Such a long-range perspective will embrace the implication of having the ideal design "out there" on the horizon and moving toward it. We called this phenomenon the *"law of the moving horizon."* This law emerges from the realization that as we move toward the horizon the horizon moves ahead of us. The landscape of our evolutionary journey changes. The law of the moving horizon best applies to evolutionary design. As time goes by in our journey toward the future, the ideal model will most likely change, as we might "remake" it. As we move toward the realization of the ideal, the environment in which our system lives and the situational context in which our system operates will also change. *We shall reexamine and possibly reshape our ideal model,* based on our commitment to coevolve with our continuously changing and transforming societal environment. But it is not only what is "out there" that leads to changes in the ideal but also what is within us. Our perceptions, insights, and ideas also change as time goes by, changing our vision of the ideal. *Evolutionary design is a journey that never ends.* An evolutionary design community is like the hunters of antiquity who pursued the miraculous stag with the hope that it would lead them to the promised land. Furthermore, the design journey does not end with us. We should also design systems that offer learning opportunities, arrangements, and resources for evolutionary learning by which future generations will be enabled and empowered to attain evolutionary competence so that *they can envision their own evolutionary ideal,* shape their own future, and continue their journey toward their promised land.

12.3.5.2.e Authentic Participation. Evolutionary design is authentic only if it is created by members of the evolutionary design community. It is from their dreams, ideas, aspirations, and preferences that the "ideal should" emerges as a collective definition and design of the EGSs. They engage in design because they genuinely and deeply care about the future and want to shape it. Thus, the design of EGS is always participative.

12.3.2.5.f Evolutionary Design by Consensus. Evolutionary design engages the creative potential of all who participate in it. Such engagement makes possible their meaningful contribution to the design of their EGS. At the same time, the inquiry provides them with unique learning experiences about how to engage in design, what the characteristics of their emerging EGS are, and what will be their individual and collective role in the EGS. They will also *learn how to generate consensus among themselves*. They will continuously apply consensus building methods in making design decisions. Because the design is their own creation, they will take part more effectively and with a greater level of commitment in the implementation of the EGS and the ES. *Authentic participation is empowering and design is always empowered by it.*

12.3.2.5.g Evolutionary Design Provides for Individual and Collective Learning. By engaging in design, people learn individually and collectively as a community. As individuals, they will gain a genuine understanding of what their EGS and ES are about and how they work. They will also realize how their individual performance affects the performance of the whole. As a community, they will learn how to examine and continuously reexamine their purposes, perspectives, values, and modes of operation. Thus, they can *collectively develop new insight and knowledge*, based on which they can redesign their evolutionary systems and make continuous contributions to its life and to its future development.

Activity 46
Your task in this activity is to: (1) review the discourse of this section and list core ideas of the ideal systems notion that you consider salient and relevant to the design of EGSs and ESs. (2) Select a set of core ideas you identified and explore how the set would enhance the design of the EGS that you have worked on earlier with your group. Enter your findings in your workbook.

12.3.3 Creativity and Evolutionary Design

As a point of departure to our continuing journey on the landscape of evolutionary design, we consider general definitions of creativity and also connect creativity with the ideal systems approach to evolutionary design. Then, I explore the realms of creativity and address such aspects as its (1) nature and characteristics, (2) process, (3) internal and external conditions, and (4) barriers and ways of removing them. In closing, I explore the evolutionary imperative of creativity. As you work with the various parts of the text you should make note of the core ideas of creativity and ask yourself: What do these core ideas mean to me? What are their implications for evolutionary design? The activity at the end of the section further elaborates these tasks.

12.3.3.1 Creativity: Definitions

As an overall context of a definition of creativity, Whitehead (1968) suggests that *creativity is the actualization of potentiality*, and the process of actualization is an occasion of experiencing creativity. Viewed in conjunction, they carry the creative act that drives the world. "Creativity expresses the notion that each event is a process issuing in novelty" (p. 236). Creativity (Barron, 1988, p. 80) is "essentially the ability to bring something new into existence purposefully." The defining properties of creative product and processes are their originality, validity, adequacy to meet needs/aspirations, and their aesthetic quality or elegance. "The emphasis is on whatever is fresh, novel, unusual, ingenious, divergent, clever, and apt" (p. 20). Creativity implies novelty and innovation (Harman and Rheingold, 1984). In human experience there is tension between maintaining equilibrium, security, and stability and seeking new possibilities. This tension is manifested in such dualities as rationality and intuition, conventional and unconventional, complexity and simplicity, certainty and uncertainty, and convergence and divergence. As we saw earlier, in evolutionary design both divergence and convergence operate.

Convergent thinking tends to use rationality to focus down to a single goal. Divergent thinking, drawing upon a richness of creative ideas and original thinking, is characterized by moving away from set patterns and goals. "While both convergent and divergent thinking are involved in creative activity, it is divergent thinking that especially characterizes that which is most widely recognized as creative" (p. 51). Rogers (1961) holds that the mainspring of creativity appears to be man's tendency to actualize himself, to become his potentialities. In human life and in the society there is a tendency toward an urge to expand, extend, develop, mature, and to express and activate all our capacities. This tendency may be deeply buried under encrusted defenses of the familiar. Still, it exists in every person. *Creativity awaits the proper conditions for its release and expression.*

The experience of creativity (Csikszentmihalyi, 1993, p, 175) "stretches our skills in new directions as we recognize and master new challenges. Every human being has this creative urge as his or her birthright. It can be squelched and corrupted, but it cannot be completely extinguished." The enjoyment of creativity comes from such experiences as surpassing ourselves and mastering new obstacles. These experiences involve concentration, absorption, deep involvement, joy, a sense of discovery, and accomplishment" (p. 177).

Throughout this work we have learned to appreciate the central role of creativity in evolution. At each evolutionary generation, creativity, coupled with collective consciousness, is the main force of bringing forth an EGS and an ES.

As I repeatedly said, design creates novelty. It creates new forms and processes. Design is a manifestation of creativity. *Design is creation*. These phrases underline the central and dominant role of creativity in evolutionary design. *We cannot understand design unless we understand creativity.* I will now connect the ideal evolutionary design notion with creativity. To do this, I reintroduce an earlier image here as Figure 12.2. The image helps to make the connection between creativity and evolutionary design.

The arrow on the top stands for the act of creating/designing the ideal evolutionary system, the EGS. The EGS can be conceived only by a collective, creative envisioning of the future by the evolutionary design community. We cannot design the EGS by staying within the boundaries of our existing evolutionary state and working out from it. We have to transcend, leap out from what exists, and, animated by new collective consciousness, create ideal EGSs. Then, we work back from the ideal to define and design the ES, which we can bring to life today.

12.3.3.2 The Realms of Creativity

The various realms of creativity are explored next, including the meaning and characteristics of creativity, the creative process, conditions of

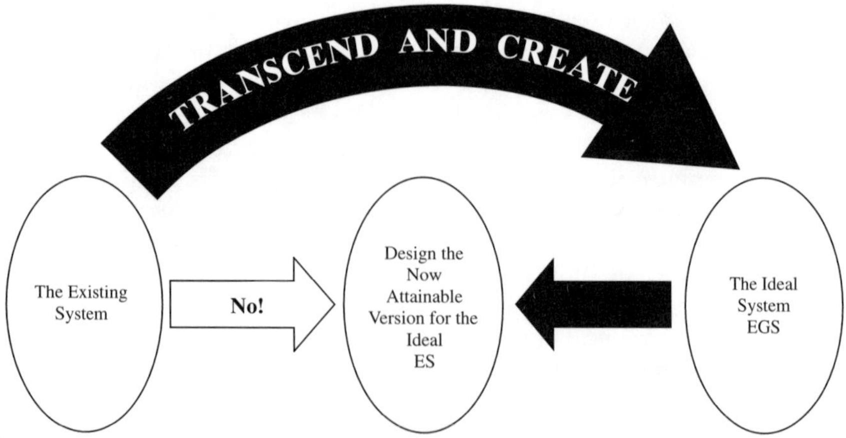

FIGURE 12.2 Creativity Is the Core Process of Design

creativity, the barriers and blocks to creativity, and misconceptions about creativity.

12.3.3.2.a Nature and Characteristics. New creative ideas emerge from us as a play of the mind (Bohm and Peat, 1987). Failure to appreciate the creating role of this play is a major block of releasing creativity. "Within the act of creative play, fresh perceptions occur which enable the person to propose a new idea that can be put forward for exploration. As the implications of this idea are unfolded, they are composed or put together with other ideas. Eventually the person supposes that these ideas are correct" (p. 48). This quote appropriately represents work in the evolutionary design space. Bohm further suggests that communication is as essential to the creative act. Creation arises from the flow of ideas between people. It flows out of a free and open communication. Discussing creativity in the whole of life, Bohm challenges the assumption that creativity is necessary only in some specialized fields. Assuming a restricted nature of creativity has very serious consequences for the society as a whole.

Bohm highlights best the role of creativity in evolutionary inquiry as he says that a free and open creative communication is the most effective way of addressing the (evolutionary) crisis that faces society. A free exchange of ideas is of fundamental relevance for transforming culture by liberating creativity. If we restrict the spirit of free play of ideas, all the problems that have plagued civilization will surface again and overwhelm us. Thus, Bohm calls for a creative surge that will involve all phases of human life. A radical (evolutionary) transformation must take place that embraces all fields of inquiry and develops a "*new view of humanity, culture, and society.* What is needed today is a new surge that is similar to the energy generated during the Renaissance but even deeper and more extensive" (p. 265).

Creativity is a function of interaction of knowledge and imagination (Parness, 1972). Without knowledge, there can be no creativity. Creativity is the innovative, novelty-creating use of knowledge. The kaleidoscope is a good analogy. The more pieces we have in it the more patterns will show. The more knowledge we use in creativity the more novel ideas we can produce. That is why we need an extensive knowledge base of the various experiential domains of EGSs. But in the kaleidoscope, merely having lots of pieces without revolving the drum does not produce patterns. Just having lots of knowledge available, without working with it, juxtaposing, combining, synthesizing it, will not allow any creative ideas to emerge. Therefore, *without knowledge, and without its dynamic use in the course of the creative process of evolutionary inquiry, our efforts will not be productive.*

12.3.3.2.b The Creative Process. The creative process is at the very core of transformation, moving from what exists toward a desired future novel state. The process involves preparation, incubation, illumination, and verification (Wallace, 1926).

Preparation. After trying to use prevailing conventional approaches in addressing issues of interest and recognizing that they do not work, we are becoming aware that we have to set aside our existing assumptions which seem to block rather than help our inquiry. As we discussed earlier, we realize that we must transcend the existing evolutionary state and be prepared to engage in the design of our EGS by acquiring evolutionary competence.

Incubation is the next stage. We open up ourselves for attaining new collective consciousness, new ways of thinking, new perceptions and assumptions. The term incubation suggests contemplation, avoiding making deliberate attempts to force insights. "Cease striving: then there will be self-transformation" (Chuang Tzu, Book XI).

Illumination is the moment of creative insight. It occurs with vivid clarity as a period of realization of new relationships with new creative ideas that emerge following the incubation period. It enables us to see our evolutionary future in a completely new way.

Verification is the process that validates the EGS we designed and brings it to fruition in its environmental context. We have to anticipate that the EGS we created might face the resistance of those who protect the status quo.

12.3.3.2.c Conditions of Fostering Evolutionary Creativity. One of the conditions is openness to new experiences, openness to novelty, as opposed of defensiveness, in which creation is prevented from coming into awareness. If we are open we welcome experiences that fall outside the usual categories. It means lack of rigidity and the permeability of boundaries in concepts, beliefs, and perceptions. We are able to receive, even welcome conflicting information; we have an "extensional orientation." *Complete openness is an essential condition of constructive creativity* in evolutionary inquiry.

For Rogers (1961) openness means the ability to play spontaneously with ideas and relationships, make impossible juxtapositions, make the given problematic, translate one form to another, transform into improbable equivalents. "It is from this spontaneous toying and exploration that there arises the hunch, the *creative seeing of life in a* new significant way" (p. 355). It is from such a seemingly wasteful spawning of a host of possibilities that one or two creative forms or process will emerge with unique and novel qualities that give them true value.

Another condition is *psychological safety.* It means unconditional acceptance of ideas and propositions that members of the designing community offer in the course of the inquiry. Acceptance conveys the feeling that each member of the community has worth in his or her own right. In evolutionary design we are to provide a climate in which *external evaluation is absent*. When we cease to form judgments of others from our own locus of evaluation, we are fostering creativity. Such an evaluation/judgment-free atmosphere is liberating and freeing. *Psychological freedom* fosters the individual's freedom of symbolic expression and creativity. It nurtures freedom to think and to feel. Rogers (1961) suggests that such freedom "fosters the openness, and the playful and spontaneous juggling of perceptions, concepts, and meaning, which are part of creativity" (p. 358).

Now that we have explored some of the conditions of creativity, we should also examine conditions that hinder or block creativity so that we can employ ways that might remove those blocks in our evolutionary inquiry.

12.3.3.2.d Conditions that Block Creativity. Bohm and Peat (1987) identify a range of conditions that block creativity. One is the common tendency toward the unconscious defense of ideas "which are assumed to be necessary to the mind's habitual state of comfortable equilibrium" (p. 50). There is a tendency to impose and cling to familiar ideas, even when there is evidence that they may be false. This then creates the illusion that no fundamental change is needed. This condition sets up the primary line of defense against conscious evolution. To cling to familiar ideas maintains a habitual sense of security and comfort and blocks the mind from engaging in the future creating evolutionary inquiry.

12.3.3.2.e Misconceptions about Creativity. Various misconceptions also hinder and block creativity. One is the widely touted assumption that creativity is necessary only in specialized fields (Bohm and Peat, 1987), such as art, literature, music, etc. This restricted assumption about "the nature of creativity is obviously of serious consequence for it clearly predetermines any program that is designed to clear up the misinformation within society" (p. 239). This in fact suggests that ordinary people, groups or organizations collectively, and society in general, cannot be creative. It is the belief that ordinary people just do not have the talent, the necessary passion, and the courage to act in a truly creative way. It holds that creativity is the privilege of the genius. Another commonly held belief is that "one must strain to try to solve the problem with the conscious mind first, and absorb a great deal of information about it before the behind-the-scene creative mind should go to work" (Harman and Rheingold, 1984, p. 79). This belief holds that one should struggle to find solutions with the rational, analytical mind, and turn to creativity only in desperation.

Otherwise, we are removing the ego-mind from its position as a gatekeeper, which, of course, will threaten its domination.

In closing, I am reminded of Toynbee's (1964) notion of the value and the crucial role and function of creativity in society. He declares that to *give a fair chance to creativity is a matter of life and death for any society. Creative ability is mankind's ultimate asset.* He warns that potential creative ability can be stifled, stunted, and stultified by the prevalence in the society of adverse attitudes of mind and habits of behavior. Therefore, *society has a moral duty to ensure that the individual's potential ability is given free play.*

Activity 47

The completion of the tasks in this activity calls for your imagination and creativity. The first task is to review each and every paragraph in the text and search for and identify "core ideas" of creativity. Write these down. The second task is to ask yourself: What are the implications of the noted core ideas for the design of my systems of interest? (The EGS you designed with your group.) The last task is to think about and formulate means and methods that you would use to remove blocks and hindrances of creativity and overcome misconceptions about it. Enter your findings in your workbook.

Reflections

Design creates novelty. Thus, creativity is central in design. Novelty cannot be produced by analysis of what is known or by associating aspects within the current frame of reference. The literature of social systems design does not question the role of creativity in design but very little of the literature emphasizes the significance of its role. It seems that design scholars take the role of creativity for granted and focus on the technical aspects of approaches and methods. The sources used in this section begin to hint at the potential powerful role and contribution that creativity might offer in evolutionary inquiry. However, it comes through that whatever has been written about the function of creativity and its function in the design of social systems or its role in conscious evolution is far from being an adequate exploration of creativity in social and evolutionary design. We must devote far more attention to creativity if we want to engage seriously in the design of evolutionary systems.

12.3.4 Evolutionary Conversation

Evolutionary design is a process that carries a stream of shared meaning by a free flow of discourse among members of an evolutionary design community who seek to create an EGS. To understand the critical nature of this communication function, the various modes of social discourse are explored. This makes it possible to search for the mode that is the most appropriate to conscious evolution.

In recent literature the notion of "dialogue" has gained prominence as the most viable form of collective social discourse. It is defined in the social discourse scholarship as "generative dialogue." It is not aimed at pursuing a specific task, such as evolutionary design. It aims to "generate" a common frame of reference, a shared view of the world among the parties of the dialogue. As such, generative dialogue that aims at creating a common ground among members of the evolutionary community is an important front part of evolutionary inquiry. More recently the notion of "strategic dialogue" has emerged as communication for collective decision-oriented inquiry. For use in the context of evolutionary inquiry, *the combination of the two forms of dialogue is proposed here as the preferred mode, which we call "evolutionary conversation."*

To tie in creativity with conversation in design, I turn to Bohm and Peat (1987), who proposed that a form of free flow of ideas, beliefs, and meaning among members of groups "may well be the most effective way of investigating the crisis that faces society, and indeed the whole of human nature and consciousness today. Such a form of free exchange of ideas and information is of fundamental relevance for transforming culture and freeing it of destructive misinformation, so that creativity can be liberated" (p. 240).

12.3.4.1 Dialogue as Social Discourse

In the course of the last several years we have seen a surge of attention devoted to "dialogue" or "generative dialogue." This mode is applied to generate a common frame of thinking in a group. Its primary architect is David Bohm (1985, 1987, 1990, 1991, 1996). Bohm and Peat (1987) make a sharp distinction between dialogue and ordinary discussion. In discussions people hold relatively fixed positions and argue their views in trying to convince each other. At best, Bohm says, this form of discourse may produce some compromise, but does not give rise to anything creative. In a discussion when something fundamental is involved, positions often tend to be non-negotiable and confrontational. This leads to either no solution or to a polite avoidance of the issue.

12.3.4.1.a The Process. The word dialogue is derived from the Greek "dia," meaning "through" and "logos," standing for the "meaning of a word." So, Bohm considers dialogue to be a free flow of meaning between people in a communication situation. In dialogue, people may prefer a certain position, but they are willing to suspend it, willing to listen to others in order to understand the meaning of their position. They are ready to change their point of view and blend it with others. In dialogue, people are able to face disagreement without confrontation and are willing to explore points of view to which they do not subscribe personally. "They will find that no fixed position is so important that it is worth holding at the expense to destroying the dialogue itself" (p. 242).

Everything that happens in the course of dialogue serves as an opportunity for learning. We learn how thoughts and feelings weave together, both collectively and individually. If members of the group are able to hold all their assumptions in suspension, they will generate shared consciousness. (The root meaning of consciousness is "knowing it all together.") In a dialogue the individual's and the group's "knowing it all together" form a subtle higher unity and come together in a harmonious way. In the dialogue event people are able to be honest and straight with each other, they level with each other, they share ideas freely. They develop a common mind, a shared mind, and can *think together in a new and creative way*. They awaken their collective intelligence and feelings of genuine participation, mutual trust, fellowship, and friendship. They can think and talk together. Shared meaning and understanding flow freely in the group. However, they can do none of this if hierarchy or authority is represented in the group.

12.3.4.1.b Dialogue Is Culture Creating. The organizing principle of dialogue implies a change of how the mind works. In true dialogue a new form of consensual mind emerges, generating a rich, creative order between the individual and the social as a more powerful force than is the individual mind alone. This creative order "arises from a spirit of friendship dedicated to clarity and the ultimate perception of what is true" (Bohm and Peat, 1987, p. 247). People who learn the potential power of such a dialogue will be able to transfer the spirit of dialogue into their activities and social relationships and into the systems and communities in which they live. *Dialogue, therefore, may create a new culture* in the evolutionary design community and, furthermore, "members of the community can explore the possibility of extending the transformation of the mind into a broader socio-cultural context" (p. 247).

Schein (1994) suggests that dialogue is a vehicle for understanding cultures and subcultures in organizations. And organizational learning depends upon such cultural understanding. It *facilitates the development of a common language and collective mental models.* Thus, the ability to engage in dialogue becomes one of the most fundamental and most needed human capabilities. Dialogue becomes a central component of any model of evolutionary transformation.

12.3.4.1.c Dimensions of Dialogue. Isaacs (1993) proposes that dialogue has three dimensions. (1) It is *collective learning* and action. The level of learning that can take place in a collective setting can never be mastered individually. In dialogue the group *creates a pool of common meaning* and new levels of coordinated action. (2) *Paradigm exploration* in dialogue enables people to step back from the context of specific problems, reflect upon what lies beneath them, and *learn a new way of seeing* and attaining a new kind of consciousness. (3) Dialogue enables us to *create a bridge between diverse cultural differences*. We can reach back into our shared cultural background, and create a common flow of meaning. (4) Dialogue *fosters the power of collective creation*. As we suspend our assumptions and begin to listen to each other in a deep way, new creative insights and new levels of wisdom emerge. We not only transform existing patterns of thought but also transmute them and create new levels of consciousness.

12.3.4.1.d Strategic Dialogue. Beck (1994) contrasts Generative Dialogue with Strategic Dialogue. According to Beck, Strategic Dialogue focuses on addressing specific tasks and is applied in finding specific solutions in decision-oriented inquiry. The dialogue scholarship community has not yet addressed substantively "strategic dialogue." Still, Schein (1994) suggests that the test of the importance of generative dialogue will be "whether or not difficult, conflict-ridden problems can be handled in a dialogue mode. Our experience with strategic dialogue the last two decades is that indeed it can. The combination of generative and strategic dialogue as a mode of communication is the answer to evolutionary design.

12.3.4.2 Evolutionary Conversation

It has been suggested that we can connect the two types of dialogue into a new communication mode called *conversation*. I will define it next and then report on its application.

12.3.4.2.a Generative + Strategic Dialogue = Evolutionary Conversation. It is proposed that the combination of generative dialogue and strategic dialogue comprises a comprehensive method of social discourse that is the most viable to use in an evolutionary designing community. We call this method *evolutionary (design) conversation*. The root meaning of conversation is "to turn to one another." Members of an evolutionary design community turn to one another without reserve and in truth and openness, accepting and honoring each other. Before the design community engages in the substantive task of evolutionary design, it involves itself in generative dialogue. This involvement will lead to the creation of a "common ground" as the community focuses on the thoughts, values, and world views of its members and creates a flow of shared meaning, shared perceptions, a shared world view, in a social milieu of friendship and fellowship. At this point the community is prepared to move on and engage in the strategic dialogue of evolutionary design inquiry.

12.3.4.2.b The Evolution of a Conversation Program. Earlier, I discussed the genesis of the design conversation program of the International Systems Institute. For a while, these conversations were the strategic dialogue type. But quickly, we became aware of how an up-front, *generative type of dialogue can enhance the potential and power of our conversations*. In fact, in the course of the last several years some of our design teams focused on the design of conversation itself.

The conversation is developed in yearly cycles. Research teams of around seven (plus or minus one or two) define themselves by a topic a year before the event and set forth a preparation program for developing the knowledge base for their selected topics. A set of resources are identified for collective study and others are selected by individuals. During the next several months members exchange ideas and findings, communicating on the Internet. Approximately two months before the conversation event, members exchange "input papers" and comment on each other's papers. During the five days of the conversation event they explore and develop their selected theme in-depth, following the disciplined inquiry process of consensus building conversations. The last day, the teams present a brief report and the conversation community prepares for the next event. Following the conversation, the teams develop their comprehensive reports, which are then published. Some teams work for several years on the same topic. Some of the recent topics include: *Designing Evolutionary Learning Communities, Designing Authentic and Healthy Communities, The Design of Evolutionary Guidance Systems, Metaphors for Evolutionary Guidance, Societal Design, The Systemic Design of Education.*

As the work of our international conversation community proceeds, we not only address significant themes but also learn about the conversation process itself. We make true progress toward ever more meaningful, genuine, authentic, and ethical conversation experiences. The work of our research teams demonstrates the power of conversation to tap into the collective intelligence of groups, create communities with shared meaning and a shared view of the world, and generate collective wisdom and capacity to engage in purposeful evolutionary design.

Activity 48
Accomplishing the next two tasks will help you to construct your own meaning of the role of conversation in evolutionary design and bring that role to life.

Task 1. Review the text of this section and mark the core ideas that represent the intent and the process of conversation. Map the core ideas in a relational arrangement by asking: What experiences should the group have that would enable it to attain shared meaning, a feeling of a caring community, and a collective view of the world? Based on this exploration and the experience of working with your group with the activities in this chapter, write a "briefing paper" (and enter it in your workbook) that will introduce to a novice group the meaning, the purpose, and the process of conversations.

Task 2. Use the briefing with a small group of your choice and have a conversation with them about the purpose and nature of conversations. Describe the experience in your workbook.

Reflections

Conversation, as described in this section, appears to be appropriate to use as a mode of communication in groups that engage in evolutionary design. Here, I make an observation on the significance of engaging in generative dialogue. In design groups there is an understandable tendency and temptation to "jump into" a strategic dialogue: into a design-task-focused-conversation right away. However, such rushing into functional tasks does not allow us to explore the assumptions, beliefs, values, and the implicit ideas that underlie the cognitive and affective characteristics of members of the evolutionary design community even though assumptions, values, and beliefs are the substantive bases of forming judgments and making design decisions. Unless we provide adequate opportunity for surfacing and openly addressing these up-front issues in a "generative dialogue" mode, they will sure bubble up later in the course of our "strategic dialogue." At that time, their earlier neglect will bog down the discourse and lead to recurring disagreements. At that point, most likely we shall dismiss or suppress them as being unwanted and interfering with carrying out design tasks. Eventually, we will wind up having debates and arguments. We might be able to coerce agreements, take votes, and may even call the outcome a "shared decision." But such a decision stands on a very shaky ground, on a shallow collective base. That which we acceded to, we have very little to care for, and we will have very little ownership or responsibility toward it. Thus, we will pay a high penalty for neglecting the use of the "generative dialogue" side of our conversation program by jumping into a "strategic dialogue."

12.3.5 *The Evolutionary Designing Community*

In Webster (1979) we find two categories of definition of a community: (1) a unified body of individuals and (2) society at large. Under (1), eight subcategories are denoted. Of those, the one that best describes the designing community is: "the people with common interest." But people in an evolutionary designing community have not only a common interest but also a common purpose, which is to engage in conscious evolution by design. Furthermore, the degree to which they can be considered to constitute a designing community is marked by the degree of effort they devote to attain their purpose, the degree of their commitment to it, and the degree of their commitment to each other. An evolutionary designing community is a self-defining, self-organizing, self-directing, self-maintaining, self-renewing, and self-governing entity.

There is an awakening aspiration and a maturing resolve in the society at large and in many segments of the human community toward self-determination, toward claiming our right and assuming the responsibility for shaping our own future, and to empowering ourselves to design our own lives and shape our own destiny. Consequently, there is general disillusionment with the way representative democracy works, when others are making decisions for us. The way democracy works today is more rhetoric and label than substance.

12.3.5.1 What Is a Community?

By reviewing the community and organizational literature one can find many different perspectives and views on the notion of community. Here, I review a few that are deemed to be appropriate to a characterization of evolutionary designing communities. MacCallum (1970) describes the community as an association of *people having a feeling of solidarity within their group that achieves coordinated action* "by virtue of shared enthusiasm and common dedication to a purpose which transcends immediate self interest" (p. 80).

During recent years we have seen the development of the Communitarian Movement (Etzioni, 1993) as a new moral, social, and public venture that reaffirms *shared values and aims to build a responsive and responsible civic society* in which we are each other's keepers. The movement aims "to bring about the changes in values, habits and public policies that will allow us to do for the society what the environmental movement seeks to do for nature" (p. 3). The establishment of moral coherence within a community is the moral foundation of the common good. The notion of common good has a dynamic element: "the community and individuals working toward a telos: a common purpose or goal" (p. 132).

Etzioni's vision is reflective of what we call here the evolutionary designing community that has the purpose of creating its future system as a *moral community, serving the shared purpose and the common good.* MacIntyer's (1984) conception of a moral community reflects Aristotle's vision of civic virtue, according to which people aim to build their moral community, serving the collective "telos," the common good, by engaging in coherent shared activities. The search for the common good cannot be an individual enterprise. For such enterprise, the community provides the only legitimate context.

A community (Peck, 1993) is a state of "being together with both individual authenticity and interpersonal harmony so that people become able to function with a collective energy even greater than the sum of their collective energies" (p. 472). Such a community has the capacity to evolve in wisdom, in effectiveness, and in maturity. *Group consciousness becomes a way of life.* Members of the community speak their minds honestly and openly. Issues are aired fairly; members feel that they are heard. They feel that everyone has equal power and responsibility.

In presenting an array of strategies for future search conferences, Weisbord (1992) proposed a set of core values that are pertinent to designing communities. He suggests that in design situations: (1) the *knowledge of ordinary* people is an extraordinary source of information; (2) people can *create their own future*; (3) they want to have opportunities to *engage their heads and hearts and participate in the creative process*; (4) *everyone is equal in the design situation*; (5) given the opportunity, *people learn to cooperate*; (6) the design process empowers people to feel more knowledgeable and in *control of their future*; and (7) *diversity is highly valued* and appreciated.

12.3.5.2 Making Use of the Collective Intelligence of the Community

Making use of the collective intelligence of designers is one of the key characteristics of evolutionary inquiry. On this line of thought, Pinchot and Pinchot (1993) suggest that the use of *collective intelligence is a key imperative* in organizations. The challenge of intelligent organizations is to establish strong communities so that everyone can contribute. The community serves as the vessel of vision and values, and the collectively developed intelligence of all members. The community recognizes the value of the intelligence of each individual. Everyone is responsible for adding value, use long-time-frame thinking; innovating, taking risk, and self-directing. What it all means to us is that evolutionary systems that are strong communities have the potential for involving everyone in the inquiry of shaping the future.

12.3.5.3 Equity and Diversity

In an egalitarian spirit, evolutionary communities treat *everyone as being of equal value*, as a high-potential member of the community. At the same time, unique individuality is nurtured. The combined honoring and encouraging of these two qualities in designing communities enables designers to see each design issue from many different viewpoints. It enables members of the evolutionary design community to learn from each other and generate creativity. Equity creates the climate that allows diversity to flourish and generate the richness and robustness needed in the design of evolutionary systems. The power of creative conversation in evolutionary design is determined a great deal by the variety of viewpoints and the diversity of participants. The greater the diversity the greater the likelihood of finding evolutionary design solutions. Reflecting on the effectiveness and richness of design, Churchman (1971) calls for the sweeping-in of the greatest possible variety of viewpoints and positions from the greatest possible variety of fields and disciplines.

Tsivacou (1990) suggests that many of the existing design methods, "when driven to rational consent, tend to suppress differences in the name of effectiveness" (p. 547). Such methods reduce the power of design when applied in the context of complex social systems in which conflicting views exist, especially in situations of deep change. And the evolutionary change we speak about is indeed deep change. Evolutionary inquiry has to satisfy two demands: the creation of a climate of conversation which nurtures the free expression of ideas as well the expression of differences and the enactment of criteria for the legitimation of choices in case of disagreement (p. 547).

One of the core principles of evolution is *requisite diversity*, which is an essential condition of the robustness of all life forms. Monocultures create barren spaces in nature as well as in the ecology of society. One brittle way to achieve a sense of community is by building on similarities. But narrowly defined communities are ineffective. They lack the diversity of viewpoints that are necessary to solve issues that cut across disciplines and cultures.

12.3.5.4 Stewardship Transcends Leadership

The search for authentic design solutions questions our current notions of leadership, which are associated with taking initiative, controlling, and knowing what is best for others. The act of a person "leading" cultural and organizational change by determining the desired future and defining the path to get there is *alien to the notion of people taking charge of their lives* and shaping their own future as members of a designing community. Block (1993) noted that search for strong leadership means that we place responsibility for our systems in the hands of others, which then reduces our opportunity to determine our own future. "The attraction of the idea of leadership is that it includes a vision of the future, some transforming quality that we yearn for" (p. 12). That is what much of the current organizational literature is promoting by calling for "buying-into" the vision of the leader. No wonder that leaders believe their "key task is to recreate themselves down through the organization as they wonder: How do I instill in others the same vision and behaviors that have worked for me?" (p. 13).

The *alternative to leadership is stewardship.* Stewardship asks each one of us to be responsible and accountable for the outcomes of the system we design. *Stewardship is shared accountability,* which is fueled by a shared commitment to service. What we have called leadership becomes stewardship which is empowerment, distributed among all participants of the evolutionary design community.

"There is pride in leadership, it evokes images of direction," says Block (1993, p. 41) and "there is humility in stewardship, it evokes images of service. *Service is central to stewardship.* When we serve, we build capability in others by supporting their ownership and empowerment and their right to participate. The kind of stewardship described here is the only viable approach to self-governance in designing communities. Yes: stewardship transcends leadership.

12.3.5.5 Designing Communities and Their Health

A community, as a designing social system, is a *shared space, a shared identity and character* (Nelson, 1989). In a community we not only share values but also share ways by which to discard old values and take on new ones. (These functions are critical to evolutionary design.) An evolutionary designing community is in search of purpose that gives direction to it. In that community the individual has "the power to participate in common choice while still retaining the sense of being unique and valuable" (p. 361). A designing community is not an organization of formally related identical individuals. It is a *social structure of unique individuals* who play specific roles by which they contribute to the overall design task.

The shared vision of the evolutionary designers provides the energy that is required to induce transcendental change, *leading to rebirth by establishing an evolutionary system.* Having a shared vision is essential to the health of an evolutionary community. However, the community faces two dangers. The first is that a quest for vision does not guarantee a safe journey. The evolutionary journey is always laden with risk. The second is that of facing the consequences of introducing disequilibrium by the pursuit of a new vision, against the conventional perception that equilibrium and balance are the desired healthy conditions. "Good health in communities does not consist of natural states of health. Good health is comprised of unnatural states which require constant energy and risk. This means that health is not something that can be achieved once and for all, requiring no further investment in human energy and intention" (p. 366). Only healthy designing communities can make judgments about how to best use their collective intelligence and resources as a means of maintaining their health.

12.3.5.6 Not the Expert but the Community

Evolutionary design carried out in the user-designer mode is not directed by an expert but *emerges from the intensive, creative, and dynamic interaction of members of the evolutionary designing community.*

Rowland's (1992) musical metaphor—the orchestra vs. the jazz ensemble—well represent the contrasting modes of the community as designers and the expert-driven mode. The orchestra is the metaphor for expert-driven design. The conductor makes all the decisions that the large number of players follow. In contrast, in the jazz ensemble the designing community decides what to play and how to play. They improvise around a basic plan, react to each other as they play, and challenge each other with new ideas. They explore opportunities and design new patterns. The play implies a high degree of interaction and cooperation. These are also the kinds of behavior that creative and interactive design implies. If we engage an evolutionary community in design then we need to learn to play design in the mode in which the jazz ensemble plays. The expert role in evolutionary design is to develop learning resources and opportunities for people in all walks of life, at all levels of education so that they can learn how to design.

12.3.5.7 An Example of a Designing Community

Our exploration of the idea of a designing community now concludes by presenting an example of a self-definition of a designing community: that of the International Systems Institute (ISI). The Institute is a not-for-profit research and educational organization, dedicated to the improvement of the human condition. Its purpose is to conduct research and development and create learning programs and resources for the development of authentic evolutionary designing communities. The ISI community defined itself as follows.

Our challenge is to learn to become an authentic community of scholarly practitioners and practicing scholars, to apply what we learn in all aspects of our lives, and help others to learn to develop their own authentic communities.

An authentic community is a group of individuals who have developed a deep and significant commitment to each other and to a shared vision and purpose. Members of the community:

- feel that they belong together and believe that they can make a difference in the world by pursuing their shared vision and purpose;
- communicate with each other openly, honestly, respecting and honoring each other's ideas and contributions;
- organize themselves with total absence of hierarchy and bureaucracy as equal partners in service and in mutual assistance; they govern themselves by shared stewardship;
- apply maximum flexibility in their shared work, taking full advantage of their unique individual and collective potential, knowledge, skills, creativity, and intuition;
- take responsibility for the continuing development of their individual and collective capabilities;
- nurture and practice genuine and authentic participation in achieving their shared purpose and in creating the common future of their community; and
- become bonded, knowing that they can rely on each other; they trust, honor, and support each other; they share values, aspirations, hopes, and they live by a collectively defined code of ethics.

The foregoing statement was formulated some time ago. It reflects some but not all of the ideas about a designing community developed in this section. The task of making use of the ideas introduced in this section now becomes your task as suggested in the activity below.

Activity 49

Review the text of this section and select core ideas that best define community in general and a designing community in particular. Organize the core ideas in clusters. The second task is to use the core ideas you selected and develop a statement of guidance for the creation of an authentic designing community. This community can be your family, a group you work with, or any system of your choice. Introduce your findings in your workbook.

12.3.6 The Democratic Agenda for the New Agoras

There is one domain in our societal way of life which—although mentioned repeatedly in this work—has not yet been addressed comprehensively. This domain is democracy. The question I propose is: What kind of democracy do we seek to realize in our evolutionary systems? In the first part of this chapter I gave some hints that we expect that the Agoras of the twenty-first century will recapture the true meaning of democracy and will build a genuine democratic culture and a vibrant cultural democracy.

> *We have not yet tried democracy. . . . We do not even have a conception of what democracy means.* (Mary Parker Follett, 1965)

This quote is an appropriate introduction to the discourse that follows. First an original definition of democracy is presented. Then, we listen to observers of the current state of democracy and consider their ideas as we contemplate the role of democracy in conscious evolution and in the lives of evolutionary designing communities.

12.3.6.1 The Original Meaning of Democracy

There are three Greek words we must keep in mind if we want to understand the original meaning of democracy in the classical Greeks' way of life. These are: "*democracy*," "*sizitisis*," and "*demosophia*." *Democracy* means "the power of the people." *Sizitisis* stands for "searching together." *Demosophia* is "the wisdom of the people." If we integrate these three we obtain the true meaning of democracy: People have the power to make decisions about issues affecting their lives. These decisions are made by searching together, by engaging in conversations. (Socrates said that the only way to arrive at the truth and attain wisdom is by searching together.) To attain wisdom is a prerequisite to exercising the power of the people. This type of democracy was practiced by citizens of the Athenian Republic as they gathered at "Agora" and engaged in conversation on various issues of shared interest.

12.3.6.2 Democratic Culture and Cultural Democracy

The New Agoras of the twenty-first century will manifest in their lives, in their shared culture the core ideas and precepts of democracy. They will create a *democratic culture* as their evolutionary culture. People of the New Agoras will organize their lives and build institutions that reflect the core ideas of democracy. So, they will build a *cultural democracy* that will manifest a democratic culture.

Follett (1965) holds that cultural democracy is far more comprehensive in organizing human relationships than the token institutions we now have. True democracy is not taking votes periodically and designating others to make decisions for us. Democracy should be what Follett calls "self-creating coherence," to be created by those most affected by a particular decision. We are now beginning to grasp that there is order in democracy, not the order left from our many authoritarian institutions, but the order that comes from coherence that is self-created, the product of many different individuals interacting and making collective decisions in their groups and in their communities. Much of self-creating democracy has to do with its open endedness. It is to be in a permanent state of reinvention. We do not know where the road to an emergent democracy leads if it is allowed to become its fullest. Democracy is always on the move, says Follett. It thrives on uncertainty and unpredictability. There is no formula for democracy, no more than there is a formula for any other kind of creativity. Democracy is self-creating and self-governing. Authority and responsibility for making decisions are vested in those who are affected by the decision. "Democracy is fueled by the willingness of individuals to involve themselves in directly confronting the issues that concern them. Democracy satisfies one of the strongest needs humans possess, the need to be useful. For democracy is on a permanent talent search: it finds uses in everyone and for everyone" (p. 188).

Reflections

We have made an important differentiation between *cultural democracy* and a *democratic culture*. *A democratic culture* is a culture that is democratic in its philosophy and in its practice. It is therefore an idea that is lived by as a way of life by people in a culture. It is shared by members of a culture and put into practice in their social interactions and as they engage in collective decision making on issues that affect and interest them. A *cultural democracy* is a set of arrangement of programs and institutions that are established and animated by democratic ideas and ideals. These arrangements provide ongoing opportunities, programs, and resources for learning a democratic way of life and create and maintain societal arrangements that empower members of a culture to competently practice cultural democracy, participate in collective decision making, and build institutions that manifest in their practice the culture of democracy.

In evolutionary inquiry, in self-guided conscious evolution, in designing our evolutionary systems in the New Agoras, *it is our task to create both a democratic culture and cultural democracy*. The coin of democracy has these two sides. It is a far cry from the current limited institutions of representative democracy.

The two-sided coin of democracy requires a transformation of our individual and collective consciousness. Rephrasing Einstein, we cannot bring forth a self-creating democratic culture from the same consciousness that brought forth our current democratic system. We have to think anew. We have to attain new collective consciousness, transcend what we have now, envision new images of democracy, and, based on those images, *transform our society by evolutionary design*.

So much has been written about the current state of democracy and the desired state. For the emerging Agoras, a deep consideration of the kind of democracy to aspire to will be a crucial issue. There is one more issue that should be addressed, as it is expected to play a key role in conscious, self-guided evolution: the role of high technology.

12.3.6.3 Teledemocracy

I have made repeated references to the power and use of technologies of the information age that enable citizen participation. Teledemocracy (TD) is such a means that can establish direct democracy through the use of communication media. Reporting on thirteen TD projects, Arterton (1987) identifies eleven institutional characteristics of citizen participation. (1) *Access* tells us the range of participation. (2) *Reach* accounts for the percentage of citizens available to participate and who actually do become involved. (3) *Effectiveness* is a measure of the direct influence of participation on public policy. (4) *Agenda setting* reveals the extent to which citizens have an influence on what issues are decided. (5) *Diversity* of paths accounts for the ways citizens can learn about a TD project. (6) *Duration* defines the length of time and number of events over which participation lasts. (7) *Individual or group-based mode* is another dimension of participation. (8) *Initiative* is the degree to which participants generate the TD opportunity themselves. (9) *Cost* is the burden that participants carry for their involvement. (10) The *educative value* reveals the degree to which participants learn about the issues addressed. (11) *Participating competence* reveals the skills of participation and the confidence attained by becoming politically active.

The idea of teledemocracy is only one of many possible approaches that the Agoras of the twenty-first century will consider as they develop their evolutionary designing communities, create their design inquiry, engage in the design of their EGSs and ESs, and begin to build the Agora webs of the evolutionary design movement. While up to now technology directed its own development, it is expected that in the era of conscious evolution the evolutionary movement will place demand on technology so that the Internet or new forms of it will serve evolutionary inquiry.

Activity 50
Review the preceding section and note core ideas of democracy. You might use other sources that address democracy, its current state, and/or its desired state. Reflect upon these and right a brief essay: "My ideas of what kind of democracy we should have." Enter your essay in your workbook.

Reflections

In this final chapter the Athenian Agora captured our imagination as a public space that we could reconstitute as the evolutionary design space of the Fourth Generation HSS. We explored some potential examples of how these New Agoras might work and the kinds of qualities they would seek to realize in their design inquiry and in their evolutionary systems. In the second part of the chapter a set of characteristics were proposed that not only would add value to our evolutionary inquiry but also establish a set of internally consistent benchmarks for the work of evolutionary designing communities. It seems to me that working through the chapter, we have connected the past to the present and the present to the future. It is now the future that becomes our creating space.

Summary and Conclusion

Our evolutionary learning journey has now reached an end, which is only the end of a new beginning. We have painted a sketch of the evolutionary saga of our species; we have an understanding of how evolution has worked and how we can make it work. I hope that in the course of your learning journey you have developed an interest in conscious evolution and even a will to enter the design space of evolution. I hope you accept the challenge to build evolutionary learning and designing communities. I hope that you will dedicate yourselves to taking part in the creation of evolutionary guidance systems. Each of us belongs to several systems, starting with our family, our kinship group, followed by the systems in which we live and work, the interests and nurturing groups in our communities, the volunteer agencies to which we belong, our places of worship, our schools, and others. Each of these are potential evolutionary design groups, places of New Agoras. How exciting, how rewarding will it be to search for common ground in—and within—these groups. To create collective value systems, to envision ideal images, and design evolutionary guidance systems! It could be the only sure way to build an internally consistent life space and life experience for us.

We are entering an age when there is a call for what is best in us, when we are invited to offer our individual and collective intelligence, when we are invited to use our greatest gift, the gift of creativity, when we are on a spiritual quest of building harmony, and above all, when we are invited to enter into the evolutionary design space, and collectively create our future.

We know now that we have a choice. We can continue to defend the status quo, resist change, and continue to struggle to save the collapsing Third Generation HSS by trying to fix it. Or, we can accept the challenge to join the many who believe that we can consciously and purposefully learn to guide our evolution and take responsibility for creating it.

Activity 51

This activity is your "free space." This is your *creating space*. It is your opportunity to enter the evolutionary design space and write an essay that explains your own evolutionary philosophy, one that underlies your choice between two paths. One is the well-traveled path of those who doubt if we have a role to play in conscious evolution; the other commits us to take the less traveled road and engage in conscious, self-guided evolution, and by so doing, take charge of and shape our destiny.

13

A Closing Thought

Engaging in the kind of evolutionary inquiry proposed here will require an immense effort and commitment of individuals, families, groups, systems, institutions, and the society as a whole. I invite you to contemplate what role you might play in this great venture, what contributions you might make to the task of building New Agoras and joining others in a harmonious way in the creation of a desired shared future.

It has been a long journey, as we traveled together on the terrain of evolutionary inquiry. I use the "we" pointedly. I was your guide on this journey and introduced to you what I have learned from exploring the work of others, from the many evolutionary conversations we had over the years, from my students and colleagues, and from my own design experiences. My own reward and satisfaction comes from the hope that I have been a useful guide to you.

You have done a great deal of work as you have tried to make sense from the material, reflected upon it, constructed your own meaning from it, defined core ideas of evolutionary inquiry, and formulated perspectives that helped you to be prepared for evolutionary design actions.

How much you have learned, the depth and the breath of your knowledge and competence in evolutionary inquiry depended upon the intensity of your interest and the effort you invested in working with the activities. Thus, you may feel that you have now a solid orientation in conscious evolution or that you have gone way beyond it, acquired a firm grounding in it, and are ready to engage in it.

We can celebrate your commencement into a continuing journey of evolutionary learning and evolutionary inquiry as you join the future creators. Because it is our destiny to create.

In closing, I offer you a quotation that always inspires me:

> I must create a system,
> Or be enslaved by another Man's;
> I will not Reason and Compare,
> My business is to Create.
>
> William Blake, *Jerusalem*, pl. 10, 1.20

References

Ackoff, R. L., 1981, *Creating the Corporate Future*, John Wiley & Sons, New York.
Ackoff, R. L., 1994, *The Democratic Corporation*, Oxford University Press, New York.
Ackoff, R. L., 1995, Wholeing the Part and Righting the Wrongs, *Systems Research*, 12:43–46.
Arendt, H., 1956, *The Human Condition*, University of Chicago Press, Chicago, IL.
Argyris, C. and Schon, D., 1978, *Organizational Learning*, Addison-Wesley, Reading, MA.
Arterton, C. F., 1987, *Teledemocracy*, Sage Library of Social Research, Sage, Newbury Park, CA.
Artigiani, R., 1996, Societal Computation and the Emergence of Mind, *Evolution and Cognition*, 2:2–15.
Bach, J., 1993, Evolutionary Guidance System in Organizational Design, *World Futures*, 36:107–128.
Baker, M., 1991, *Spirituality*, New Thought, Duluth, CA.
Banathy, H. B., 1973, *Developing a Systems View of Education: A Systems Models Approach*, Fearon, Palo Alto, CA.
Banathy, B. H., 1979, The Dynamics of Integrative Design, in *General Systems Research: A Science, A Methodology, A Technology* (Erickson, R., ed.), Society of General Systems Research, Louisville, KY, pp. 191–197.
Banathy, B. H., 1987a, The Characteristics and Acquisition of Evolutionary Competence in *World Futures*, 23:123–144.
Banathy, B. H., 1987b, *The Design of the Far West Laboratory*, Far West Laboratory, San Francisco, CA.
Banathy, B. H., 1989, The Design of Evolutionary Guidance Systems, *Systems Research*, 6:193–212.
Banathy, B. H., 1991, *Systems Design of Education*, Educational Technology Publications, Englewood Cliffs, NJ.
Banathy, B. H., 1992, *A Systems View of Education*, Educational Technology Publications, Englewoods Cliffs, NJ.
Banathy, B. H., 1993a, The Cognitive Mapping of Social Systems, in *The Evolution of Cognitive Maps* (Laszlo, E., ed.), Gordon and Breach, New York, pp. 205–219.
Banathy, B. H., 1993b, From Evolutionary Consciousness to Conscious Evolution, *World Futures*, 36:73–79.
Banathy, B. H., 1996, *Designing Social Systems in a Changing World*, Plenum Press, New York.
Banathy, B. H., 1998, Evolution Guided by Design, *Systems Research and Behavioral Science*, 15:161–172.
Banathy, B. H. and Johnson, D., 1977, Cooperative Group Interaction Curriculum, in *Curriculum Handbook* (Rubin, L., ed.), Allyn & Bacon, Boston, MA, pp. 570–577.
Barron, F., 1988, Putting Creativity to Work, in *The Nature of Creativity* (Sternberg, R., ed.), Columbia University Press, New York.

Beck, M., 1994, *The Concept of Dialogue*, unpublished manuscript.
Bell, D., 1976, *The Coming of the Post-Industrial Society*, Basic Books, New York.
Bergson, H., 1907/1983, *Creative Evolution* (Mitchell, A., trans.), University Press of America, Luham, MD.
Biatch, M., 1995, An Idealized Design of a Design to Guide the Growth and Evolution of a Synagogue Community, *Progress*, 4(Summer):41–58.
Blais, S., 1995, From Work Life to Life Work: Creating the Journey, *Progress*, (Summer):140–160.
Block, P., 1993, *Stewardship*, Barrett-Koehler, San Francisco, CA.
Bohm, D., 1983, *Wholeness and the Implicate Order*, Routledge and Kegan, London.
Bohm, D., 1985, *Unfolding Meaning*, Ark Paperback, London.
Bohm, D., 1990, *On Dialogue*, David Bohm Seminars, Ojai, CA.
Bohm, D., 1996, *On Dialogue*, Routledge, New York.
Bohm, D. and Edwards, M., 1991, *Changing Consciousness*, HarperCollins, San Francisco, CA.
Bohm, D. and Peat, D., 1987, *Science, Order, and Creativity*, Bantam Books, New York.
Botnik, J. W. and Maltiza, M., 1979, *No Limits to Learning*, Pergamon Press, Oxford.
Boulding, E., 1981, Evolutionary Visions, Sociology and the Human Life Span, in *Evolutionary Vision* (Jantsch, E., ed.), Westwiew Press, Boulder, CO.
Boulding, K., 1956, *The Image*, The University of Michigan Press, Ann Arbor, MI.
Boulding, K., 1978, *Ecodynamics: A Theory of Social Evolution*, Sage, Beverly Hills, CA.
Boulding, K., 1980, General Systems Theory—The Skeleton of Science in *The General Theory of Systems, Applied to Management and Organizations*, Vol I (Jamieson, G., Chen, L., Schkade, L., and Smith, C. H., eds.), Intersystems, Salinas, CA.
Boulding, K., 1981, *Ecodynamics: A New Theory of Societal Evolution*, Sage, Beverly Hills, CA.
Boulding, K., 1985, *Human Betterment*, Sage, Beverly Hills, CA.
Breuil, H., 1906, *L'Evolution de la Peinture et de la Gravure sur Murailles dans les Caverness Ornees de l'Age du Renne*, Congress Prehistorique, Paris.
Bridges, W., 1991, *Managing Transitions*, Addison-Wesley, Reading, MA.
Brin, D., 1998, *The Transparent Society*, Addison-Wesley, Reading, MA.
Burn, A. and Burn, M., 1980, *The Living Past of Greece*, Little Brown and Company, Boston, MA.
Campbell, B., 1974, *Human Evolution*, 2nd edit., Aldine, Chicago, IL.
Campbell, J., 1959, *The Mask of Good*, Viking Press, New York.
Chaisson, E., 1981, *Cosmic Dawn*, Berkeley Books, New York.
Chaisson, E., 1987, *The Life Era*, The Atlantic Monthly Press, New York.
Chaisson, E., 1988, *Universe*, Prentice-Hall, Englewood Cliffs, NJ.
Checkland, P., 1981, *Systems Thinking, Systems Practice*, John Wiley & Sons, New York.
Checkland, P. and Scholes, J., 1990, *Soft Systems Methodology*, John Wiley & Sons, New York.
Christakis, A., 1966, A People Science, The Cogniscope Systems Approach, *Journal of Transdisciplinary Systems Sciences*, 1:16–19.
Christakis, A. and Conaway, N., 1995, *Building High Performance Project Teams, Learning About the Cogniscope*, CWA, Berwyn, PA.
Christakis, A. and Conaway, N., 1996, *Designing the Ideal Societal Image and Designing a Design Inquiry Program for Designing It*, CWA, Berwyn, PA.
Churchman, W. C., 1968, *The Systems Approach*, Delacorte Press, New York.
Churchman, W. C., 1971, *The Design of Inquiring Systems*, Basic Books, New York.
Churchman, W. C., 1979, *The Systems Approach and Its Enemies*, Basic Books, New York.
Churchman, W. C., 1982, *Thought and Wisdom*, Intersystems, Salinas CA.
Combs, A., 1996, *The Radiance of Being*: Complexity, Chaos, and the Evolution of Consciousness, Paragon House, St. Paul, MN.

Cronbach, L. J. and Suppes, P., 1969, *Research for Tomorrow's Schools: Disciplined Inquiry in Education*, Macmillan, New York.
Cross, N., 1990, The Nature and Nurture of Designability, *Design Studies*, 2(3): 127–140.
Csanyi, V., 1989, *Evolutionary Systems and Society*, Duke University Press, Durham, NC.
Csikszentmihalyi, M., 1993, *The Evolving Self: A Psychology for the Third Millenium*, Harper Perennial, New York.
Curtis, R. K., 1982, *Evolution or Extinction*, Pergamon Press, New York.
Darwin, C., 1859, *The Origin of Species by Means of Natural Selection*, Murray, London.
Darwin, C., 1871, *The Descent of Man*, Modern Library, New York.
Davis, P., 1989, *The Cosmic Blueprint*, Touchstone, New York.
Davis, P., 1992, *The Mind of God*, Simon & Schuster, New York.
Dawkins, R., 1989, *The Selfish Gene*, Oxford University Press, New York.
Deacon, T., 1997, *The Symbolic Species*, W.W. Norton, New York.
Dennett, D., 1996, *Darwin's Dangerous Idea. Evolution and the Meanings of Life*, Touchstone, New York.
De La Cal, M., Hawthorn, P., Owen, E., and Robinson, S., 1999, How We Became Human, *Time*, Internet Addition, 154:1–5.
Dils, J., 1995, Idealized Design of a Senior Support Group, *Progress*, 4 (Summer):91–112.
Diamond, J., 1998, *Guns, Germs, and Steel: The Fates of Human Societies*, W.W. Norton & Company, New York, London.
Drucker, P., 1989, *The New Realities*, Harper & Row, New York.
Eiseley, L., 1979, *Darwin and the Mysterious Mr. X*, E. P. Dutton, New York.
Elgin, D., 1993, *Awakening Earth*, William Morrow and Company, New York.
Etzioni, A., 1993, *The Spirit of Community*, Crown, New York.
Feurstein, C., 1987, *Structures of Consciousness*, Integral, LoverLake, CA.
Flood, R. L. and Jackson, M., 1991, *Creative Problem Solving*, Plenum Press, New York.
Follett, M. P., 1965, *The New State: Group Organization: The Solution of Popular Government*, Peter Smith, Gloucester, MA.
Frankfort, H., 1946, *The Intellectual Adventure of Ancient Man*, University of Chicago Press, Chicago, IL.
Frantz, T. G., 1995, Evolutionary Competence in the Postmodern Family: An Idealized Design Approach, in *Progress* (Summer):20–40.
Frantz, T. G. and Miller, C., 1993, An Idealized Design Approach, in *World Futures*, 36:83–106.
Fukuyama, F., 1999a, The Great Disruption: Human Nature & the Reconstruction Social Oder, *The Atlantic Monthly*, May, pp. 53–80.
Fukuyama, F., 1999b, *The Great Disruption*, Free Press, New York.
Gebser, J., 1949/1986, *The Ever Present Origin*, Ohio University Press, Athens, OH.
Goerner, S., 1999, *After the Clockwork Universe*, Floris Books, Edinburgh.
Goodwin, B. C. 1994, *How the Leopard Got its Spots: The Evolution of Complexity*, C. Sribner's Sons, New York.
Gould, S., 1977, *Ever Since Darwin*, W.W. Norton, New York.
Gould, S., 1996, *Full House*, Three River Press, New York.
Gould, S., 1998, *Leonardo's Mountain of Clams and the Diet of Worms*, Harmony Books, New York.
Hall, R., 1973, *The Silent Language*, Anchor Books, New York.
Hammer, M. and Champy, J., 1993, *Reengineering the Corporation*, HarperCollins, New York.
Harman, W. and Horman, J., 1990, *Creative Work*, Knowledge Systems, Indianapolis, IN.
Harman, W. and Rheingold, H., 1984, *Higher Creativity*, Jeremy P. Tarcher, Los Angeles, CA.
Havel, V., 1990, *Address*, Presented at a Joint Session of the US Congress.
Hubbard, B., 1998, *Conscious Evolution*, New World Library, Novato, CA.

Humphrey, N., 1986, *The Inner Eye*, Faber & Faber, New York.

Huxley, A., 1945, *The Perennial Philosophy*, Harper Brothers, New York.

Huxley, A., 1946, *Essays of a Humanist*, Harper of Row, New York.

Isaacs, W., 1992, *Dimensions of Generative Dialogue*, The Dialogue Project, The MIT Press, Cambridge, MA.

Isaacs, W., 1993, Dialogue, Collective Thinking, and Organizational Learning, *Organizational Dynamics*, Fall.

Jackson, M., 1992, *Systems Methodology for the Management Sciences*, Plenum Press, New York.

Jackson, M., 1995, Beyond the Fads: Systems Thinking for Managers, *Systems Research*, 12:25–42.

Jantsch, E., 1975, *Design for Evolution*, Braziller. New York.

Jantsch, E., 1980, *The Self-Organizing Universe*, Pergamon Press, Oxford.

Jantsch, E., 1981, *The Evolutionary Vision*, Westview Press, Boulder, CO.

Jantsch, E. and Waddington, C. H., 1976, *Evolution and Consciousness*, Addison-Wesley, Reading, MA, pp. 37–67.

Johanson, D., Johanson, L., and Edward, B., 1994, *Ancestors: In Search of Human Origin*, Villard Books, New York.

Jones, C. J., 1984, *Essays in Design*, John Wiley & Sons, New York.

Judson, H., 1979, *The Eight Days of Creation*, Simon & Schuster, New York.

Kavanaugh, J., 1989, *Participatory Design of a Citizens Group for Social Empowerment*, A Journey of Trust, Unpublished Ph.D. Dissertation.

Koestler, A., 1964, *The Act of Creation*, Arkana, London.

Krippner, S., 1997, Integrating Individual and Community Approach to Healing, Wellness: The Shamanic Approach to Healing, *Alternative and Complementary Therapies*, 3:281–287, 359–368.

Krippner, S., 1998, Epistemology in Shamanic States of Consciousness, Paper given at Tucson IV: Toward a Science of Consciousness, Tucson, AZ.

Laszlo, E., 1972, *The Systems View of the World*, Braziller, New York.

Laszlo, E., 1987, *Evolution: A Grand Synthesis*, New Science Library, Boston, MA.

Laszlo, E., 1996, *Evolution: The General Theory*, Hampton Press, Cresskill, NJ.

Leakey, R. and Lewin, R., 1977, *Origin*, Dutton, New York.

Leakey, R. and Lewin, R., 1993, *Origin Reconsidered*, Anchor Books, New York.

Leroi-Gourhan, A., 1967, *Treasures of Prehistoric Art*, H. N. Abrahams, N.Y.

Lewin, R., 1998, *The Origin of Modern Humans*, Scientific American Library, NY.

Linstone, H. and Mitroff, I., 1994, *The Challenge of the 21st Century*, State University Press, New York.

Lorenz, K., 1977, *Behind the Mirror: A Search for The Natural History of Human Knowledge*, Harcourt, Brace, Jovanovich, New York.

MacIntyre, A., 1997, *After Virtue, 2nd Ed.*, University of Notre Dame Press, Notre Dame, IN.

MacCallum, S., 1970, *The Art of Community*, Institute for Human Studies, Menlo Park, CA.

Margulis, L., 1999, *Symbiotic Planet: A New Look at Evolution*, Basic Books, New York.

Markely, O. W., 1978, Human Consciousness in Transformation,in *Evolution and Consciousness* (Jantsch, E. and Waddington, C. H., eds.), Addison-Wesley, Reading, MA.

Markley, O. W. and Harman, W., 1982, *Changing Images of Man*, Pergamon Press, London.

Martin, D., 1999, *The Spirit of Dialogue*, International Communication for the Renewal on Earth, Cross River, NY.

Martin, D., Banathy, B. H., and Ward, L., 1999, *Healthy Community Project: A Collaborative Systems Approach*, Cross River Connections, Cross River, NY.

Mayr, E., 1942, *Systematics and the Origin of Species*, Columbia University Press, New York.

McGee, D., 1993, A Medical Practice as a Vehicle for Evolutionary Learning, *World Futures*, 36:129–140.

Mitroff, I. and Linstone, H., 1993, *The Unbounded Mind*, Oxford University Press, New York.

Moore, C. M., 1987, *Group Techniques for Idea Building*, Sage, Newbury Park, CA.

Mumford, L., 1967, *The Myth of the Machine: Techniques of Human Development*, Harcourt, Brace and World, New York.

Nadler, G., 1981, *The Planning and Design Approach*, John Wiley & Sons, New York.

Nadler, G. and Hibino, S., 1990, *Breakthrough Thinking*, Prima, Rocklin, CA.

Nelson, H., 1989, Unnatural States of Health in Natural Organizations, in *The Well-Being of Organizations* (Churchman, W., ed.), Intersystems, Salinas, CA.

Newell, A. and Simon, H., 1972, *Human Problem Solving*, Prentice-Hall, Englewoods Cliffs, NJ.

Nickols, S., 1974, Personal Communication.

Ornstein, R., 1991, *The Evolution of Consciousness*, Prentice-Hall Englewood Cliffs, NJ.

Osgood, C., 1959, *Ingalik Mental Culture*, Yale University Press, New Haven, CT.

Page, J., 1966, *Conference Report*, Ministry of Public Building and Works, London.

Pankow, W., 1976, Openness as Self-Consciousness, in *Evolution and Consciousness* (Jantsch, E. and Waddington, C. H., eds.), Addison-Wesley, Reading, MA.

Papanek, V., 1972, *Design for the Real World*, Bantam Books, New York.

Parness, S., 1972, Programming Creative Behavior, in *Climate for Creativity* (Taylor, C., ed.), Pergamon Press, New York, pp. 193–228.

Pastorino, R., 1993, A "User-Designer" Mediation Approach: Fostering Evolutionary Consciousness and Competence, *World Futures*, 36:151–166.

Peccei, A., 1977, *The Human Quality*, Pergamon Press, Oxford.

Peck, S. M., 1993, *A World Waiting to Be Born*, Bantam Books, New York.

Pinchot, G. and Pinchot, E., 1993, *The End of Bureaucracy & the Rise of the Intelligent Organization*, Berret-Koehler, San Francisco, CA.

Popper, K. R., 1974, Autobiography of Karl Popper, in *The Philosophy of Karl Popper* (Schilpp, P. A., ed.), Open Court, La Salle, Il, pp. 3–181.

Potts, R., 1997, *Human Descent*, Avon Books, New York.

Prigogine, I. and Stengers, I., 1984, *Order Out of Chaos*, Bantam Books, New York.

Ray, P., 1996, *The Integral Cultural Survey*, Research Report 96-A, Institute of Noetic Sciences, Sausalito, CA.

Rittel, H. and Webber, H., 1984, Planning Problems Are Wicked Problems, in: *Developments in Design Methodology* (Cross, N., ed.), John Wiley & Sons, New York.

Rogers, C. R., 1961, *On Becoming a Person*, Houghton Mifflin, Boston, MA.

Rowland, G., 1992, Do You Play Jazz? *Performance Instruction*, November/December, pp. 19–25.

Salk, J., 1983, *Anatomy of Reality: Merging of Intuition and Reason*, Columbia University Press, New York.

Sallstrom, P., 1992, The Possibility of the Impossible, *Cybernetics and Human Knowing*, 1:49–52.

Sapiro, M., 1999, *The North-End Agora*, Idaho Systems Institute, Boise, ID.

Sass, S., 1998, *The Substance of Civilization*, Arcade, New York.

Schein, E., 1994, The Process of Dialogue: Creating Effective Communication, *The Systems Thinker*, 5:1–4.

Shreeve, J., *The Neanderthal Enigma*, Avon Books, New York.

Singer, A. A., 1948, *In Search of a Way of Life*, Columbia University Press, New York.

Spencer, H., 1874, *Principles of Sociology*, 2nd edit., Williams and Northgate, London.

Spencer, H., 1898, *Principles of Sociology*, 3rd edit., Williams and Northgate, London.

Tattersall, I., 1998, *Becoming Human*, Harcourt Brace & Co., New York.
Teilhard de Chardin, 1975, *The Phenomenon of Man*, Harper & Row, New York.
Toynbee, A., 1947, *A Study of History*, Oxford University Press, New York.
Toynbee, A., 1959, *Hellenism*, Oxford University Press, New York.
Toynbee, A., 1964, Is America Neglecting Her Creative Minority? in: *Widening Horizons in Creativity* (Taylor, C. W., ed.), Wiley, New York, pp. 3–9.
Toynbee, A., 1981, *The Greeks and Their Heritage*, Oxford University Press, New York.
Tsivacou, I., 1990, An Evolution of Design Methodology, *Systems Practice*, 3:545–560.
Ulrich, W., 1983, *Critical Heuristics in Social Planning*, Haupt, Bern.
Vickers, G., 1981a, Some Implications of Systems Thinking, in *Systems Behavior* (Open University Systems Group, ed.), Harper & Row, London, pp. 19–25.
Vickers, G., 1981b, Social Ethics, in *Changing Images of Man* (Markely, O. and Harman, W., eds.) Pergamon Press, London.
Wailand, C., 1993, Evolutionary/Systemic Management of Organizations, *World Futures*, 36:141–154.
Wallace, G., 1926, *The Art of Thought*, Harcourt, New York.
Walsh, N., 1990, *The Spirit of Shamanism*, Jeremy P. Tarcher, Los Angeles,
Warfield, J., 1987, Developing Design Culture in Higher Education, In: *Design Inquiry* (B. H. Banathy, ed.), International Systems Institute, Carmel, CA, pp. 14–18.
Warfield, J., 1990, *A Science of General Design*, Intersystems, Salinas, CA.
Warfield, J. and Cardenas, R., 1994, *A Handbook of Interactive Management*, Iowa State University Press, Arnold, IA.
Webster, 1979, *New Collegiate Dictionary*, G & C. Merriam, Springfield, MA.
Weisbord, M., 1989, *Productive Workplaces; Organizing and Managing for Dignity, Meaning, and Community*, Berret-Koehler, San Francisco, CA.
Weisbord, M., 1992, *Discovering Common Ground*, Berrett-Koehler, San Francisco, CA.
Wesson, R., 1997, *Beyond Natural Selection*, The MIT Press, Cambridge, MA.
Whitehead, A. N., 1968, *Modes of Thought*, Free Press, New York.
Wilber, K., 1981a, *Up from Eden*, Shambhala, Boulder, CO.
Wilber, K., 1981b, *No Boundary*, New Science Library, Shambhala, Boulder, CO.
Wilson, A. and Cann, R., 1992, The Recent African Genesis of Humans, *Scientific American*, April, 68–73.
World Futures, 1993, *Evolutionary Consciousness* (Banathy, B. H., ed.), 36:75–166.
Zeldin, T., 1994, *Intimate History of Humanity*, HarperCollins, New York.

Index